Boron in Plants and Agriculture

Boron in Plants and Agriculture

Exploring the Physiology of Boron and Its Impact on Plant Growth

Edited by

TARIQ AFTAB
Aligarh Muslim University, Aligarh, India

MARCO LANDI
University of Pisa, Pisa, Italy

IOANNIS E. PAPADAKIS
Agricultural University of Athens, Athens, Greece

FABRIZIO ARANITI
University of Milan, Milan, Italy

PATRICK H. BROWN
University of California-Davis, Davis, CA, United States

ELSEVIER

ACADEMIC PRESS
An imprint of Elsevier

Academic Press is an imprint of Elsevier
125 London Wall, London EC2Y 5AS, United Kingdom
525 B Street, Suite 1650, San Diego, CA 92101, United States
50 Hampshire Street, 5th Floor, Cambridge, MA 02139, United States
The Boulevard, Langford Lane, Kidlington, Oxford OX5 1GB, United Kingdom

Notices

Knowledge and best practice in this field are constantly changing. As new research and experience broaden our understanding, changes in research methods, professional practices, or medical treatment may become necessary.

Practitioners and researchers must always rely on their own experience and knowledge in evaluating and using any information, methods, compounds, or experiments described herein. In using such information or methods they should be mindful of their own safety and the safety of others, including parties for whom they have a professional responsibility.

To the fullest extent of the law, neither the Publisher nor the authors, contributors, or editors, assume any liability for any injury and/or damage to persons or property as a matter of products liability, negligence or otherwise, or from any use or operation of any methods, products, instructions, or ideas contained in the material herein.

ISBN: 978-0-323-90857-3

For Information on all Academic Press publications
visit our website at https://www.elsevier.com/books-and-journals

Publisher: Nikki Levy
Acquisitions Editor: Nancy J. Maragioglio
Editorial Project Manager: Kathrine Esten
Production Project Manager: Punithavathy Govindaradjane
Cover Designer: Greg Harris

Typeset by MPS Limited, Chennai, India

Contents

List of contributors *xiii*
Preface *xvii*

1. Essentiality of boron in higher plants **1**

Seyed Majid Mousavi and Tahereh Raiesi

1.1 Introduction 1
1.2 Importance of boron in agriculture 2
1.3 Importance of boron in higher plants during the vegetative phase 4
1.3.1 Structural functions of boron 4
1.3.2 Biochemical/hormonal functions of boron 6
1.4 Importance of boron in higher plants during the reproductive phase 8
1.5 Boron dynamics and transport inside higher plants 9
1.6 B translocation and distribution in higher plants 10
1.7 Sensitive and tolerant plants to different concentrations of boron 11
1.8 Boron deficiency in higher plants 12
1.8.1 Management of boron deficiency in higher plants 15
1.9 Boron toxicity in higher plants 16
1.9.1 Management of boron toxicity in higher plants 18
1.10 Conclusion and future perspectives 19
References 20

2. Boron in fruit crops: plant physiology, deficiency, toxicity, and sources for fertilization **29**

Renato Vasconcelos Botelho, Marcelo Marques Lopes Müller, Renan Caldas Umburanas, James Matheus Ossacz Laconski and Maurilo Monteiro Terra

2.1 Introduction 29
2.2 Boron roles in physiology and transport 31
2.3 Boron toxicity and deficiency symptoms 32
2.4 Boron availability to plants, fertilization, and boron sources 34
2.5 Boron mineral sources and fertilization 37
Acknowledgements 46
References 46

3. Boron deficiency and toxicity symptoms in plants **51**

Monireh Mones Sardrodi, Zahra Dehghanian, Khashayar Habibi, Arash Hemati, Behnam Asgari Lajayer, Janhvi Pandey and Tess Astatkie

3.1 Introduction 51
3.2 Symptoms of boron deficiency 53
3.3 The deficiency of boron 55
3.3.1 Membranes, cytoskeleton, and cell wall 55
3.3.2 Nitrate uptake and fixation 57
3.3.3 Oxidative stress and secondary metabolism 58
3.4 The boron toxicity 59
3.4.1 Boron application for abiotic stress relief 59
3.4.2 Tolerance to boron toxicity 61
3.5 Gene expression and boron 64
3.6 Conclusion and future prospect 64
References 65

4. Molecular regulatory mechanisms in plants that underlie phenotypic adaptations to low boron levels **71**

Michaela S. Matthes

4.1 Introduction 71
4.2 Cessation of plant growth in boron deficient conditions: cell elongation or cell division 75
4.3 Interactions of boron with phytohormones 76
4.3.1 Boron—auxin interactions 76
4.3.2 Boron—cytokinin interactions 79
4.3.3 Boron—ethylene interactions 80
4.3.4 Boron—abscisic acid (ABA) interactions 80
4.3.5 Boron—brassinosteroid interactions 81
4.3.6 Boron—jasmonic acid interactions 81
4.4 Boron deficiency and transcript level changes 82
4.5 Conclusions and outlook 83
References 84

5. From outside to inside: mechanisms modulating plant responses to boron stress **91**

Greice Leal Pereira, João Antonio Siqueira, Paula da Fonseca-Pereira, Rebeca P. Omena-Garcia, Vitor L. Nascimento, Adriano Nunes-Nesi and Wagner L. Araújo

5.1 Introduction 91
5.2 Architectural adaptation of the root system to boron availability 94

5.3 Regulation of boron uptake: a central process for homeostasis 97
5.4 The role of boron distribution and redistribution in plant adaptation 100
5.5 Biochemical and physiological changes regulating stress tolerance by boron 104
5.6 Molecular effects underlying the effects of boron in plants 108
5.7 Adaptive mechanisms to boron stress in the reproductive phase 110
5.8 Concluding remarks 113
Funding 114
References 114

6. Physiological and biochemical mechanisms and adaptation strategies of plants under boron deficiency conditions 127

Milan Kumar Lal, Rahul Kumar Tiwari, Arvind Jaiswal, Biswaranjan Behera, Aalok Shiv, Awadhesh Kumar, Sudhir Kumar, Vijay Paul, Madan Pal Singh, Brajesh Singh and Prakash Kumar Jha

6.1 Introduction 127
6.2 Impact of boron stress on crop productivity 129
6.3 Physiological response of crops under boron stress 130
6.3.1 Photosynthesis 132
6.3.2 Carbon partitioning and source-sink relationship 133
6.3.3 Nitrogen metabolism 134
6.3.4 Interaction other nutrients 134
6.3.5 Transpiration 135
6.3.6 Pollen tube formation 136
6.4 Biochemical response of crops under boron stress 136
6.4.1 Cell wall and membrane permeability 136
6.4.2 Boron transporters 137
6.4.3 Reactive oxygen species and antioxidant system 138
6.4.4 Carbohydrate metabolism 139
6.5 Conclusion 139
References 140

7. Role of physical and chemical agents in plants for tolerance to boron nutrition 147

Shiva Aliyari Rad, Zahra Dehghanian, Behnam Asgari Lajayer, Muazzez Gürgan and Tess Astatkie

7.1 Introduction 147
7.2 Public attributes and chemistry of boron 148
7.3 Higher plants require boron as a micronutrient 148
7.4 Toxicity of boron in plants and boron nutrition 149

7.5 Boron pptake and transport mechanisms 150
7.6 Boric acid channels, functions and regulation 151
7.7 The borate exporters: physiological functions 152
7.8 Boron roles in plants 154
7.9 Roles of boron in plant metabolism 155
7.10 Plant tolerance to boron 157
7.11 Primary considerations 159
7.12 Revisiting tolerant mechanisms 159
7.13 Plant genetic modifications for boron susceptibility and resilience 161
7.14 Conclusions 162
References 163

8. Impact of boron and its toxicity on photosynthetic capacity of plants 169

Chrysovalantou Antonopoulou and Christos Chatzissavvidis

8.1 Introduction 169
8.2 Boron toxicity and photosynthesis 174
8.3 Conclusion 179
References 180

9. Comprehensive analyses of gene expression and identification of metabolites for boron stress tolerance 187

Firat Kurt and Ertugrul Filiz

9.1 Introduction 187
9.2 BOR1 homologs in plants 188
9.3 The metabolites for boron stress tolerance 192
9.4 Gene regulation under boron-deficient conditions 193
9.5 Gene regulation under conditions of excessive boron 194
9.6 Conclusion 200
References 201

10. Transcription factors and target genes involved in plant responses to high boron adaptation 207

Emre Aksoy and Ceyhun Kayıhan

10.1 Introduction 207
10.2 Transcription factors identified in *Arabidopsis thaliana* under boron toxicity 209
10.3 Transcription factors identified in barley under boron toxicity 210
10.4 Transcription factors identified in poplar under boron toxicity 212

10.5 Transcription factors identified in rice under boron toxicity 213
10.6 Transcription factors identified in wheat under boron toxicity 214
10.7 Transcription factors identified in *Puccinellia distans* under boron toxicity 214
10.8 miRNAs involved in post-transcriptional control under boron toxicity 215
10.9 Long noncoding RNAs involved in post-transcriptional control under B toxicity 217
10.10 Known functions of identified transcription factor families under boron toxicity 217
10.11 Conclusion 227
References 228

11. Alleviation of boron toxicity in plants by silicon: mechanisms and approaches 233

Muhammad Riaz, Muhammad Kamran, Omar Aziz and Xiurong Wang

11.1 Introduction 233
11.2 Silicon-induced alleviation of boron toxicity 235
11.2.1 Plant growth traits 235
11.2.2 Reduction of boron transport from roots to shoots 237
11.2.3 Oxidative stress and plant defense system 239
11.2.4 Silicon induced improvement in the photosynthesis under boron toxicity 242
11.3 Prospects and challenges 243
References 244

12. Agronomic aspects of boron: fertilizers, agronomical strategy, and interaction with other nutrients 249

Seyed Majid Mousavi, Seyed Ali Ghaffari Nejad, Fereydun Nourgholipour and Shirzad Abbaszadeh Zoshkey

12.1 Introduction 249
12.2 Status of boron in soils 250
12.3 Importance and functions of boron in plants 251
12.4 Importance of boron in agriculture and quality of production 252
12.5 Responses of different plants/varieties to the status of boron in the soil 255
12.5.1 Plant responses under boron deficiency condition 255
12.5.2 Plant responses under boron toxicity condition 256
12.6 Effect of soil properties on bioavailability of boron 257
12.7 Interactions of boron with other nutrients 258
12.8 Management of boron in the agricultural soils 260

12.8.1 Management of boron under deficiency condition 260
12.8.2 Management of boron under toxicity condition 262
12.9 Boron fertilizers 262
12.10 Conclusion and future perspectives 264
References 265

13. Boron, hormones and secondary metabolites in plants: a molecular point of view 271

Faiçal Brini and Marco Landi

13.1 Introduction 271
13.2 Roles of boron in plant metabolism 272
13.3 Boron transport mechanisms 274
13.4 The boron nutritional status evokes contrasting changes in plant hormones metabolism 277
13.5 Effect of boron deficiency on plant development 281
13.6 Alleviation of the effects of boron toxicity 283
13.7 Conclusions and future prospects 285
References 285

14. An overview on boron and pollen germination, tube growth and development under in vitro and in vivo conditions 293

Muhammad Aasim, Özlem Akgür and Büşra Yıldırım

14.1 Introduction 293
14.2 In vitro studies on boron and pollen 295
14.2.1 Almond (*Prunus amygdalus*; Rosaceae) 297
14.2.2 Almond (*Prunus amygdalus*; Rosaceae) and peach (*Prunus persica*; Rosaceae) 297
14.2.3 Apocynaceae family (Allamanda, Alstonia, Catharanthus, Nerium, Plumeria, Thevetia, and Tabernaemontana) 298
14.2.4 Areca palm (*Areca catechu* L.; Arecaceae) 299
14.2.5 Calabash tree (*Crescentia Cujete* L.; Bignoniaceae) 299
14.2.6 Chinese fir (*Cunninghamia lanceolata* L.; Cupressaceae) 299
14.2.7 *Jacquinia ruscifolia* Jacq. (Theophrastaceae) 300
14.2.8 Kiwifruit (*Actinidia deliciosa* Cultivar Matua; Actinidiaceae) 300
14.2.9 Henna tree (*Lawsonia inermis* Linn.; Lythraceae) 301
14.2.10 Lychee (*Litchi chinensis* Sonn.; Sapindaceae) 302
14.2.11 Maize (*Zea mays* L.; Poaceae) 302
14.2.12 *Mitragyna parvifolia* (Roxb.) Korth.; (Rubiaceae) 303
14.2.13 Olive (*Olea europaea* L.; Oleaceae) 303

14.2.14 Pistachio (*Pistacia vera* L.; Anacardiaceae) 303
14.2.15 Pomegranate (*Punica granatum*; Lythraceae) 304
14.3 In vivo studies on boron and pollen 305
14.3.1 Apple (*Malus domestica* L.; Rosaceae) 305
14.3.2 Lowbush Blueberry (*Vaccinium angustifolium* Ait.; Ericaceae) 305
14.3.3 Mango (*Mangifera indica* L. cv.; Anacardiaceae) 306
14.3.4 Peach (*Prunus persica*; Rosaceae) 307
14.3.5 Petunia and Agapanthus (*Petunia* Juss.; Solanaceae and *Agapanthus* L Herit.; Amaryllidaceae) 307
14.3.6 *Picea meyeri* (Pinaceae) 307
14.4 Conclusion 308
References 308

15. Impact of boron nutrition on pollen stigma interaction and seed quality 311

Archana, Preetam Verma and Nalini Pandey

15.1 Introduction 311
15.2 Experimental design 313
15.3 Pollen–stigma interaction 314
15.4 Steps in pollen–stigma interaction 314
15.5 Enzymes responsible for pollen–stigma interaction 315
15.5.1 Esterase 316
15.5.2 Acid phosphatase 317
15.5.3 Peroxidase (POD) 317
15.6 Impact of boron stress on floral development, pollen growth and enzymatic activities 318
15.7 Impact of boron stress on seed yield and its quality 320
15.8 Conclusion 322
References 322

Index *327*

List of contributors

Muhammad Aasim
Department of Plant Protection, Faculty of Agricultural Sciences and Technologies, Sivas University of Science and Technology, Sivas, Turkey

Özlem Akgür
Department of Plant Protection, Faculty of Agricultural Sciences and Technologies, Sivas University of Science and Technology, Sivas, Turkey

Emre Aksoy
Faculty of Arts and Sciences, Department of Biological Sciences, Middle East Technical University, Ankara, Turkey

Shiva Aliyari Rad
Department of Biotechnology, Faculty of Agriculture, Azarbaijan Shahid Madani University, Tabriz, Iran

Chrysovalantou Antonopoulou
Department of Agricultural Development, School of Agricultural and Forestry Sciences, Democritus University of Thrace, Orestiada, Greece

Wagner L. Araújo
Departamento de Biologia Vegetal, Universidade Federal de Viçosa, Viçosa, Minas Gerais, Brazil

Archana
Plant Nutrition and Stress Physiology Laboratory, Department of Botany, University of Lucknow, Lucknow, India

Behnam Asgari Lajayer
Department of Soil Science, Faculty of Agriculture, University of Tabriz, Tabriz, Iran

Tess Astatkie
Faculty of Agriculture, Dalhousie University, Truro, NS, Canada

Omar Aziz
Department of Soil & Environmental Sciences, University of Agriculture Faisalabad Sub Campus Depalpur, Okara, Pakistan

Biswaranjan Behera
ICAR-Indian Institute of Water Management, Bhubaneswar, India

Renato Vasconcelos Botelho
Department of Agronomy, State University of Midwestern of Paraná, Guarapuava, Paraná, Brazil

Faiçal Brini
Laboratory of Biotechnology and Plant Improvement, Center of Biotechnology of Sfax, Sfax, Tunisia

Christos Chatzissavvidis
Department of Agricultural Development, School of Agricultural and Forestry Sciences, Democritus University of Thrace, Orestiada, Greece

Paula da Fonseca-Pereira
Departamento de Biologia Vegetal, Universidade Federal de Viçosa, Viçosa, Minas Gerais, Brazil

Zahra Dehghanian
Department of Biotechnology, Faculty of Agriculture, Azarbaijan Shahid Madani University, Tabriz, Iran

Ertugrul Filiz
Cilimli Vocational School, Duzce University, Cilimli, Duzce, Turkey

Muazzez Gürgan
Department of Biology, Tekirdağ Namık Kemal University, Tekirdağ, Turkey

Khashayar Habibi
Department of Biotechnology, College of Agriculture, Isfahan University of Technology, Isfahan, Iran

Arash Hemati
Department of Soil Science, Faculty of Agriculture, University of Tabriz, Tabriz, Iran

Arvind Jaiswal
ICAR-Central Potato Research Station, Jalandhar, India

Prakash Kumar Jha
Feed the Future Sustainable Intensification Innovation Lab, Department of Agronomy, Kansas State University, Manhattan, KS, United States

Muhammad Kamran
School of Agriculture, Food and Wine, The University of Adelaide, Adelaide, SA, Australia

Ceyhun Kayıhan
Faculty of Sciences and Letters, Department of Molecular Biology and Genetics, Başkent University, Ankara, Turkey

Awadhesh Kumar
ICAR-National Rice Research Institute, Cuttack, India

Sudhir Kumar
ICAR-Indian Agricultural Research Institute, New Delhi, India

Firat Kurt
Faculty of Applied Sciences, Department of Plant Production and Technologies Muş Alparslan University, Muş, Turkey

James Matheus Ossacz Laconski
Department of Agronomy, State University of Midwestern of Paraná, Guarapuava, Paraná, Brazil

Milan Kumar Lal
ICAR-Central Potato Research Institute, Shimla, India; ICAR-Indian Agricultural Research Institute, New Delhi, India

Marco Landi
Department of Agriculture, Food and Environment, University of Pisa, Pisa, Italy; Centre for Climatic Change Impact, University of Pisa, Pisa, Italy

Michaela S. Matthes
Institute for Crop Science and Resource Conservation, Crop Functional Genomics, University of Bonn, Bonn, Germany

Monireh Mones Sardrodi
Department of Plant Breeding and Biotechnology, Faculty of Agriculture, University of Tabriz, Tabriz, Iran

Seyed Majid Mousavi
Agricultural Research Education and Extension Organization (AREEO), Soil and Water Research Institute (SWRI), Department of Soil Fertility and Plant Nutrition, Karaj, Iran

Marcelo Marques Lopes Müller
Department of Agronomy, State University of Midwestern of Paraná, Guarapuava, Paraná, Brazil

Vitor L. Nascimento
Setor de Fisiologia Vegetal - Departamento de Biologia, Universidade Federal de Lavras, Lavras, Minas Gerais, Brazil

Seyed Ali Ghaffari Nejad
Agricultural Research Education and Extension Organization (AREEO), Soil and Water Research Institute (SWRI), Department of Soil Fertility and Plant Nutrition, Karaj, Iran

Fereydun Nourgholipour
Agricultural Research Education and Extension Organization (AREEO), Soil and Water Research Institute (SWRI), Department of Soil Fertility and Plant Nutrition, Karaj, Iran

Adriano Nunes-Nesi
Departamento de Biologia Vegetal, Universidade Federal de Viçosa, Viçosa, Minas Gerais, Brazil

Rebeca P. Omena-Garcia
Departamento de Biologia Vegetal, Universidade Federal de Viçosa, Viçosa, Minas Gerais, Brazil

Janhvi Pandey
Division of Agronomy and Soil Science, CSIR-Central Institute of Medicinal and Aromatic Plants, Lucknow, India

Nalini Pandey
Plant Nutrition and Stress Physiology Laboratory, Department of Botany, University of Lucknow, Lucknow, India

Vijay Paul
ICAR-Indian Agricultural Research Institute, New Delhi, India

Greice Leal Pereira
Departamento de Biologia Vegetal, Universidade Federal de Viçosa, Viçosa, Minas Gerais, Brazil

Tahereh Raiesi
Citrus and Subtropical Fruit Research Centre, Horticultural Science Research Institute, Agricultural Research, Education and Extension Organization (AREEO), Ramsar, Iran

Muhammad Riaz
State Key Laboratory for Conservation and Utilization of Subtropical Agro-bioresources, Root Biology Center, College of Natural Resources and Environment, South China Agricultural University, Guangzhou, China; College of Resources and Environment, Zhongkai University of Agriculture and Engineering, Guangzhou, China

Aalok Shiv
ICAR-Indian Institute of Sugarcane Research, Lucknow, India

Brajesh Singh
ICAR-Central Potato Research Institute, Shimla, India

Madan Pal Singh
ICAR-Indian Agricultural Research Institute, New Delhi, India

João Antonio Siqueira
Departamento de Biologia Vegetal, Universidade Federal de Viçosa, Viçosa, Minas Gerais, Brazil

Maurilo Monteiro Terra
Agronomic Institute of Campinas, Campinas, São Paulo, Brazil

Rahul Kumar Tiwari
ICAR-Central Potato Research Institute, Shimla, India; ICAR-Indian Agricultural Research Institute, New Delhi, India

Renan Caldas Umburanas
Department of Agronomy, State University of Midwestern of Paraná, Guarapuava, Paraná, Brazil

Preetam Verma
Government Polytechnic, Department of Biotechnology, Ayodhya, India

Xiurong Wang
State Key Laboratory for Conservation and Utilization of Subtropical Agro-bioresources, Root Biology Center, College of Natural Resources and Environment, South China Agricultural University, Guangzhou, China

Büşra Yıldırım
Department of Plant Protection, Faculty of Agricultural Sciences and Technologies, Sivas University of Science and Technology, Sivas, Turkey

Shirzad Abbaszadeh Zoshkey
Department of Agronomy and Plant Breeding, College of Agricultural and Natural Resources, The University of Tehran, Tehran, Iran

Preface

Boron (B) deficiency is one of the most widespread deficiencies among plant micronutrients in agriculture while B toxicity is a major constraint in many temperate and arid irrigated regions of the world. B is an essential micronutrient and plays a crucial role in the growth and development of plants. The primary function of B is as a structural component of the plant cell wall which is critical for the formation of dimers of rhamnogalacturonan. A deficiency of B results in a wide range of metabolic disruptions including reduced lignification, altered auxin metabolism, compromised membrane integrity, altered sugar and carbohydrate metabolism, reduced photosynthesis, altered nucleic acid metabolism, and reductions in the activity of several enzyme activities. It has not been determined if these effects are a consequence of cell wall disruption or are evidence of a direct functional requirement for B in these processes. B mobility within the plant body varies with the plant species; in plant species that produce polyalcohols, B can easily translocate via phloem as a B-polyalcohol complex, whereas B is largely phloem-immobile in species that do not make polyalcohols. High B levels in the soil, particularly in arid and semiarid regions, inhibit plant growth and decrease crop yield and quality. In species with poor phloem mobility, B toxicity causes chlorosis and necrosis in the marginal regions of mature leaves. In species with phloem B mobility, symptoms are expressed in young tissues and occur as a result of phloem blockage. B toxicity also causes structural damage to thylakoids, thus inhibiting photosynthesis and limiting CO_2 uptake. B toxicity also disturbs electron transport leading to the overproduction of reactive oxygen species such as superoxide (O_2^-), hydroxyl (OH^-) radicles, which causes membrane damage, that eventually leads to cell death. Tolerance to excess B is a result of within-plant localization and the activity efflux transporters that reduce B accumulation in both roots and shoots.

As a consequence of its primary role in cell wall formation, B deficiency causes rapid and diverse growth defects, mainly in young rapidly growing tissues. Symptoms include the death of the apical meristem, cessation of root elongation, reduced leaf expansion, the loss of pollen viability, and reduced reproductive success. Boron deficiency reduces not only yield but also the quality of crops due to vegetative deformation, damage to fruit and storage organ structure, and reduced storage life.

In "*Boron in Plants and Agriculture: Exploring the Physiology of Boron and Its Impact on Plant Growth*", we provide a comprehensive update on the significant roles of boron in plant physiology and highlight the emerging techniques and applications that are currently being used in plant–boron interaction studies and its prospects. We are confident that this book will cater to the need of all those working in this field and will benefit scientists, research personnel, and graduate students involved in such studies.

Tariq Aftab[1], Marco Landi[2], Ioannis E. Papadakis[3], Fabrizio Araniti[4] and Patrick H. Brown[5]

[1]*Aligarh Muslim University, Aligarh, India*
[2]*University of Pisa, Pisa, Italy*
[3]*Agricultural University of Athens, Athens, Greece*
[4]*University of Milan, Milan, Italy*
[5]*University of California-Davis, Davis, CA, United States*

CHAPTER 1

Essentiality of boron in higher plants

Seyed Majid Mousavi[1] **and Tahereh Raiesi**[2]
[1]Agricultural Research Education and Extension Organization (AREEO), Soil and Water Research Institute (SWRI), Department of Soil Fertility and Plant Nutrition, Karaj, Iran
[2]Citrus and Subtropical Fruit Research Centre, Horticultural Science Research Institute, Agricultural Research, Education and Extension Organization (AREEO), Ramsar, Iran

1.1 Introduction

The necessity of boron (B) for the growth and development of plants was first proven at the beginning of the 20th century and, its essentiality has been reported not only in vascular plants but also in cyanobacteria, algae, fungi, and animals (Camacho-Cristóbal et al., 2018). Presently, in spite of the very well-shown essentiality of B for all vascular plants, the understanding of its functions remains unfinished (Herrera-Rodríguez et al., 2010).

Boron has vital functions both in the vegetative and reproductive phases and has impairments in numerous metabolic and anatomical processes induced by its deficiency or toxicity. Some scientists have reported that the essentiality of B is due to its ability to stabilize the molecules with cis−diol groups (Bolaños et al., 2004). Because the two existing forms of B in plants (i.e., boric acid and borate anion) can react easily to produce complexes with a range of compounds containing cis diol groups producing stable cyclic borate esters. Accordingly, some researchers reported that the plasma membrane integrity and function may be affected by B (Cakmak and Römheld, 1997), most likely due to the formation of B-complexes with membrane constituents containing cis−diol groups (Brown et al., 2002). Therefore, it is well known that the most usual role of B is its structural function in the cell wall, and its deficiency is often related to abnormal walls that have changed the physical and structural properties and do not develop normally (Brown et al., 2002; Ahmad et al., 2009, 2012). Boron also plays a vital role in development, cell division, fruit and seed development, sugar transport, and hormone development (Mousavi and Motesharezadeh, 2020). This chapter tries to evaluate different aspects of B essentiality for higher plants during vegetative and

Boron in Plants and Agriculture
DOI: https://doi.org/10.1016/B978-0-323-90857-3.00008-4

reproductives growth phases. Also, by classifying plant species in respect to response to B, it tries to investigate the structural and biochemical/hormonal functions of B under deficiency and toxicity conditions and offers management strategies.

1.2 Importance of boron in agriculture

B is an essential micronutrient for optimum growth and development and quality of crops (Marschner, 2011). Also, of all essential nutrients for plant growth, only plant requirements for molybdenum and copper are lower than the requirements for this nutrient (Chatzissavvidis and Antonopoulou, 2020). B is a nonmetal among micronutrients. It performs many crucial functions in plants and is mainly involved in several metabolic pathways and functions such as sugar transport, cell wall synthesis, and structure, metabolism of carbohydrate and RNA, synthesis of DNA, lignification, phenol metabolism, respiration, the integrity of plasma membranes and regulation of metal-activated enzymes (Dordas and Brown, 2000; Marschner, 2011). It has been reported that B deficiency affects the performance of 132 crops from more than 80 countries, affecting at least 8 million hectares worldwide (Das and Purkait, 2020).

From the standpoint of plant nutrition, B is unique among the trace elements in that the range between the concentration of B essential for normal crop production and the concentration causing injury is narrow (Gupta et al., 1985). The boron requirement varies among different plant species, as well as the genotype within the species, and it is possible that the available B in one soil is in the deficient range for one species but exhibits effects of toxicity for another (Brdar-Jokanović, 2020). it is essential for plants' growth, development, and yield that there is a proper amount of available B in the soil (Shah et al., 2017).

The amount of total B in most of the world's agricultural soils is approximately from 5 to 30 ppm (Brdar-Jokanović, 2020); however, only a small fraction of the total B (5%–10%) is hydro-soluble and, therefore, available to plants as a nutrient (Atique-ur et al., 2018). B in the soil is found in a chemical pool (Padbhushan and Kumar, 2017). The pH, texture, moisture, temperature, organic matter, and clay mineralogy are the most important soil factors affecting the availability of B to plants (Das and Purkait, 2020; Goldberg, 1997). Hot water-soluble soil B is mainly considered as an indication of plant-available B on a wide array of crops worldwide (Johnson and Fixen, 1990). The critical level of hot water-soluble soil B ranges from <0.1 to 1.0 ppm in most soils for different crops (Berger, 1949; Colak et al., 2013; Debnath and Ghosh, 2014; Gupta et al., 1985; Mathew et al., 2018; Sarkar et al., 2008) but the predictive value of this test is not always satisfactory. Note that, Plant tissue analysis is reliable for confirming B deficiency in plants.

B concentration can be roughly categorized by climatic zone. B deficiency is most likely in coarse-textured soils in humid regions. Temporary deficiency of B can occur by liming acid soils due to increased B adsorption at higher soil pH (Johnson and Fixen, 1990). In fact, insufficient levels of available B in soils adversely affect the metabolism and growth of crops (Atique-ur et al., 2018; Bolaños et al., 2004), fruit yield and quality (Davarpanah et al., 2016; Ekbic et al., 2018), the shelf life of fruits (Vajari et al., 2018) and increase the susceptibility of crops to diseases (Wang and Ko, 1975).

B requirement can be satisfied in soils with low B availability for the plant by incorporating B in the soil or foliar spray of B-containing fertilizers. However, frequent fertilizer applications can result not only in environmental pollution but also can disrupt soil fertility and result in B toxicity (Shireen et al., 2018). Using several useful and eco-friendly techniques, the concentration of B in the soil solution can be maintained within the optimal range.

The selection of genotypes with high efficiency of nutrient acquisition has been considered as a cost-effective and environmentally-friendly technology to minimize inefficient use of inputs and environmental pollution (Bilal et al., 2018; Raiesi and Moradi, 2021). In a number of studies, genotypes within the species have been screened for B uptake efficiency (Atique-ur et al., 2018; Mei et al., 2016; Rerkasem and Jamjod, 1997; Zhou et al., 2014). For instance, Zhou et al. (2014) reported that the different performances of citrus rootstocks indicated that Carrizo citrange was the most tolerant while Trifoliate orange was the most sensitive to B-deficiency. Another way to cope with B deficiency is the use of arbuscular mycorrhiza fungi. It has been suggested that ectomycorrhizas may either increase B uptake or sequester B in forms unavailable for plants (Ruuhola and Lehto, 2014). Some studies report reduced (Clark et al., 1999), unaffected (Lu and Miller, 1989), and enhanced (Kothari et al., 1990; Lehto et al., 2004) acquisition of B in shoots of MF-inoculated plants, but the exact role of mycorrhiza in the uptake of B is still poorly known and requires further investigation.

B deficiency does not usually occur in regions of the world where the climate is arid and semiarid, salt-affected soils (Naz et al., 2016) and where saline irrigation water (Ben-Gal and Shani, 2002) is prevalent. In these regions, B toxicity is a much more probable occurrence than deficiency. Hot water-soluble B concentration of 5 mg/kg or more in the soil is considered toxic for most plants (Bañuelos et al., 1999). Excess and B deficiency inhibit plant growth (Bañuelos et al., 1999; Mesquita et al., 2016; Paull et al., 1988) and reduce crop yield (Yau et al., 1995).

1.3 Importance of boron in higher plants during the vegetative phase

It is known for almost a century that B is an essential micronutrient; however, because there is no radioactive isotope is available (Shaaban, 2010), most of the known information about its role in plant physiology is still being debated (Lewis, 2019; Wimmer et al., 2019).

Recently, research has progressed to the point where it is possible to demonstrate B participation in some main aspects of plant physiology, including (1) a structural role for B in cell walls, (2) a role for B in the maintenance of plasma membrane functions, (3) B involvement in the biosynthesis of some metabolic compounds, (Landi et al., 2019; Pereira et al., 2021; Shireen et al., 2018), (4) B role in nitrogen fixation and sugar transport, and (5) B involvement in the division of meristem cells and the formation of flower buds and leaf buds (Goldbach et al., 2001; Shireen et al., 2018).

1.3.1 Structural functions of boron

The primary cell wall of higher plants is an important factor influencing the structural integrity of the cells and determining the cells' shape, size, rigidity, and strength during plant development (Blevins and Lukaszewski, 1998; Somerville et al., 2004). Cross-links between their major components, cellulosic polymers, and matrix polymers such as hemicellulosic and pectic polysaccharides can modify the mechanical properties of growing cell walls (Voiniciuc et al., 2018). According to reports, there is a close relationship between B nutrition and the primary cell walls (Blevins and Lukaszewski, 1998). B was reported to be complex with many cell wall constituents (Marschner, 2011; Matoh, 1997; Wu et al., 2017). In plants, two B species, including boric acid ($B(OH)_3$) and borate anion ($B(OH)_4^-$), exist in equilibrium with each other. Both chemical species can react quickly to form complexes with various compounds containing cis–diol groups generating stable cyclic borate esters (Camacho-Cristóbal et al., 2018). Based on this property, it has been demonstrated that the best-known physiological function of B is its structural role in the cell wall through the formation of borate diol diester linkage with apiose residues of rhamnogalacturonan II (RG-II) (Camacho-Cristóbal et al., 2002; O'Neill et al., 2004) that stabilizes the pectin network and regulates the tensile strength and porosity of the cell wall (Fleischer et al., 1999; Ryden et al., 2003). Additionally, The RG-II-B complex is decisive for the cell elongation process (De Cnodder et al., 2005). According to reports, the majority of cellular B has been localized in the cell wall fraction especially when B supply is limited. The amount of B located in the wall depends on the plant species, organ, and B supply (Goldbach and Wimmer, 2007). Under B-limiting conditions in tobacco (*Nicotiana tabacum* L.) and squash (*Cucurbita pepo* L.) plants, up to 95% of the cellular B has been localized in the cell wall fraction of leaves and most of the cell wall B ($>70\%$) was associated with pectin (Hu and Brown, 1994). Another research found that cell walls of citrange, compared to other citrus rootstocks, have more RG-II to maintain a stronger ability to combine with B under B deficiency. In fact, B role in the formation of RGII-B dimer could explain many of the characteristics of B-deficient plants. According to Wang et al. (2015), vital visible symptoms of B-deficient in trees, including depression of growing points and deformity of organs, may be related to the functional role of B in the structure of cell wall and membrane.

Although the quantities of B in membranes were not significant, especially compared with those in cell wall fractions, it plays an important role in both the structural and functional integrity of plasma membranes (Martini and Thellier, 1993; Pollard et al., 1977). However, the fundamental mechanisms of B deficiency's effect on the structure and function of plasma membranes are still unknown (Goldbach and Wimmer, 2007). According to reports, B probably forms cis–diol complex compounds with some membrane molecules, including hydroxylated ligands such as glycoproteins and glycolipids (Goldbach and Wimmer, 2007), thus maintaining its integrity and function (Cakmak and Römheld, 1997). Although, so far, the happening of these B complexes has not been verified yet. Moreover, B deficiency reduces H^+-ATPase activity in roots (Ferrol et al., 1993; Lawrence et al., 1995), and optimum B concentration stimulates ATP hydrolysis and H^+ transport activity and control membrane voltage charging. In B-deficient plants, plasma membranes are highly leaky and lose their functional integrity (Cheng and Rerkasem, 1993). Despite this, experimental evidence showed that B deficiency reduced the transport of ions, metabolites, and enzymes (Marschner, 2011; Schon et al., 1990) across the plasma membrane (Wang et al., 1999) due to the depression of ATPase activity and alters the membrane potential (Ferreira et al., 2021).

Recently, Ferreira et al. (2021) indicate that effects of B supply on PM H^+-ATPase activity explain distinct nutrient uptake (N-NH_4^+, K^+, Ca^{+2}, Fe^{+2}, and Cu^{+2}) patterns by citrus rootstocks, which supports fine-tuning fertilization guidelines of citrus considering rootstock varieties. Also, Zhou et al. (2014) indicated that In addition to B, Mg, Fe, and Mn concentrations in plant tissues are also affected by B deficiency. Mei et al. (2016) reported that absorption and transportation inhibition of other mineral nutrients might be completely related to the changes in root morphology and vessel element anatomical characteristics under B deficiency conditions.

1.3.2 Biochemical/hormonal functions of boron

In addition to the above scenarios concerning the effects of B deficiency in the cell wall and plasma membrane function, this mineral deficiency causes numerous biochemical, physiological, and anatomic changes (Blevins and Lukaszewski, 1998; Camacho-Cristóbal et al., 2002). Among them, B involvement in the nucleic acids metabolism and synthesis of protein, carbohydrate metabolism and transport, synthesis and metabolism of phenolic compounds, plant enzymes and hormonal regulation, and B role in nitrogen fixation has been widely documented (Shaaban, 2010). These are processes that determine vegetative and reproductive growth (quality of pollen, pollen tube growth).

B deficiency affects the photosynthetic capacity and transport of the photosynthetic products (Brdar-Jokanović, 2020; Li et al., 2017; Shah et al., 2017; Wang et al., 2015). According to Lu et al. (2014), B deficiency can lead to the excessive accumulation of nonstructural carbohydrates and then downregulation of photosynthesis in leaves. A decrease in stomatal conductance and leaf gas exchange and intercellular CO_2 was found in B-deficient Citrus sinensis leaves. Also, A decrease in chlorophyll and carotenoid was observed in B-deficient Citrus sinensis leaves (Han et al., 2008). The results of another study showed that the chlorophyll and carotenoid contents in the leaves of Moringa oleifera were improved following soil drenching and or foliar spray of B.

Accumulation of phenolics accurses under short-term B deprivation as a consequence of the stimulation of the enzyme phenylalanine ammonium-lyase (Camacho-Cristóbal et al., 2002). In fact, B contributes to phenolic metabolism and accumulation (Camacho-Cristóbal et al., 2018). B–sugar cis–diol complex formation is crucial to reducing the potential toxicity of phenols. However, due to the metabolic shift from glycolysis to pentose phosphate pathway in B-deficient leaves, plants fail to form this complex and resulting in phenolic compounds being synthesized and accumulated (Mengel and Kirkby, 2012).

B deficiency activates polyphenol oxidase enzymes that can be oxidized to phenolic compounds (Camacho-Cristóbal et al., 2002), resulting in elevated polyphenol oxidase and quinine concentrations and ultimately reactive oxygen species (ROS) (Cakmak and Römheld, 1997), which are hazardous for plant growth and development (Hajiboland et al., 2013). The ascorbate/glutathione cycle is one of the most important mechanisms contributing to ROS detoxification in cells. Many studies have shown that both ascorbate and glutathione levels decrease in roots and leaves under B deficiency (Cakmak and Römheld, 1997). It has been suggested that phenolics result in reversible changes in membrane permeability, and this effect of phenolics occurs during their passage through the membranes (Glass, 1973; Goldbach et al., 1990).

There is no agreement on the role of B in enzymatic activity. Marschner (2011) reported that B has no enzymatic function in plants. According to many others reports, however, this element can change the activity of some enzymes within the plant cells (Han et al., 2008). A previous study has also highlighted traits such as antioxidant activity and total phenol content of olive cv. "Zard" was improved by the application of nano-B fertilizer (Vishekaii et al., 2019). According to Han et al. (2008), even though under B deficiency, antioxidant enzymes remain active in citrus leaf; however, the system as a whole is not strong enough to protect from oxidative damage. Along this line, Mukhopadhyay et al. (2013) reported that B deficiency actives ascorbate peroxidase, catalase, peroxidase, and superoxide dismutase in tea leaves. By and large, the results suggest that the generation of ROS is intensified under boron deficiency, but the antioxidant system is unable to adequately protect the plant from oxidative damage.

B is also hypothesized to suppress IAA oxidase activity; therefore, B deficiency leads to the production and accumulation of IAA (Bohnsack and Albert, 1977). Also, B deficiency induces vital changes in nitrogen metabolism as it leads to reduction of nitrate reductase activity and nitrate accumulation (Shen et al., 1993). In many instances, it is proved that B is essential in rhizobial N_2 fixation, actinomycete symbiosis, and Cyanophyceae heterocyst formation in leguminous crops (Bolaños et al., 2004).

1.4 Importance of boron in higher plants during the reproductive phase

Based on the latest research, B deficiency induces significant changes in reproductive growth, including decreased pollen viability, pollen, and ovule infertility, reduced seed quality, reduced flower production or premature flowering, as well as fruit downfall (Dell and Huang, 1997; Marschner, 2011). Among reproductive structures, pollen and, particularly, pollen tube development show a high sensibility to B availability, probably due to the role of this element in the cell wall structure and function of the pollen tubes (Camacho-Cristóbal et al., 2018). Recently reported B might take an active part in promoting pollen tube growth by affecting $[Ca^{2+}]$ cyt, actin filament assembly, pectins, and AGP deposition at the pollen tube in apple (*Malus Domestica*) (Fang et al., 2019). Along this line, Quiles-Pando et al. (2019) reported that B deficiency increases cytosolic Ca^{2+} levels mainly via Ca^{2+} influx from the apoplast in arabidopsis thaliana roots.

According to reports, in most plant species, the B requirement for reproductive growth is much higher than for vegetative growth (Miwa and Fujiwara, 2010; Shireen et al., 2018; Wojcik et al., 2003). This is especially true for graminaceous species. B deficiency decreases grain set production and grain yield while having little effect on their vegetative growth (Lordkaew et al., 2013). In wheat, for instance, B deficiency depresses grain set and grain yield due to poor development of anthers and pollen (Rerkasem and Jamjod, 1997). Al-Amery et al. (2011) reported that B supply led to decreased empty seed percentage of sunflower plants, thus reducing yield. Rashid et al. (1994) observed that the grain yield of rice cultivars increases as a consequence of reduced panicle sterility after B application. Lordkaew et al. (2013) reported that B deficiency depressed rice grain yield through adverse effects on reproductive growth. Hussain et al. (2012) found that B can be applied as a soil application at the flowering stage to improve rice grains yield and net economic income.

The results of another research indicate that B application at an appropriate concentration (the optimum dose of foliar spray and soil drench were 1 g/L and 0.9 g/plant, respectively) could increase the fruit-setting of Moringa oleifera due to improving the vitality of pollen and (Zheng et al., 2019).

1.5 Boron dynamics and transport inside higher plants

As was stated before, the amount of total B in the majority of the world's agricultural soils is approximately from 5 to 30 ppm (Brdar-Jokanović, 2020); however, only a small fraction of the total B (5%–10%) is hydro-soluble and, therefore, available to plants as a nutrient (Atique-ur et al., 2018). B in the soil is found in a chemical pool (Padbhushan and Kumar, 2017). B uptake by plants depends upon the existing form and content of available B in soil (Sun et al., 2019). pH, texture, moisture, temperature, organic matter and clay mineralogy, fertility levels, soil biological activity, and management practices are the most important soil factors affecting the dynamics and transformations of B in soils and then the availability of B to plants (Ahmadi and Souri, 2018; Das and Purkait, 2020; Dhaliwal et al., 2019; Goldberg, 1997; Souri and Hatamian, 2019). Boron-containing minerals are mainly found in the soil in the form of tourmaline (insoluble) or hydrated minerals (soluble) and generally do not control the solubility of B in the soil solution (Goldberg et al., 1993).

Therefore, the level of B in the soil solution is controlled by the adsorbed and desorbed reactions of B pools in the soil. B occurs predominantly in the soil solution as H_3BO_3 and partially as $B(OH)_4^-$ (Marschner, 2011). Boric acid is a very weak, monobasic acid (pKa = 9.25) that acts as a Lewis acid by accepting a hydroxyl ion to form the borate anion. Equilibrium concentrations of these two species in the soil solution strongly depend on the soil pH (Klochko et al., 2006). In pH = 9.25, the concentration of these two species is equal, and in neutral or slightly acid soils, B is located mainly in the form of undissociated boric acid (Tabata et al., 2003). Boric acid has a high permeability through the lipid bilayer of biological membranes (Dordas and Brown, 2000). Both metabolic and nonmetabolic processes predominantly regulate the B uptake mechanisms. To date, three main ways are recognized for the entering and transport of B in plant cells: (1) passive diffusion through lipid-bilayer, which has a high permeability to B and under sufficient or excessive B availability conditions; (2) facilitated transport via selective or nonselective channels, such as the nodulin 26-like intrinsic proteins; and (3) high-affinity active transport reconciled by borate transporters persuaded under insufficient B conditions which a great number of B extrusion transporters has been identified for many plants and different tissues (Brown et al., 2002; Miwa and Fujiwara, 2010; Shireen et al., 2018; Takano et al., 2008).

1.6 B translocation and distribution in higher plants

Following B uptake by the roots, B moves through xylem cells with the transpiration force and accumulates in currently active sink regions (Marschner, 2011). By analyzing B efficiency, beyond the intrinsic regulation mediated by transporters in distinctly efficient species (Miwa and Fujiwara, 2010; Pereira et al., 2021), it was possible to verify that plant differences in sensitivity to B deficiency may be associated with B mobility (Bellaloui et al., 2003; Stangoulis et al., 2010) or/and the pattern of B distribution within the plant (Nachiangmai et al., 2004; Wang et al., 2014). Wang et al. (2014) suggested that two ratios, including (1) the ratio of B concentration in the rootstock stem to the scion stem and (2) the ratio in the buds to the leaves, could consider as indicators for evaluating B efficiency between the citrus scions.

According to reports, B is always considered an immobile nutrient in the phloem of many plant species (Brown and Shelp, 1997). However, studies demonstrate that species that utilize cis-polyols (complex sugars) as primary photosynthetic metabolites are likely to allow B to be transported through the phloem (Brown and Hu, 1996; Shaaban, 2010). In species transported carbon as polyols, such as loquat olive, apple, and peach, the poly-B complex is formed in the leaves and is transported in the phloem to currently active sink regions (Bieleski, 1982; Brown and Shelp, 1997; García-Sánchez et al., 2020; Landi et al., 2019; Oikonomou et al., 2019; Papadakis et al., 2018; Tsiantas et al., 2019). In some species, cis-polyols are not synthesized (Boaretto et al., 2008). In these species, B mobility in the phloem is probably related to the formation of *bis*-*N*-acetyl-borate and *bis*-Suc borate complexes in the phloem (Stangoulis et al., 2010). Along this line, Du et al. (2020) indicated that B present in the leaves can be long-distance transported from these tissues to roots via the phloem, at least by forming a B-sucrose complex in citrus plants.

1.7 Sensitive and tolerant plants to different concentrations of boron

As was stated before, B is an essential nutrient to plants, and very little concentration is needed. Plant species differ in their capacity to take up B, even when they are grown in the same growth medium. These differences generally reflect different B requirements for growth. Plants in general use less than 5% of B in the soils (Uygan and Çetin, 2004). The efficient concentration of B for the growth and development of plants varies between 0.5 and 2.0 mg/L. In general, the soils with less than 0.5 mg/L of B cannot supply the needed B to plants, and its deficiency can happen in the plants. On the contrary, in the soil where its concentration is further 2.0 mg/L, there is B contamination and it can reduce growth and yield (Taiz et al., 2015).

Generally, based on the reports of Bergmann (1988) and Marschner (2011), dicots (cotton and leguminous plants) have 4–7 times higher B requirement (20–70 mg B/kg) than monocots (Gramineae family), 5–10 mg B/kg. Numerous researches have reported certain concentrations of B are required for plant development in different aspects of biochemical, physiological, and morphological (Hale and Orcutt, 1987). There is a very slight range between toxicity and deficiency of B as in the conditions in which concentrations of available B are more than 5.0 mg/L, toxication can happen to many crops. Production of forage legumes and some vegetable crops is often limited by lack of B, but recovery happens by supplying B. On the contrary, excess of B also, through affecting plants' physiological and biochemical properties, causes defects in the fruits and leaves of the plants (Hartmann et al., 1981).

The reported results for sunflower and bean fields confirm the importance of B in the optimum growth and yield of plants. At the concentration of 0.5 mg/L, the yield of sunflower is high (418 kg/ 1000 m^2), but with increasing the concentration of B up to 16 mg/L, the yield decreases down to 306 kg/1000 m^2. A similar status happened for the bean plant. A concentration of 0.5 mg/L has measured a yield of about 180 kg/1000 m^2 and was gown down to 73 kg/1000 m^2 at 16 mg/L (Sener and Özkara, 1989). Therefore, it is vital to maintain B in the growth media within an optimum range for maximum growth and production.

There is an overlap of the beneficial and injurious effects of B between plant species. Ayvaz (2002) has reported three broad categories of tolerance (sensitive, semitolerant, and tolerant). The sensitive species can tolerate 0.5 mg/L of B, but tolerant species can tolerate up to 4 mg/L (Batar et al., 2009). The tolerant plants endure a wide range of B concentrations with little effect, and the sensitive plants exhibit a strong reaction to either too much or too little B. In Table 1.1, different plants have been classified based on their sensitivity to B.

1.8 Boron deficiency in higher plants

Sufficient nutrition of B is essential both for high yields and the high quality of plants. As the most important roles of B in plants are thought to be its structural role in cell wall development and stimulation or inhibition of specific metabolism pathways (Ahmad et al., 2009, 2012), thereby, differences in the B demand of plant species are probably related to the differences in their cell wall composition, and cis–diol configuration in the cell walls, such as pectic substances.

Boron deficiency is a nutritional disorder that differs from high B soils and high rainfall areas throughout the world (Marschner, 2011). Plants that are cultivated in B deficient conditions must display B deficiency symptoms in the first stages, such as unusual growth of the apical growing section (Mousavi and Motesharezadeh, 2020). B deficiency disturbs cell wall plasticity and inhibits root meristem functioning, and consequently affects root growth (Herrera-Rodríguez et al., 2010). Some of the main morpho-anatomical alterations caused by B deficiency at the root level, stem level, and leaf-level are summarized in Table 1.2, along with its biochemical alterations in different plants (Table 1.3). In the previous study by Mousavi and Motesharezadeh (2020) deficiency of B in fruit crops was investigated.

Table 1.1 Classification of different plants based on their sensitivity to B (Marschner, 2011; Ahmad et al., 2012).

Tolerant species	Moderately tolerant species	Moderately sensitive species	Sensitive species
Most Vegetables, Alfalfa, Beet, Cotton, Grain, Sorghum, Oat, Sugar beet, Tomato, Barley	Barley, Cabbage, Celery, Corn, Squash, Sweet clover, and Turnip	Tubers, Cereals, Broccoli, Carrot, Cucumber, Pea, Pepper, Potato and Radish	Citrus (Grapefruit, Lemon, Orange), Stone fruits, Nut trees Avocado, Bean, Grape, and Wheat

Table 1.2 Some of the main morpho-anatomical alterations caused by B deficiency in different parts of plants.

Root	Stem	Leaf
Inhibition of root elongation Marschner (2011)	The thickness of the wall in xylem parenchyma cells and collenchyma tissue. Hajiboland et al. (2012)	Inhibition of leaf expansion. Hajiboland et al. (2012)
Reduction of cell division in root meristematic regions Dell and Huang (1997)	Malformation of xylem. Robertson and Loughman (1974)	Discoloration of Leaves depending on the plant species. Hajiboland et al. (2012)
Increase of lateral root initiation and higher-order lateral root branching. Hajiboland et al. (2012)	Shortened, thickened, cracked, or showing external corkiness (celery). Internal cavities (cauliflower) leading to breakdown and rotting. Robertson and Loughman (1974)	Reduction of chlorophyll content. Hajiboland and Farhanghi (2010)
Reduction of apical dominance. Hajiboland et al. (2012)		Decline and increase in the anthocyanin content of leaves. Hajiboland and Farhanghi (2011)
Storage organs develop brown internal areas (brown heart of turnips)or cracks, which may lead to rotting (heart rot of sugar beet)		Reduction of cell wall thickness. Kouchi and Kumazawa (1976)

Unlike nitrogen (N), B is not mobile in the plant; therefore, its deficiency symptoms first happened in the shoot apical meristem and in fruits. Deficiency of B may observe on soil with sufficient content of B due to over liming, dry or wet soil conditions, and low partial pressure of oxygen in the soil (Wojcik et al., 2008). Loomis and Durst (1992) reported that B has a critical role in generative processes affecting

Table 1.3 The biochemical and molecular responses of plants under B deficiency conditions.

Plant	The biochemical alteration	Reference
Pea (*Pisum sativum*)	Decrease in the nodule weight and N2 fixation capacity of legumes	Bolaños et al. (2004)
Bean (*Phaseolus vulcans*)	Reduction in the levels of ascorbic acid, SH-compounds, and H_2O_2 scavenging enzymes.	Bolaños et al. (2004)
-	Increase in the accumulation of phenolic compounds in the plant	Cakmak and Römheld (1997)
The higher plants	Disturbance of sugar transport, cell wall synthesis, signification, carbohydrate metabolism, RNA metabolism, respiration, indole acetic acid (IAA) metabolism, phenol metabolism.	Pollard et al. (1977)
Tobacco	occurring of enzymatic or nonenzymatic oxidation; generation of toxic quinones and destructive O2 species	Juan et al. (2002)
Maize	An increase in the level of actin and tubulin proteins and their polymerization patterns change in the plant cytoskeleton	Yu et al. (2003)
Tobacco	Causing the unbalanced oxidation and reduction in the plant tissues	Kobayashi et al. (2004)
-	Inhibition of the activity of enzymes effective in cell wall formation	Camacho-Cristóbal et al. (2008)
The vascular plants	Increase the oxidative stress in the plant	Herrera-Rodríguez et al. (2010)
Rapeseed	Reduction in shoot IAA concentrations	Zhou et al. (2016)
Oilseed rape	Increase of indole-3-acetonitrile (IAN); Accumulation of auxin in the shoot, which is highly associated with suppressed shoot growth as well as an enhanced IAA oxidation rate; Increase in ABA concentrations	Eggert and von Wirén (2017)
Squash, *Arabidopsis*	Reduction of ascorbic acid concentration in roots and increases NADPH oxidase activity in Roots which could be additional factors that facilitate the formation of ROS.	Camacho-Cristóbal et al. (2015, 2018)

another development, pollen germination, and pollen tube growth. Therefore, in the B-sensitive plants, abortion of flower initials and a poor set of fruit or seeds are recorded under deficiency of B (Mozafar, 1989). In Chapter 3, the symptoms of B deficiency in plants will be carefully discussed, as well.

1.8.1 Management of boron deficiency in higher plants

Soils deficient in Boron can be amended with boron fertilizer such as Borax, boric acid, and Solubor (Table 1.4), based on soil tests and crop requirements. As was stated before, due to the narrow margin between B deficiency and toxicity, an excess rate can easily happen and injure the plant growth (Marschner, 2011). Therefore, high care is required to apply the correct rate of B fertilizer. As shown in Table 1.5, B application rates generally range from 0.3 to 4.0 kg B/ha, depending on crop requirement and application method. Higher rates are required for broadcast applications than for banded soil applications or foliar sprays. In high pH soils, foliar applications are preferred. Once symptoms of B deficiency are observed, it is usually too late to apply B. As was stated before, different plants vary in their susceptibility to B deficiency, and when B deficiency symptoms occur, B is recommended at application rates determined by the plant. Because B is immobile in plants, its deficiency in the plants growing in soils with marginal B levels can happen during peak growing periods (vegetative, flowering, and seed development stages), so a steady supply of B throughout the growing season is

Table 1.4 List of boron-containing chemical fertilizers.

Fertilizer	Formula	% B content
Borax	$Na_2B_4O_7.10H_2O$	11
Boric acid	H_3BO_3	17
Solubor	$Na_2B_8O_{13}.4H_2O$	20
Sodium borate	Na2B4O7.5H2O	20
Sodium tetraborate	Na2B4O2.5H2O	14
Colemanite	Ca2B6O11.5H2O	10
B frits	-	2–6

*Borax, solubor, sodium borate, and sodium tetraborate have been most commonly used for soil application. Boric acid, colemanite, and B frits are considered to be more promising on highly leached sandy soils as well as for long-duration field crops, including perennial forages and fruit plants, owing to their low solubility, and slow release of B. Boronated superphosphate has also been tried to correct B deficiency in crops (Patil et al., 1987).

Table 1.5 Application rates of boron recommended for correcting deficiency of B by the higher plants (Ahmad et al., 2012).

Plant	Rate of application (kg/ha)	Plant	Rate of application (kg/ha)
Alfalfa	2–4	Potato	0.6–1
Apple	0.3–1.4	Clovers	0.6–2.3
Tomato	0.6–1.7	Corn	0.6–1
Sweet potato	0.6–1.7	Pea	0.9–1.2
Rice	1.7–2	Wheat	1.7–2
Cabbage	1–4.5	Peanut	0.3–0.5
Soybean	2.8	Cotton	0.6–1
Carrot	1–1.7	Grapes	0.6–1
Strawberry	0.6–1		

needed for the optimum growth and yield. Foliar application is also an efficient method to supply B to plants, particularly when root activity is restricted and B deficiency in crop appears under dry soil conditions in the growing season.

1.9 Boron toxicity in higher plants

Boron toxicity often happens naturally in alkaline and saline soils together with a little rainfall and very scarce leaching, in agricultural lands close to the coastal areas or in areas with persistent geothermal activities (Landi et al., 2019). B toxicity happens when the B concentration in the plant tissue is more than 200 mg/kg. However, nonuniform B distribution in different plant species makes it difficult to define the critical level (Gupta et al., 1985). Extra B in soil results in growth damage and change of plant metabolism, resulting in advanced necrosis in leaf, stem, and deformities of fruits (Reid et al., 2004). Also, B toxicity has been shown to affect root growth in plants. In Table 1.6, some of the main morpho-anatomical caused by B excess at the root level, stem level, and leaf level and also, its biochemical alterations are summarized.

Boron tends to accumulate in old leaves, especially at the margin of leaves, Because B is translocated along the transpiration streams and concentrates at the end of the transpiration stream. Excess B also decreases plant yield (Yau et al., 1995). Boron toxicity is a usual challenge in the

Table 1.6 Some of the main morpho-anatomical alterations caused by boron excess in different parts of the plant.

Root	Stem	Leaf
Alterations of root meristems. CHOI et al. (2007)	Development of cork and collenchyma cells. Papadakis et al. (2018)	Depending on the plant species, both reduction and increase of leaf thickness. Papadakis et al. (2004, 2018)
Deposition of suberin in cortical cells and thickening of hypodermis and lignification of root. Ghanati et al. (2002)	Increase of cell wall thickness. Papadakis et al. (2018)	Reduction of transpiration rate due to severe stomata limitations. Macho-Rivero et al. (2018)
Alteration of root meristem and thereby reduction of root cells number, increase their length and width. CHOI et al. (2007)	Increase in thickness of the cell wall of phloem cells. Huang et al. (2014)	Alterations of photosynthetic pigment content and its relative proportion. Kayıhan et al. (2017)
		Reduction of chloroplast size, thylakoid density, and increase of the relative volume of plastoglobuli. Papadakis et al. (2004)
		The appearance of chlorosis and Necrosis in leaves. Roessner et al. (2006)

arid and semiarid environments throughout the world that can limit plant growth. The usual symptom of B toxicity in plants is chlorosis (yellowing) of the leaf tip, progressing along the leaf margin and into the blade. Necrosis of the chlorotic tissue occurs, followed by leaf abscission. Necrosis of the leaf tissue results in a loss of photosynthetic capacity, which reduces plant productivity (Lovatt and Dugger, 1984). Pollen germination and pollen tube growth may also be inhibited (Versar Inc, 1975). In chapter 3 entitled "The symptoms of Boron deficiency and toxicity in plants," the symptoms of B toxicity in plants will be carefully discussed, as well (Table 1.7).

Table 1.7 Some of the usual biochemical and molecular responses of plants under boron toxicity conditions.

Plant	The biochemical alteration	Reference
Triticum aestivum L.	Reduction of nitrate reductase activity and increase the activity of glutamate dehydrogenase in roots and leaves	Mahboobi et al. (2002)
-	Disruption of gene splicing, osmotic imbalance, and reduction of the plant cell capacity to cope with oxidative damage	Reid (2007)
Tomato, Pepper, Lettuce	Increase of fat peroxidation and proline accumulation	Eraslan et al. (2007a,b)
Chickpea	Prevention of photosynthesis through introducing damage to the assembly of thylakoids	Ardıc et al. (2009)
The vascular plants	Affecting the enzymes associated with N metabolism, including nitrate reductase and asparagine synthase	Herrera-Rodríguez et al. (2010)
Solanum tuberosum L.	Accumulation of ROS such as superoxide radicals, hydroxides, and hydrogen peroxide (H2O2) in the plant	Ayvaz et al. (2016)
Triticum aestivum L.	Reduction of leaf water potential and increase of Chlorophylle a/b ratio	Kayıhan et al. (2017)
-	Increase of lipid peroxidation and H_2O_2 followed by genotoxic effects	Çatav et al. (2018)
Portulaca oleracea L.	Increase in concentrations of K, Mg, and B in the shoot and also increase in the activity of CAT, APX, and content of proline MDA and H_2O_2 content	Samet and Çıkılı (2019)

1.9.1 Management of boron toxicity in higher plants

Management of B toxicity has its special condition, and it is entirely different from the management of B deficiency. Alleviation of B toxicity in the agricultural soils can be done through leaching, applying of different amendments, or growing tolerant plants (Savić and Marjanović-Jeromela, 2013; Goldberg and Suarez, 2012, 2011; Dobermann, 2000; Mousavi et al., 2018a,b,c; Mousavi, 2021). The relationships between B and other nutrients (for example, Ca, S, Zn, Si, and Al) were considered for developing amendments that are recommended either for improving the soil properties or for ameliorating the toxic effects of B on higher plants

(Zakir Hossain et al., 2005; Inal et al., 2009; Turan et al., 2009). In addition to the application of different soil conditioners to the contaminated soils, different researchers (Hilal et al., 2011; Shaffer et al., 2012; Wolska and Bryjak, 2013) have suggested that developing and using cost-effective strategies for B removal from irrigation water can solve the challenge of the most usual anthropogenic sources of contaminants. In general, the most usual methods for management of B toxicity can be summarized as follows:

- Plow when the soil is dry so that B accumulates in the topsoil. Leach with water containing a small amount of B;
- Use surface water with a low B content for irrigation;
- Applying soil conditioners like lime, gypsum, etc., in order to change the soil pH and subsequently change in the B bioavailability depending on the initial pH of the soil;
- Applying organic fertilizers in the soil, because organic matter is an important component that adsorbs B and affects the concentration of B and its availability;
- Selecting of cultivars and genotypes tolerant to high levels of B.

1.10 Conclusion and future perspectives

Boron is a micronutrient that has a vital role in the health and quality of the growth and yield of higher plants. The most usual function of B is its role in the strength and development of cell walls which consequently increases plant tolerance to different biotic and abiotic stresses. By careful investigation of the specific functions of B, it is found that B plays a vital role in the life cycle of higher plants, both in the vegetative and reproductive phases. Although the higher plants have completely different reactions to the status of B in the soil, scientists have divided different plants into 4 groups in terms of reaction to B including tolerant, moderately tolerant, moderately sensitive, and sensitive species. The plants show physiological, molecular/biochemical responses under deficiency and toxicity conditions of B, which in some plants, such as blackheart rot in sugar beet, these responses are used as an efficient index for determining the real status of B in the growth media. There are different strategies for management of B nutrition in the higher plants, which due to the narrow margin between deficiency and toxicity of B, it is necessary to act carefully, especially under the deficiency condition. The essentiality of B in the higher plants has been evaluated in various aspects so far, and most researchers have emphasized its essentiality. However, this was not confirmed by some researchers (Lewis, 2019; Matthes et al., 2020). Therefore, it is necessary to evaluate more carefully the behavior of B in the life cycles of the higher plants. Hopefully, by identifying the unknown aspects of the behavior of this element, effective steps can be taken to achieve sustainable production and food security.

References

Ahmad, W., Niaz, A., Kanwal, S., Rahmatullah, Rasheed, M.K., 2009. Role of boron in plant growth: a review. J. Agric. Res. 47 (3), 329–338.

Ahmad, W., Zia, M.H., Malhi, S.S., Niaz, A., Ullah, S., 2012. Boron Deficiency in soils and crops: a review. Crop. Plant. 2012, 65–97.

Ahmadi, M., Souri, M.K., 2018. Growth and mineral content of coriander (*Coriandrum sativum* L.) plants under mild salinity with different salts. Acta Physiologiae Plant. 40 (11), 1–8.

Al-Amery, M.M., Hamza, J.H., Fuller, M.P., 2011. Effect of boron foliar application on reproductive growth of sunflower (*Helianthus annuus* L.). Int. J. Agron. 2011, 1–5.

Ardıc, M., Sekmen, A.H., Tokur, S., Ozdemir, F., Turkan, I., 2009. Antioxidant responses of chickpea plants subjected to boron toxicity. Plant. Biol. 11 (3), 328–338.

Ayvaz, M., 2002. Bazı arpa çeşitlerinde borun büyüme ve gelişme üzerine etkileri (Doctoral dissertation, Yüksek Lisans Tezi, Biyoloji Anabilim Dalı).

Atique-ur, R., Farooq, M., Rashid, A., Nadeem, F., Stuerz, S., Asch, F., Bell, R.W., Siddique, K.H. M., 2018. Boron nutrition of rice in different production systems. A review. Agronomy for Sustainable Development 38 (3), 25. Available from: https://doi.org/10.1007/s13593-018-0504-8.

Ayvaz, M., Guven, A., Blokhina, O., Fagerstedt, K.V., 2016. Boron stress, oxidative damage and antioxidant protection in potato cultivars (*Solanum tuberosum* L.). Acta Agriculturae Scandinavica Sect. B—Soil & Plant. Sci. 66 (4), 302–316.

Bañuelos, G.S., Ajwa, H.A., Caceres, L., Dyer, D., 1999. Germination Responses and Boron Accumulation in Germplasm from Chile and the United States Grown with Boron-Enriched Water. Ecotoxicology and Environmental Safety 43 (1), 62–67. Available from: https://doi.org/10.1006/eesa.1999.1765.

Batar, T., Koksal, N.S., Yersel, S.E., 2009. Production and characterization of wall plaster with borax and paper wastes and perlite additives. Ekoloji 18 (72), 45–53.

Bellaloui, N., Yadavc, R.C., Chern, M.S., Hu, H., Gillen, A.M., Greve, C., et al., 2003. Transgenically enhanced sorbitol synthesis facilitates phloem-boron mobility in rice. Physiologia Plant. 117 (1), 79–84.

Ben-Gal, A., Shani, U., 2002. Yield, transpiration and growth of tomatoes under combined excess boron and salinity stress. Plant. Soil. 247 (2), 211–221.

Berger, K.C., 1949. Boron in soils and crops. Adv. Agron. 1, 321–351.

Bergmann, W., 1988. Ernährungsstörungen bei Kulturpflanzen (Nutrient Flow in Crop Plants). VEB Gustav Fischer Verlag, Jena (in German).

Bieleski, R.L., 1982. Sugar alcohols. In Plant Carbohydrates I. Springer, Berlin, Heidelberg, pp. 158–192.

Bilal, H.M., Aziz, T., Maqsood, M.A., Farooq, M., Yan, G., 2018. Categorization of wheat genotypes for phosphorus efficiency. PLoS one 13 (10), e0205471.

Blevins, D.G., Lukaszewski, K.M., 1998. Boron in plant structure and function. Annu. Rev. Plant. Biol. 49 (1), 481–500.

Boaretto, R.M., Quaggio, J.A., de Assis Alves Mourão Filho, F., Giné, M.F., Boaretto, A. E., 2008. Absorption and mobility of boron in young citrus plants. Commun. Soil. Sci. Plant. Anal. 39 (17–18), 2501–2514.

Bohnsack, C.W., Albert, L.S., 1977. Early effects of boron deficiency on indoleacetic acid oxidase levels of squash root tips. Plant. Physiol. 59 (6), 1047–1050.

Bolaños, L., Lukaszewski, K., Bonilla, I., Blevins, D., 2004. Why boron? Plant. Physiol. Biochem. 42 (11), 907–912.

Brdar-Jokanović, M., 2020. Boron toxicity and deficiency in agricultural plants. Int. J. Mol. Sci. 21 (4), 1424.

Brown, P.H., Bellaloui, N., Wimmer, M.A., Bassil, E.S., Ruiz, J., Hu, H., et al., 2002. Boron in plant biology. Plant. Biol. 4 (02), 205–223.

Brown, P.H., Hu, H., 1996. Phloem mobility of boron is species dependent: evidence for phloem mobility in sorbitol-rich species. Ann. Botany 77 (5), 497–506.

Brown, P.H., Shelp, B.J., 1997. Boron mobility in plants. Plant. Soil. 193 (1), 85–101.

Cakmak, I., Römheld, V., 1997. Boron deficiency-induced impairments of cellular functions in plants. Plant. Soil. 193 (1), 71–83.

Camacho-Cristóbal, J.J., Anzellotti, D., González-Fontes, A., 2002. Changes in phenolic metabolism of tobacco plants during short-term boron deficiency. Plant. Physiol. Biochem. 40 (12), 997–1002.

Camacho-Cristóbal, J.J., Martín-Rejano, E.M., Herrera-Rodríguez, M.B., Navarro-Gochicoa, M.T., Rexach, J., González-Fontes, A., 2015. Boron deficiency inhibits root cell elongation via an ethylene/auxin/ROS-dependent pathway in Arabidopsis seedlings. J. Exp. botany 66 (13), 3831–3840.

Camacho-Cristóbal, J.J., Navarro-Gochicoa, M.T., Rexach, J., González-Fontes, A., Herrera-Rodríguez, M.B., 2018. Plant response to boron deficiency and boron use efficiency in crop plants. Plant. Micronutrition Use Efficiency 109–121. Academic Press.

Camacho-Cristóbal, J.J., Rexach, J., González-Fontes, A., 2008. Boron in plants: deficiency and toxicity. J. Integr. Plant. Biol. 50 (10), 1247–1255.

Çatav, Ş.S., Genç, T.O., Oktay, M.K., Küçükakyüz, K., 2018. Effect of boron toxicity on oxidative stress and genotoxicity in wheat (*Triticum aestivum* L.). Bull. Environ. contamination Toxicol. 100 (4), 502–508.

Chatzissavvidis, C., Antonopoulou, C., 2020. Boron toxicity in fruit crops: agronomic and physiological implications. Fruit. Crop. 211–221. Elsevier.

Cheng, C., Rerkasem, B., 1993. Effects of boron on pollen viability in wheat. Plant. Soil. 155 (1), 313–315.

Choi, E.Y., Kolesik, P., McNeill, A.N.N., Collins, H., Zhang, Q., Huynh, B.L., et al., 2007. The mechanism of boron tolerance for maintenance of root growth in barley (*Hordeum vulgare* L.). Plant Cell & Environ. 30 (8), 984–993.

Clark, R.B., Zobel, R.W., Zeto, S.K., 1999. Effects of mycorrhizal fungus isolates on mineral acquisition by *Panicum virgatum* in acidic soil. Mycorrhiza 9 (3), 167–176.

Colak, B., Korkmaz, A., Horuz, A., 2013. Boron status, boron fractions and its availability in sugar beet grown soils. Anadolu Tarım Bilimleri Derg. 28 (3), 157–167.

Das, A.K., Purkait, A., 2020. Boron dynamics in soil: classification, sources, factors, fractions, and kinetics. Commun. Soil. Sci. Plant. Anal. 51 (22), 2778–2790.

Davarpanah, S., Tehranifar, A., Davarynejad, G., Abadía, J., Khorasani, R., 2016. Effects of foliar applications of zinc and boron nano-fertilizers on pomegranate (*Punica granatum* cv. Ardestani) fruit yield and quality. Sci. Horticulturae 210, 57–64.

De Cnodder, T., Vissenberg, K., Van Der Straeten, D., Verbelen, J.P., 2005. Regulation of cell length in the Arabidopsis thaliana root by the ethylene precursor 1-aminocyclopropane-1-carboxylic acid: a matter of apoplastic reactions. N. Phytol 168 (3), 541–550.

Debnath, P., Ghosh, S.K., 2014. Assessment of critical limit of available boron for pea in acidic alfisols of East Sikkim. India. Legume Res-An Int. J. 37 (5), 508–514.
Dell, B., Huang, L., 1997. Physiological response of plants to low boron. Plant. Soil. 193 (1), 103–120.
Dhaliwal, S.S., Naresh, R.K., Mandal, A., Singh, R., Dhaliwal, M.K., 2019. Dynamics and transformations of micronutrients in agricultural soils as influenced by organic matter build-up: A review. Environ. Sustainab Indic. 1, 100007.
Dobermann, A., 2000. Rice: Nutrient disorders & nutrient management. Int. Rice Res. Inst.
Dordas, C., Brown, P.H., 2000. Permeability of boric acid across lipid bilayers and factors affecting it. J. Membr. Biol. 175 (2), 95–105.
Du, W., Pan, Z.Y., Hussain, S.B., Han, Z.X., Peng, S.A., Liu, Y.Z., 2020. Foliar supplied boron can be transported to roots as a boron-sucrose complex via phloem in citrus trees. Front. Plant. Sci. 11, 250.
Eggert, K., von Wirén, N., 2017. Response of the plant hormone network to boron deficiency. N. Phytol. 216 (3), 868–881.
Ekbic, H.B., Gokdemir, N., Erdem, H., 2018. Effects of boron on yield, quality and leaf nutrients of Isabella (*Vitis labrusca* L.) grape cultivar. ActaScientiarumPolonorum-HortorumCultus 17 (1), 149–157.
Eraslan, F., Inal, A., Gunes, A., Alpaslan, M., 2007a. Boron toxicity alters nitrate reductase activity, proline accumulation, membrane permeability, and mineral constituents of tomato and pepper plants. J. Plant. Nutr. 30 (6), 981–994.
Eraslan, F., Inal, A., Savasturk, O., Gunes, A., 2007b. Changes in antioxidative system and membrane damage of lettuce in response to salinity and boron toxicity. Sci. Horticulturae 114 (1), 5–10.
Fang, K.F., Du, B.S., Zhang, Q., Xing, Y., Cao, Q.Q., Qin, L., 2019. Boron deficiency alters cytosolic Ca2 + concentration and affects the cell wall components of pollen tubes in *Malus domestica*. Plant. Biol. 21 (2), 343–351.
Ferreira, G.A., Hippler, F.W., Prado, L.A.D.S., Rima, J.A., Boaretto, R.M., Quaggio, J. A., et al., 2021. Boron modulates the plasma membrane H + -ATPase activity affecting nutrient uptake of citrus trees. Ann. Appl. Biol. 178 (2), 293–303.
Ferrol, N., Belver, A., Roldan, M., Rodriguez-Rosales, M.P., Donaire, J.P., 1993. Effects of boron on proton transport and membrane properties of sunflower (*Helianthus annuus* L.) cell microsomes. Plant. Physiol. 103 (3), 763–769.
Fleischer, A., O'Neill, M.A., Ehwald, R., 1999. The pore size of non-graminaceous plant cell walls is rapidly decreased by borate ester cross-linking of the pectic polysaccharide rhamnogalacturonan II. Plant. Physiol. 121 (3), 829–838.
García-Sánchez, F., Simón-Grao, S., Martínez-Nicolás, J.J., Alfosea-Simón, M., Liu, C., Chatzissavvidis, C., et al., 2020. Multiple stresses occurring with boron toxicity and deficiency in plants. J. Hazard. Mater. 397, 122713.
Ghanati, F., Morita, A., Yokota, H., 2002. Induction of suberin and increase of lignin content by excess boron in tobacco cells. Soil. Sci. Plant. Nutr. 48 (3), 357–364.
Glass, A.D., 1973. Influence of phenolic acids on ion uptake: I. Inhibition of phosphate uptake. Plant. Physiol. 51 (6), 1037–1041.
Goldbach, H.E., Hartmann, D., Rötzer, T., 1990. Boron is required for the stimulation of the ferricyanide-induced proton release by auxins in suspension-cultured cells of *Daucus carota* and *Lycopersicon esculentum*. Physiologia Plant. 80 (1), 114–118.
Goldbach, H.E., Wimmer, M.A., 2007. Boron in plants and animals: is there a role beyond cell-wall structure? J. Plant. Nutr. Soil. Sci. 170 (1), 39–48.
Goldbach, H.E., Yu, Q., Wingender, R., Schulz, M., Wimmer, M., Findeklee, P., et al., 2001. Rapid response reactions of roots to boron deprivation. J. Plant. Nutr. Soil. Sci. 164 (2), 173–181.

Goldberg, S., 1997. Reactions of boron with soils. Plant. Soil. 193 (1), 35–48.
Goldberg, S., Forster, H.S., Heick, E.L., 1993. Boron adsorption mechanisms on oxides, clay minerals, and soils inferred from ionic strength effects. Soil. Sci. Soc. Am. J. 57 (3), 704–708.
Goldberg, S., Suarez, D.L., 2011. Distinguishing boron desorption from mineral dissolution in arid-zone soils. Soil. Sci. Soc. Am. J. 75 (4), 1347–1353.
Goldberg, S., Suarez, D.L., 2012. Role of organic matter on boron adsorption-desorption hysteresis of soils. Soil. Sci. 177 (7), 417–423.
Gupta, U.C., Jame, Y.W., Campbell, C.A., Leyshon, A.J., Nicholaichuk, W., 1985. Boron toxicity and deficiency: a review. Can. J. Soil. Sci. 65 (3), 381–409.
Hajiboland, R., Bahrami-Rad, S., Bastani, S., 2013. Phenolics metabolism in boron-deficient tea [Camellia sinensis (L.) O. Kuntze] plants. Acta Biologica Hungarica 64 (2), 196–206.
Hajiboland, R., Farhanghi, F., 2010. Remobilization of boron, photosynthesis, phenolic metabolism and anti-oxidant defense capacity in boron-deficient turnip (*Brassica rapa* L.) plants. Soil. Sci. & Plant. Nutr. 56 (3), 427–437.
Hajiboland, R., Farhanghi, F., 2011. Effect of low boron supply in turnip plants under drought stress. Biologia Plant. 55 (4), 775.
Hajiboland, R., Farhanghi, F., Aliasgharpour, M., 2012. Morphological and anatomical modifications in leaf, stem and roots of four plant species under boron deficiency conditions. An. de. Biología 34, 15–29. Servicio de Publicaciones de la Universidad de Murcia.
Hale, M.G., Orcutt, D.M., 1987. The Physiology of Plants Under Stress. John Wiley & Sons.
Han, S., Chen, L.S., Jiang, H.X., Smith, B.R., Yang, L.T., Xie, C.Y., 2008. Boron deficiency decreases growth and photosynthesis, and increases starch and hexoses in leaves of citrus seedlings. J. Plant. Physiol. 165 (13), 1331–1341.
Hartmann, H.D., Flocker, W.J., Kofranek, A.M., 1981. Plant science. Growth, Development and Utilization of Cultivated Plants. Prentice-Hall Inc.
Herrera-Rodríguez, M.B., González-Fontes, A., Rexach, J., Camacho-Cristobal, J.J., Maldonado, J.M., Navarro-Gochicoa, M.T., 2010. Role of boron in vascular plants and response mechanisms to boron stresses. Plant. Stress. 4 (2), 115–122.
Hilal, N., Kim, G.J., Somerfield, C., 2011. Boron removal from saline water: a comprehensive review. Desalination 273 (1), 23–35.
Hu, H., Brown, P.H., 1994. Localization of boron in cell walls of squash and tobacco and its association with pectin (evidence for a structural role of boron in the cell wall). Plant. Physiol. 105 (2), 681–689.
Huang, J.H., Cai, Z.J., Wen, S.X., Guo, P., Ye, X., Lin, G.Z., et al., 2014. Effects of boron toxicity on root and leaf anatomy in two citrus species differing in boron tolerance. Trees 28 (6), 1653–1666.
Hussain, M., Khan, M.A., Khan, M.B., Farooq, M., Farooq, S., 2012. Boron application improves growth, yield and net economic return of rice. Rice Sci. 19 (3), 259–262.
Inal, A., Pilbeam, D.J., Gunes, A., 2009. Silicon increases tolerance to boron toxicity and reduces oxidative damage in barley. J. Plant. Nutr. 32 (1), 112–128.
Johnson, G.V., Fixen, P.E., 1990. Testing soils for sulfur, boron, molybdenum, and chlorine. Soil. Test. Plant. Anal. 3, 265–273.
Juan, J., Camacho-Cristóbal, D. Anzellotti and A. González-Fontes. 2002. Changes in phenolic metabolism of tobacco plants during short-term boron deficiency. J. Pl. Physiol. Biochem. 40:997-1002.
Kayıhan, C., Öz, M.T., Eyidoğan, F., Yücel, M., Öktem, H.A., 2017. Physiological, biochemical, and transcriptomic responses to boron toxicity in leaf and root tissues of contrasting wheat cultivars. Plant. Mol. Biol. Report. 35 (1), 97–109.

Klochko, K., Kaufman, A.J., Yao, W., Byrne, R.H., Tossell, J.A., 2006. Experimental measurement of boron isotope fractionation in seawater. Earth Planet. Sci. Lett. 248 (1–2), 276–285.

Kobayashi, M., Mutoh, T., Matoh, T., 2004. Boron nutrition of cultured tobacco BY-2 cells. IV. Genes. Induc. Low. Boron Supply. J. Exp. Botany 55 (401), 1441–1443.

Kothari, S.K., Marschner, H., Römheld, V., 1990. Direct and indirect effects of VA mycorrhizal fungi and rhizosphere microorganisms on acquisition of mineral nutrients by maize (*Zea mays* L.) in a calcareous soil. N. Phytol. 116 (4), 637–645.

Kouchi, H., Kumazawa, K., 1976. Anatomical responses of root tips to boron deficiency: III. Effect of boron deficiency on sub-cellular structure of root tips, particularly on morphology of cell wall and its related organelles. Soil. Sci. Plant. Nutr. 22 (1), 53–71.

Landi, M., Margaritopoulou, T., Papadakis, I.E., Araniti, F., 2019. Boron toxicity in higher plants: an update. Planta 250 (4), 1011–1032.

Lawrence, K., Bhalla, P., Misra, P.C., 1995. Changes in NAD (P) H-dependent redox activities in plasmalemma-enriched vesicles isolated from boron-and zinc-deficient chick pea roots. J. Plant. Physiol. 146 (5–6), 652–657.

Lehto, T., Lavola, A., Kallio, E., Aphalo, P.J., 2004. Boron uptake by ectomycorrhizas of silver birch. Mycorrhiza 14 (3), 209–212.

Lewis, D.H., 2019. Boron: the essential element for vascular plants that never was. New Phytol. 221 (4), 1685–1690.

Li, M., Zhao, Z., Zhang, Z., Zhang, W., Zhou, J., Xu, F., et al., 2017. Effect of boron deficiency on anatomical structure and chemical composition of petioles and photosynthesis of leaves in cotton (*Gossypium hirsutum* L.). Sci. Rep. 7 (1), 1–9.

Loomis, W.D., Durst, R.W., 1992. Chemistry and biology of boron. BioFactors (Oxford, Engl.) 3 (4), 229–239.

Lordkaew, S., Konsaeng, S., Jongjaidee, J., Dell, B., Rerkasem, B., Jamjod, S., 2013. Variation in responses to boron in rice. Plant. Soil. 363 (1), 287–295.

Lovatt, C.J., Dugger, W.M., 1984. Boron. Biochem. Essent. ultratrace Elem. 389–421.

Lu, S., Miller, M.H., 1989. The role of VA mycorrhizae in the absorption of P and Zn by maize in field and growth chamber experiments. Can. J. Soil. Sci. 69 (1), 97–109.

Lu, Y.B., Yang, L.T., Li, Y., Xu, J., Liao, T.T., Chen, Y.B., et al., 2014. Effects of boron deficiency on major metabolites, key enzymes and gas exchange in leaves and roots of citrus sinensis seedlings. Tree Physiol. 34 (6), 608–618.

Macho-Rivero, M.A., Herrera-Rodríguez, M.B., Brejcha, R., Schäffner, A.R., Tanaka, N., Fujiwara, T., et al., 2018. Boron toxicity reduces water transport from root to shoot in Arabidopsis plants. Evidence for a reduced transpiration rate and expression of major PIP aquaporin genes. Plant. Cell Physiol. 59 (4), 841–849.

Mahboobi, H., Yücel, M., Öktem, H.A., 2002. Nitrate reductase and glutamate dehydrogenase activities of resistant and sensitive cultivars of wheat and barley under boron toxicity. J. Plant. Nutr. 25 (8), 1829–1837.

Marschner, H., 2011. Marschner's Mineral Nutrition of Higher Plants. Academic press.

Martini, F., Thellier, M., 1993. Boron distribution in parenchyma cells of clover leaves. Plant. Physiol. Biochem. (Paris.) 31 (5), 777–786.

Mathew, J., Krishnakumar, V., Srinivasan, V., Bhat, R., Namboothiri, C.G., Haris, A.A., 2018. Standardization of critical boron level in soil and leaves of coconut palms grown in a tropical Entisol. J. Soil. Sci. Plant. Nutr. 18 (2), 376–387.

Matoh, T., 1997. Boron in plant cell walls. Plant. Soil. 193 (1), 59–70.

Matthes, M.S., Robil, J.M., McSteen, P., 2020. From element to development: the power of the essential micronutrient boron to shape morphological processes in plants. J. Exp. botany 71 (5), 1681–1693.

Mei, L., Li, Q., Wang, H., Sheng, O., Peng, S.A., 2016. Boron deficiency affects root vessel anatomy and mineral nutrient allocation of *Poncirus trifoliata* (L.) Raf. Acta physiologiae Plant. 38 (4), 1–8.

Mengel, K., Kirkby, E.A., 2012. Principles of Plant Nutrition. Springer Science & Business Media.

Mesquita, G.L., Zambrosi, F.C., Tanaka, F.A., Boaretto, R.M., Quaggio, J.A., Ribeiro, R.V., et al., 2016. Anatomical and physiological responses of citrus trees to varying boron availability are dependent on rootstock. Front. Plant. Sci. 7, 224.

Miwa, K., Fujiwara, T., 2010. Boron transport in plants: co-ordinated regulation of transporters. Ann. Botany 105 (7), 1103–1108.

Mousavi, S.M., 2021. Silicon and nano-silicon mediated heavy metal stress tolerance in plants. In: Etesami, H., Anwar Hossain, M., Al-Saeedi, A.H., El-Ramady, H., Fujita, M., Pessarakli, M. (Eds.), Silicon and Nano-Silicon in Environmental Stress Management And Crop Quality Improvement: Recent Progress and Future Prospects. Elsevier publications.

Mousavi, S.M., Moshiri, F., Moradi, S., 2018a. Mobility of heavy metals in sandy soil after application of composts produced from maize straw, sewage sludge and biochar: discussion of Gondek et al. (2018). J. Environ. Manag. 222, 132–134.

Mousavi, S.M., Motesharezadeh, B., 2020. Boron deficiency in fruit crops. In: Srivastava, A.K., Hu, C. (Eds.), Fruit Crops. Elsevier, pp. 191–209.

Mousavi, S.M., Motesharezadeh, B., Hosseini, H.M., Alikhani, H., Zolfaghari, A.A., 2018b. Root-induced changes of Zn and Pb dynamics in the rhizosphere of sunflower with different plant growth promoting treatments in a heavily contaminated soil. Ecotoxicol. Environ. Saf. 147, 206–216.

Mousavi, S.M., Motesharezadeh, B., Hosseini, H.M., Alikhani, H., Zolfaghari, A.A., 2018c. Geochemical fractions and phytoavailability of zinc in a contaminated calcareous soil affected by biotic and abiotic amendments. Environ. Geochem. Health 40 (4), 1221–1235.

Mozafar, A., 1989. Boron effect on mineral nutrients of maize. Agron. J. 81 (2), 285–290.

Mukhopadhyay, M., Ghosh, P.D., Mondal, T.K., 2013. Effect of boron deficiency on photosynthesis and antioxidant responses of young tea plantlets. Russian J. Plant. Physiol. 60 (5), 633–639.

Nachiangmai, D., Dell, B., Bell, R., Huang, L., Rerkasem, B., 2004. Enhanced boron transport into the ear of wheat as a mechanism for boron efficiency. Plant. Soil. 264 (1), 141–147.

Naz, T., Akhtar, J., Iqbal, M.M., ul Haq, M.A., Saqib, M., 2016. Boron toxicity in salt-affected soils and effects on plants. Soil, Science: Agricultural Environmental, Prospectives. Springer, Cham, pp. 259–286.

Oikonomou, A., Ladikou, E.V., Chatziperou, G., Margaritopoulou, T., Landi, M., Sotiropoulos, T., et al., 2019. Boron excess imbalances root/shoot allometry, photosynthetic and chlorophyll fluorescence parameters and sugar metabolism in apple plants. Agronomy 9 (11), 731.

O'Neill, M.A., Ishii, T., Albersheim, P., Darvill, A.G., 2004. Rhamnogalacturonan II: structure and function of a borate cross-linked cell wall pectic polysaccharide. Annu. Rev. Plant. Biol. 55, 109–139.

Padbhushan, R., Kumar, D., 2017. Fractions of soil boron: a review. J. Agric. Sci. 155 (7), 1023–1032.

Papadakis, I.E., Dimassi, K.N., Bosabalidis, A.M., Therios, I.N., Patakas, A., Giannakoula, A., 2004. Effects of B excess on some physiological and anatomical parameters of 'Navelina' orange plants grafted on two rootstocks. Environ. Exp. Botany 51 (3), 247–257.

Papadakis, I.E., Tsiantas, P.I., Tsaniklidis, G., Landi, M., Psychoyou, M., Fasseas, C., 2018. Changes in sugar metabolism associated to stem bark thickening partially assist young tissues of *Eriobotrya japonica* seedlings under boron stress. J. Plant. Physiol. 231, 337–345.

Patil, G.D., Patil, M.D., Patil, N.D., Adsule, R.N., 1987. Effect of boronated superphosphate, single superphosphate and borax on yield and quality of groundnut. J. Maharashtra Agricult Univ. 12, 168–170.

Paull, J.G., Cartwright, B., Rathjen, A.J., 1988. Responses of wheat and barley genotypes to toxic concentrations of soil boron. Euphytica 39 (2), 137–144.

Pereira, G.L., Siqueira, J.A., Batista-Silva, W., Cardoso, F.B., Nunes-Nesi, A., Araújo, W. L., 2021. Boron: more than an essential element for land plants? Front. Plant. Sci. 11, 2234.

Pollard, A.S., Parr, A.J., Loughman, B.C., 1977. Boron in relation to membrane function in higher plants. J. Exp. Botany 28 (4), 831–841.

Quiles-Pando, C., Navarro-Gochicoa, M.T., Herrera-Rodríguez, M.B., Camacho-Cristóbal, J.J., González-Fontes, A., Rexach, J., 2019. Boron deficiency increases cytosolic Ca2 + levels mainly via Ca2 + influx from the apoplast in *Arabidopsis thaliana* roots. Int. J. Mol. Sci. 20 (9), 2297.

Raiesi, T., Moradi, B., 2021. Young navel orange rootstock improves phosphorus absorption from poorly soluble pools through rhizosphere processes. Rhizosphere 17, 100316.

Rashid, A., Rafique, E., Bughio, N., 1994. Diagnosing boron deficiency in rapeseed and mustard by plant analysis and soil testing. Commun. Soil. Sci. Plant. Anal. 25 (17–18), 2883–2897.

Reid, R., 2007. Update on boron toxicity and tolerance in plants. Adv. Plant. Anim. Boron Nutr. 83–90.

Reid, R.J., Hayes, J.E., Post, A., Stangoulis, J.C.R., Graham, R.D., 2004. A critical analysis of the causes of boron toxicity in plants. Plant Cell Environ. 27 (11), 1405–1414.

Rerkasem, B., Jamjod, S., 1997. Genotypic variation in plant response to low boron and implications for plant breeding. Plant. Soil. 193 (1), 169–180.

Robertson, G.A., Loughman, B.C., 1974. Response to boron deficiency: a comparison with responses produced by chemical methods of retarding root elongation. N. Phytol 73 (5), 821–832.

Roessner, U., Patterson, J.H., Forbes, M.G., Fincher, G.B., Langridge, P., Bacic, A., 2006. An investigation of boron toxicity in barley using metabolomics. Plant. Physiol. 142 (3), 1087–1101.

Ruuhola, T., Lehto, T., 2014. Do ectomycorrhizas affect boron uptake in Betula pendula? Can. J. For. Res. 44 (9), 1013–1019.

Ryden, P., Sugimoto-Shirasu, K., Smith, A.C., Findlay, K., Reiter, W.D., McCann, M. C., 2003. Tensile properties of Arabidopsis cell walls depend on both a xyloglucan cross-linked microfibrillar network and rhamnogalacturonan II-borate complexes. Plant. Physiol. 132 (2), 1033–1040.

Samet, H.A.L.İ.L., Çıkılı, Y., 2019. Response of purslane (*Portulaca oleracea* L.) to excess boron and salinity: physiological approach. Russian J. Plant. Physiol. 66 (2), 316–325.

Sarkar, D., Mandal, B., Mazumdar, D., 2008. Plant availability of boron in acid soils as assessed by different extractants. J. Plant. Nutr. Soil. Sci. 171 (2), 249–254.

Savić, J., Marjanović-Jeromela, A., 2013. Glamočlija, Đ., Prodanović, S. Oilseed Rape Genotypes Response Boron Toxicity. Genetika-Belgrade 45 (2), 565–574.

Schon, M.K., Novacky, A., Blevins, D.G., 1990. Boron induces hyperpolarization of sunflower root cell membranes and increases membrane permeability to K. Plant. Physiol. 93 (2), 566–571.

Sener, S., Özkara, M., 1989. The boron pollution of Simav creek waters and its effects on the soils and agricultural srops of Balikesir region. Plants Pollut. Develop Ctries. Ege University, Bornova, Izmir.
Shaaban, M.M., 2010. Role of boron in plant nutrition and human health. Am. J. Plant. Physiol. 5 (5), 224–240.
Shaffer, D.L., Yip, N.Y., Gilron, J., Elimelech, M., 2012. Seawater desalination for agriculture by integrated forward and reverse osmosis: improved product water quality for potentially less energy. J. Membr. Sci. 415, 1–8.
Shah, A., Wu, X., Ullah, A., Fahad, S., Muhammad, R., Yan, L., et al., 2017. Deficiency and toxicity of boron: alterations in growth, oxidative damage and uptake by citrange orange plants. Ecotoxicol. Environ. Saf. 145, 575–582.
Shen, Z., Liang, Y., Shen, K., 1993. Effect of boron on the nitrate reductase activity in oilseed rape plants. J. Plant. Nutr. 16 (7), 1229–1239.
Shireen, F., Nawaz, M.A., Chen, C., Zhang, Q., Zheng, Z., Sohail, H., et al., 2018. Boron: functions and approaches to enhance its availability in plants for sustainable agriculture. Int J Mol Sci 19 (7), 1856.
Somerville, C., Bauer, S., Brininstool, G., Facette, M., Hamann, T., Milne, J., et al., 2004. Toward a systems approach to understanding plant cell walls. Science 306 (5705), 2206–2211.
Souri, M.K., Hatamian, M., 2019. Aminochelates in plant nutrition: a review. J. Plant. Nutr. 42 (1), 67–78.
Stangoulis, J., Tate, M., Graham, R., Bucknall, M., Palmer, L., Boughton, B., et al., 2010. The mechanism of boron mobility in wheat and canola phloem. Plant. Physiol. 153 (2), 876–881.
Sun, A., Gou, D., Dong, Y., Xu, Q., Cao, G., 2019. Extraction and analysis of available boron isotopes in soil using multicollector inductively coupled plasma mass spectrometry. J. Agric. food Chem. 67 (25), 7183–7189.
Tabata, S.I., Hirakimoto, T., Nishiura, M., Watanabe, M., 2003. Synthesis of a Lewis-acidic boric acid ester monomer and effect of its addition to electrolyte solutions and polymer gel electrolytes on their ion transport properties. Electrochim. Acta 48 (14–16), 2105–2112.
Taiz, L., Zeiger, E., Møller, I.M., Murphy, A., 2015. Plant. Physiol. Dev. (No. (Ed.) 6). Sinauer Associates Incorporated.
Takano, J., Miwa, K., Fujiwara, T., 2008. Boron transport mechanisms: collaboration of channels and transporters. Trends Plant. Sci. 13 (8), 451–457.
Tsiantas, P.I., Papadakis, I.E., Tsaniklidis, G., Landi, M., Psychoyou, M., 2019. Allocation pattern, nutrient partitioning, sugar metabolism, and pigment composition in hydroponically grown loquat seedlings subjected to increasing boron concentrations. J. Soil. Sci. Plant. Nutr. 19 (3), 556–564.
Turan, M.A., Taban, N., Taban, S., 2009. Effect of calcium on the alleviation of boron toxicity and localization of boron and calcium in cell wall of wheat. Not. Botanicae Horti Agrobotanici Cluj-Napoca 37 (2), 99–103.
Uygan, D., Çetin, Ö., 2004. Bor'un Tarımsal ve Çevresel Etkileri: Seydisuyu Su Toplama Havzası [Agricultural and environmental effects of boron: seydisuyu water deposit]. Uluslar. Bor. Sempozyumu 23–25.
Vajari, M.A., Eshghi, S., Fatahi Moghadam, J., 2018. Late-season foliar application of mineral compounds effects on postharvest quality of Hayward kiwifruit. J. Berry Res. 8 (2), 95–107.
Versar Inc, 1975. Preliminary investigation of effects on the environment of boron, indium, nickel, selenium, tin, vanadium and their compounds. Boron, Versar, Inc, Springfield, Virginia.

Vishekaii, Z.R., Soleimani, A., Fallahi, E., Ghasemnezhad, M., Hasani, A., 2019. The impact of foliar application of boron nano-chelated fertilizer and boric acid on fruit yield, oil content, and quality attributes in olive (*Olea europaea* L.). Sci. Horticulturae 257, 108689.

Voiniciuc, C., Pauly, M., Usadel, B., 2018. Monitoring polysaccharide dynamics in the plant cell wall. Plant. Physiol. 176 (4), 2590–2600.

Wang, D.N., Ko, W.H., 1975. Relationship between deformedfruit disease of papaya and boron deficiency. Phytopathology 65 (4), 445–447.

Wang, N., Yan, T., Fu, L., Zhou, G., Liu, Y., Peng, S.A., 2014. Differences in boron distribution and forms in four citrus scion–rootstock combinations with contrasting boron efficiency under boron-deficient conditions. Trees 28 (6), 1589–1598.

Wang, N., Yang, C., Pan, Z., Liu, Y., Peng, S.A., 2015. Boron deficiency in woody plants: various responses and tolerance mechanisms. Front. Plant. Sci. 6, 916.

Wang, Z.Y., Tang, Y.L., Zhang, F.S., Wang, H., 1999. Effect of boron and low temperature on membrane integrity of cucumber leaves. J. Plant. Nutr. 22 (3), 543–550.

Wimmer, M.A., Abreu, I., Bell, R.W., Bienert, M.D., Brown, P.H., Dell, B., et al., 2019. Boron: an essential element for vascular plants. N. Phytologist 226 (5), 1232–1237.

Wojcik, P., Wojcik, M., Klamkowski, K., 2008. Response of apple trees to boron fertilization under conditions of low soil boron availability. Sci. horticulturae 116 (1), 58–64.

Wojcik, P., Wojcik, M., Treder, W., 2003. Boron absorption and translocation in apple rootstocks under conditions of low medium boron. J. Plant. Nutr. 26 (5), 961–968.

Wolska, J., Bryjak, M., 2013. Methods for boron removal from aqueous solutions—a review. Desalination 310, 18–24.

Wu, X., Riaz, M., Yan, L., Du, C., Liu, Y., Jiang, C., 2017. Boron deficiency in trifoliate orange induces changes in pectin composition and architecture of components in root cell walls. Front. Plant. Sci. 8, 1882.

Yau, S.K., Nachit, M.M., Ryan, J., Hamblin, J., 1995. Phenotypic variation in boron-toxicity tolerance at seedling stage in durum wheat (*Triticum durum*). Euphytica 83 (3), 185–191.

Yu, Q., Baluška, F., Jasper, F., Menzel, D., Goldbach, H.E., 2003. Short-term boron deprivation enhances levels of cytoskeletal proteins in maize, but not zucchini, root apices. Physiologia Plant. 117 (2), 270–278.

Zakir Hossain, A.K.M., Asgar, M.A., Hossain, M.A., Tosaki, T., Koyama, H., Hara, T., 2005. Boron-calcium synergically alleviates aluminum toxicity in wheat plants (*Triticum aestivum* L.). Soil. Sci. Plant. Nutr. 51 (1), 43–49.

Zheng, Y., Wu, J., Peng, X., Zhang, Y., 2019. Field-grown Moringa oleifera response to boron fertilization: Yield component, chemical composition of seed-oil and physiology. Ind. Crop. Products 138, 111449.

Zhou, G.F., Peng, S.A., Liu, Y.Z., Wei, Q.J., Han, J., Islam, M.Z., 2014. The physiological and nutritional responses of seven different citrus rootstock seedlings to boron deficiency. Trees 28 (1), 295–307.

Zhou, T., Hua, Y., Huang, Y., Ding, G., Shi, L., Xu, F., 2016. Physiological and transcriptional analyses reveal differential phytohormone responses to boron deficiency in Brassica napus genotypes. Front. Plant. Sci. 7, 221.

CHAPTER 2

Boron in fruit crops: plant physiology, deficiency, toxicity, and sources for fertilization

Renato Vasconcelos Botelho[1], Marcelo Marques Lopes Müller[1], Renan Caldas Umburanas[1], James Matheus Ossacz Laconski[1] and Maurilo Monteiro Terra[2]
[1]Department of Agronomy, State University of Midwestern of Paraná, Guarapuava, Paraná, Brazil
[2]Agronomic Institute of Campinas, Campinas, São Paulo, Brazil

2.1 Introduction

Boron (B) is a plant essential micronutrient and its deficiency occurs in many agricultural regions of the world (Wang et al., 2015). To overcome its deficiency in plants, when it is necessary, its direct application to soil or foliar fertilization has been used. The effectiveness of the foliar application of B has been reported in several fruit crops, such as citrus (*Citrus* sp) (Boaretto et al., 2011; Macedo et al. 2020; Wu et al., 2020), lychee (*Litchi chinensis*) (Will et al., 2012), almond (*Prunus dulcis*) (Nyomora et al., 1997, 2000), olive (*Olea europaea*) (Perica et al., 2001), Pistachio (*Pistacia vera*) (Torun et al. 2021), grapevine (*Vitis vinifera*) (Terra, 2003), and coffee (*Coffea* sp) (Rosolem et al., 2020).

The functions of boron are related to those of calcium, that is, regulation of membrane and cell wall functioning, cell division and increase, flower retention as well as favoring the pollen grain germination (Fregoni, 1980). B improves the rigidity of the plant cell wall, its strength and shape as well as plasma membrane integrity and permeability (Broadley et al., 2012). B is also associated with the synthesis of lignin and phenolic compounds (Marschner, 2012). These attributes suggest that plants well-nourished with B have a greater defense against pathogens (Marschner, 2012; Dordas, 2008). In contrast, plants that are malnourished in B, however, presents impaired structural integrity of their cell membranes, becoming subject to cellular leakage and, therefore, releasing organic compounds (e.g., amino acids, carbohydrates, mainly) to the outside of the cell. This can serve as a substrate for the development and spread of plant pathogens (Huber et al., 2012). Some studies that reported effect of B on the mitigation of diseases in fruit crops are reported in Table 2.1.

Boron in Plants and Agriculture
DOI: https://doi.org/10.1016/B978-0-323-90857-3.00015-1

Table 2.1 Relationship between diseases in fruit crops and the presence of the nutrient boron reported in different studies.

Disease/ Physiological disorder	Causal organism	Crop	Presence of nutrient	References
Eutypa dieback	*Eutypa lata*	Grapevine	Resistance increased with B	Rolshausen and Gubler (2005)
Tomato yellow leaf curl virus	TYLCV	Tomato	Reduction in disease with B application	Graham and Webb (1991)
Blue rot of orange	*Penicillium digitatum*	Orange	Reduction with B treatment before storage	Tarabih, El-Metwelly (2014)
Fusarium wilt	*Fusarium oxysporum*	Banana	B and Fe play a reduced the severity	Dong et al. (2016)
—	*Botrytis cinerea*	—	In culture medium, B inhibited spore germination, elongation and mycelial growth	Qin et al. (2010)

The availability of B in the soil depends on several factors such as soil pH, organic matter content, iron and aluminum oxides and clay mineral contents. Regions with high precipitation associated with sandy soils are favorable for plants to present B deficiencies. The same is true for acidic soils recently submitted to liming.

In addition to the soil factors, different plant species present a wide variation in nutritional requirements of B. To a lesser extent, such a difference also occurs between genotypes within the same plant species. Apple (*Malus domestica* Borkh.) is a crop well known for its high requirement of B, and among the fruit crops, the most studied in terms of responsiveness to fertilization with B are apple and pear (Shorrocks, 1997).

The purpose of this chapter is to review some aspects related to B roles in physiology and transport, B toxicity and deficiency symptoms on fruit crops, as well as B in soils and sources for B fertilization.

2.2 Boron roles in physiology and transport

Boron participates in plant physiological processes such as sugar transport, synthesis and structure of the cell wall, plant respiration, and metabolism of carbohydrates. Also, B participates in the synthesis of RNA, indole acetic acid, phenolic compounds; in the nitrogen fixation and reduction of aluminum toxicity (Epstein and Bloom, 2006; Taiz et al., 2017).

The mobility of B depends on the plant species as well as the genotype, due to the occurrence of B-sucrose complex (Du et al., 2020) or polyols (Brown and Hu, 1996) in the phloem, whereas in plant species that is immobile in the phloem, it occurs because of complexation with pectin of the cell wall (Hu et al., 1996). Among the possible mechanisms of foliar absorption of nutrients are the cuticle, as well as its irregularities, the stomata, trichomes, veins and bundle sheath extensions (Fernández et al., 2021), but for the absorption of B this is not yet fully understood.

B acts in several stages of reproductive growth, for example, it acts in the germination of pollen grains, pollen tube growth and formation of seeds and cell walls (Nava et al., 2009). In apple leaves grown in a culture medium, B increased the activity of the enzymes peroxidase, catalase, and dismutase (Sotiropoulos et al., 2006). In a field study soil and foliar application of B in apple plants improved the activity of catalase and glutathione reductase enzymes in leaves (Wojcik et al., 2008). These enzymes are antioxidants and are related to the reduction of oxidative stress in plants.

Foliar application of B has specificities depending on the crop. For example, in lychee plants, foliar absorption of B is threefold greater in the abaxial part compared to the adaxial part of the leaf (Will et al., 2012). Also, it was observed that the absorption of B applied to coffee leaves occurs in greater intensity when compared to cotton leaves (Rosolem et al. 2020). In sweet cherry (*Prunus avium* L.), foliar application of B increased the content of photosynthetic pigments (chlorophylls and carotenoids) (Thurzó et al., 2010). In pear trees (*Pyrus pyrifolia*), foliar application of B increased pollen germination and pollinic tube length (Lee et al., 2009). The same was observed in trees of plum (*Prunus domestica* L.) (Gravite et al., 2017), apple (*Malus domestica*) (Sá et al., 2014), loquat (*Eriobotrya japonica* Lindl.) (Nogueira et al., 2015), and quince (*Cydonia oblonga* Mill.) (Zambon et al., 2014).

The greater growth of the pollen tube in response to the application of B is associated with the formation of the boron-polysaccharide complex, which migrates through membranes to the synthesis sites since the precursors of the pollinic tube are polysaccharides and glycoproteins (Lee et al., 2009).

The foliar application of boric acid associated with sugar alcohol was reported to increase foliar absorption and translocation compared to boric acid alone, in coffee and cotton plants (Rosolem et al. 2020). Although the sugars-alcohols improve the penetration of foliar application, in Lychee the B remained immobile even in this association (Will et al., 2012). B is related to the metabolism and incorporation of calcium (Ca) in the cell wall (Ganie et al., 2013), and plants malnourished with boron have less extensibility of the cell wall, making them fragile (Nyomora et al., 2000).

2.3 Boron toxicity and deficiency symptoms

B is an essential micronutrient for plants, but in high concentrations, it can be toxic to plant cells. Deficiency of B is usually apparent in reproductive tissues, which shows a relationship with its mobility in the phloem, as well as its high demand. In fruit plants, especially, B deficiency is commonly associated with low fixation of flowers and fruits, which directly affects the agricultural yield of orchards (Table 2.2).

The concentration range between deficiency and toxicity for B is narrower compared to other essential nutrients (Marschner, 2012), thus the toxicity and deficiency of B has been reported in different plant species,

Table 2.2 Symptoms of B deficiency in different fruit species.

Crop	Symptoms (physiological disorder)	References
Apple (*Malus domestica*)	Bitter pit. B absence reduces Ca translocation.	Fang et al. (2016)
Pomelo (*Citrus grandis*) and Sweet orange (*Citrus sinensis*)	Low vigor. Cracking and yellowing of leaves, as well as a reduction in leaf area and plant biomass.	Zhou et al. (2020); Han et al. (2008)
Sweet orange (*Citrus sinensis*) and acid lime (*Citrus aurantifolia* Swingle)	Low vigor. Reduces flower and fruit fixation.	Noor et al. (2019); Yadav et al. (2020)
Guava (*Psidium guajava* L.)	Reduces flower and fruit fixation.	Baranwal et al. (2017).
Sweet Passion fruit (*Passiflora alata*)	Low vigor. New leaves with curling on the edges, shorter internodes and branches length, apical bud necrosis and flower malformation.	Freitas et al. (2011)
Blackberry (*Rubus* sp.)	Low vigor. Death of the apical meristem, chlorosis in new leaves and canopy wrinkling.	Souza et al. (2015)
Walnut (*Carya illinoinensis* W.)	Low vigor. Reduces flower and fruit fixation.	Wells et al. (2008)
Peach (*Prunus persica*)	Reduces flower and fruit fixation.	Nava et al. (2009)
Red raspberry (*Rubus idaeus* L.)	Less fruiting, productivity, fruits with less soluble solids and firmness.	Wojcik (2005)
Pineapple (*Ananas comosus*)	Deformed and smaller fruits, with cortical outgrowths between the berries.	Siebeneichler et al. (2008)
Pineapple (*Ananas comosus*)	Fruits with lower mass, juice volume and vitamin C content.	Cunha et al. (2019)
Grapevine (*V. labrusca* x *V. vinifera*)	Leaf chlorosis, internerval browning of the young fruits and malformed clusters, with normal-sized berries and small ones.	Kuniyuki et al. (1985)

however, it varies widely among species and/or genotypes (Chen et al., 2012; Huang et al. 2016). According to Askew (1944) and Cook et al. (1960), boron is basically not mobile inside of the plant, therefore the first deficiency symptoms begin to be visible show on younger tissues of the vine.

In apples (*Malus domestica*), high B concentrations inhibit pollen grain germination and pollen tube growth, as well as causes abnormalities in pollen tube morphology due to its functions on structural and regulatory role with Ca^{2+}, actin cytoskeleton and components of the cell wall (Fang et al., 2016).

In grapevine, B deficiency is associated with leaf concentrations below 25 mg/kg (Fregoni, 1980), whereas the optimum range is reported to be between 25 and 41 mg/kg (Fregoni and Scienza, 1976) or 15–38 mg/kg (Fregoni, 1980) or 45–53 mg/kg (Terra et al., 2007). Possibly this small difference between these studies is due to differences between the evaluated genotypes as well as production environments, which were different. Some symptoms of B deficiency in the grapevine are presented in Fig. 2.1. According to Terra (2003), for the vine is more appropriate to use the concentration ranges for leaf analysis interpretation (Table 2.2).

In grapevine leaves with symptoms of boron toxicity, a B concentration of 424 mg/kg was reported, 7.6 times higher when compared to the concentration of leaves without toxicity symptoms, of 56 mg/kg (Terra et al., 2007) (Fig. 2.2). It was also reported toxic levels in grapevine leaves with B concentration higher than 150 mg/kg (Fregoni, 1980) and 200 mg/kg (Winkler et al., 1974).

The symptoms of toxicity in grapevines are reported to initiate by yellowing of the leaf blade and necrotic spots on the leaf margins, which coalesce to form a continuous lesion from the margins to the center of the leaf as the symptom progresses (Terra et al., 2007). Especially in grapevine leaves of the cultivar "Italia," it was characterized that B toxicity is associated with symptoms of chlorosis, blight and curling of the leaf edges (Terra et al., 2007).

2.4 Boron availability to plants, fertilization, and boron sources

There are, at least, four different forms of B in the soil. Among the minerals (I), tourmaline is the main soil mineral of B, being poorly soluble and resistant to weathering, which makes the release of B very slow and

Figure 2.1 Boron sufficiency (+B) and deficiency (−B) symptoms of in young leaves (A, B) and berries (B, C, D, E) of "Italia" grapevine. *Courtesy of Dr. Maurilo Monteiro Terra.*

leaves only a small fraction (≤ 5%) of the total B in the soil available to plants (Havlin et al., 2016). The B from organic matter (II) is gradually released with mineralization, and therefore less is lost through leaching (Kot et al., 2012), making organic matter the main source of soil B for plants.

Once released, B can be adsorbed (III) to the surface of clay minerals and iron (Fe) and aluminum (Al) oxides, as well as to the organic matter

Figure 2.2 Boron toxicity symptoms of in (A, B) banana leaves and in (C, D) young leaves of "Italia" grapevine. *For (A, B) courtesy of Dr. Edson Shigueaki Nomura; For (C, D) Dr. Maurilo Monteiro Terra.*

compounds, or it may remain in soil solution (IV) in dynamic equilibrium with the solid phase, in the form of boric acid [H_3BO_3] mainly but also as borate [$B(OH)_4^-$], both forms being subject to leaching, especially H_3BO_3 (zero net charges) (Goldberg, 2014). With soil pH $\leq$ 7.0, the predominant H_3BO_3 present relatively low affinity to the clay, whereas with soil pH > 7.0 $B(HO)_4^-$ forms increases quickly and reaches maximum adsorption around pH 9.0 (Elrashidi and O'Connor, 1982), as well as it is adsorbed to iron oxides through the bond exchange mechanism with OH^- (Goldberg, 2014).

The average levels of available B in the soil are 10–20 ppm (Woods, 1994). The critical level of B in the soil depends on the soil texture, pH, and organic matter content, as well as on the climatic conditions, which interfere in the processes of leaching and organic matter decomposition. There is no consensus, but Sims and Johnson (1991) summarized the soil critical levels of B extracted in hot water (HW-B), the most common method to predict the availability of B to the plants, between 0.1 and 2.0 mg/kg of soil.

Levels of 0.5–3.0 mg/kg of B in the soil are considered adequate depending on the crops (Bell, 1997). Wojcik and Wojcik (2003) proposes 0.45 mg/kg as a critical level of HW-B in the soil for most fruit crops in temperate climates. As hidden hunger or visual symptoms of deficiency may occur below the critical level, B fertilization should be considered when values close to 0.45 mg/kg of B occurs in soil test results. Raij et al. (1997) consider the limits of boron contents (mg/dm^3 in hot water) in soils for several crops, namely, 0 –0.20 as low, 0.21–0.60 as medium and greater than 0.60 as high.

Plants take up B from the soil solution as $B(OH)_3$ (boric acid) (Dordas et al., 2000), which in plant apoplast (pH 5.5) and cytoplasm (pH 7.5) is >99.95% and >98%, respectively, of B free forms (Marschner, 2012). Critical B concentrations in new leaves range from 2 to 20 mg/kg, and concentrations for toxicity usually vary in the range of 50–200 $mg_{[B]}$/kg depending on plant species, and Ca:B relation in leaf tissue is an important conditioning factor for B deficiency-toxicity nutritional state (Gupta, 1993).

2.5 Boron mineral sources and fertilization

There are two ways to supply B to crops: one is through the application to the soil, and the other is through foliar spraying. The soil fertilization can be done using borate fertilizers, but also by adding these B sources to mixed fertilizers, the so-called NPK formulations, which results in greater ease of application and uniformity of distribution in the field, once the rates of micronutrients are relatively low (Reetz, 2016).

Soil application is the most used method for B fertilization, but as soil-applied B may remain mostly in the roots instead of translocating to the above-ground part of the trees (Wojcik et al., 2008), most growers also use foliar sprays for B maintenance (Nagy et al., 2011).

Terra (2003) recommends, as a preventive treatment, spray application of a solution containing 1 g of boric acid per liter of water, applied three times before bloom, every 7 days, for "Niagara Rosada" grapevine. Also, it is possible to apply 10–20 g of borax per plant in the soil, right after pruning as a preventive measure.

Sources of B for B-fertilizer production are derived from mineral deposits, at arid sites with a history of volcanism or hydrothermal activity, where B occurs as borates. Deposits with ore of high B concentration and economically viable size are rare, being explored in Turkey, the United States, and some other parts of the world (Matterson, 1980).

Most common borate bearing minerals are ulexite, borax (tincal ore), natural boric acid (sassolite), colemanite, and kernite (Brdar-Jokanović, 2020). Borax and boric acid are commonly submitted to refinement and are widely used in the production of solid and liquid fertilizers for their high solubility in water. Also, both are also used for foliar spraying and in fertigation, whereas colemanite is moderately soluble in water, and ulexite is insoluble in water, and are more adequate for situations with high leaching potential, as light texture soils in rainy regions (Lopes, 1999; Woods, 1994).

The B rates for soil application varies depending on the crop, but also on the soil and climate conditions, ranging from 0.25 to 3.0 kg B/ha (Mortvedt and Woodruff, 1993) and from 0.5 to 2.0 kg B/ha (Reetz, 2016). In fruit crops, the rate of application may also vary from 2 to 3 g of B/tree (Nagy et al., 2011).

It is recommended to analyze the leaf tissue periodically (e.g., 4 years) in case of normality, and annually if nutritional symptoms are observed. Use soil analysis concurrently with tissue analysis to establish fertilizer recommendations. Leaf B sufficiency range of several fruit crops are presented in Table 2.3.

The exportation of B varies among different species, and some of them are presented in Table 2.4. Estimating long-term nutrient export contributes to maintenance fertilization strategies. The recommendation for fertilization with B for some fruit species is presented in Tables 2.5 and 2.6.

Table 2.3 Boron contents ranges (mg/kg) on complete leaf (petiole + leaf blade), on the petiole and leaf blade, collected at the bloom of more recent mature leaves of the vine productive branch.

Ranges of boron contents	Complete leaf	Petiole	Leaf blade
Deficiency	< 20	<10	<10
Rapid deficiency	20–44	10–34	10–34
Optimum	45–53	35–43	35–43
Slight excess	54–100	44–90	44–90
Excess	> 100	> 90	> 90

Table 2.4 Leaf sufficiency range of boron (B) on fruit crops reported in literature.

Crop	B (mg/kg)	Reference	Diagnostic leaves sampling description
Avocado (*Persea americana*)	50–100	Raij et al. (1997)	In early spring, collect newly expanded leaves between 5 and 7 months old, from the average height of the tree canopy.
	< 35 insufficient 40–100 normal > 150 excessive	CQFS (2004)	In early winter collect mature complete leaves, aged between 5 and 7 months, which originated from spring shoots. Collect four to eight leaves per tree, at an approximate height of 1.5–2 m from the ground in the four quadrants of the canopy.
Pineapple (*Ananas comosus*)	20–40	Raij et al. (1997)	Sample, just before floral induction, a freshly ripe leaf, which is usually the fourth leaf from the apex. Cut the leaves into pieces 1 cm wide, eliminate the basal portion without chlorophyll.
	20–40	CQFS (2004)	Collect the fourth leaf, counted from the apex, just before flower induction. Cut the leaves into pieces 1 cm wide, eliminating the basal portion without chlorophyll.
Acerola (*Malpighia emarginata*)	25–100	Raij et al. (1997)	Sampled in the four sides of the plant, fully expanded young leaves of fruitful branches.
Banana (*Musa* sp)	10–25	Raij et al. (1997)	Remove the central 5–10 cm of the third leaf from the inflorescence, eliminating the midrib and peripheral halves.
	9–75	SBCS/NEPAR (2019)	When the inflorescence has all female bunches without bracts and with two or three open male bunches, collect the third leaf below the apex, removing a central strip 10 cm wide, eliminate the midrib and the outer halves of the leaf.
	10–25	CQFS (2004)	When the leaves are at the beginning of flowering, collect the third youngest leaf, from which two pieces are cut, from the middle third of the leaf, excluding the vein and the outer half.

(Continued)

Table 2.4 (Continued)

Crop	B (mg/kg)	Reference	Diagnostic leaves sampling description
Fig (*Ficus carica*)	30–75	Raij et al. (1997)	Collect newly mature and fully expanded leaves from the middle portion of the branches three months after budding.
	30–75 normal >300 excessive	CQFS (2004)	Collect complete leaves (blade and petiole) that are newly mature and fully expanded, located in the middle portion of the branches, approximately three months after the beginning of budding.
Citrus (*Citrus* sp)	36–100 normal	Raij et al. (1997)	Collect the third leaf from the fruit, generated in spring, at 6 months of age, on branches with 2–4 cm diameter fruits. Sample four leaves per plant.
	50–100 normal	SBCS/NEPAR (2019)	Collect the third leaf from the fruit generated in the spring, after six months of age, on branches with fruits 2–4 cm in diameter. Collect four leaves per tree, one at each cardinal point.
	<36 insufficient 36–100 normal >150 excess	CQFS (2004)	In the period between late summer and early winter, collect belted leaves at seven months of age from fruiting branches that originated in spring shoots. Eight to sixteen leaves per tree, depending on leaf size, should be collected at a height of approximately 1.5 m from the ground, in the four quadrants of the canopy.
Apple (*Malus domestica*)	25–50	Raij et al. (1997)	Collect four to eight mature, fully expanded leaves.
	30–50	SBCS/NEPAR (2019)	Collect complete leaves, with blade and petiole, from the middle part of the shoots emitted in the growing season (summer), four leaves per tree in each cardinal position. Choose branches in the middle height of the canopy and on the outside of the plant, with medium growth, and avoid branches with excessive growth (thieves).
	<20 insufficient 20–29 below normal 30–50 normal 51–140 above normal >140 excessive	CQFS (2004)	Collect complete and normal leaves from the middle part of the year's shoots, in mid-summer

Macadamia (*Macadamia* sp)	25–50	Raij et al. (1997)	Collect newly mature leaves and fully expanded in the middle of the last flow of vegetation.
Papaya (*Carica papaya*)	20–30	Raij et al. (1997)	Collect 15 petioles of young, fully expanded, and mature leaves (seventeenth and twentieth leaves from the apex), with a visible flower in the armpit.
Mango (*Mangifera indica*)	50–100	Raij et al. (1997)	Collect leaves in flowering from the middle of the last stream of vegetation, from branches with flowers at the end. Samples four leaves per tree.
	24–30 (var. Tommy Atkins) 22–29 (var. Palmer)	SBCS/NEPAR (2019)	Collect four leaves per tree, with petiole, one in each cardinal position, in the middle of the last vegetation flow and in the middle third of the plant, at the beginning of full flowering.
Passion fruit (*Passiflora edulis*; P. sp)	40–60	Raij et al. (1997)	In autumn, collect the third or fourth leaf, starting from the apex of unshaded branches. Alternatively, collect the leaf with a floral bud in the armpit, about to open.
	20–57	SBCS/NEPAR (2019)	Collect four leaves per plant, fourth or fifth leaf from the apex of unshaded branches, during spring.
	40–150	CQFS (2004)	Collect the newly mature fourth leaf from the apex of the productive branches in autumn.
Peach (*Prunus persica*)	20–60	Raij et al. (1997)	Collect 26 newly mature and fully expanded leaves from the middle portion of the branches.
	15–30	SBCS/NEPAR (2019)	Collect four leaves per plant, the fourth or fifth leaf from the apex of nonshaded branches, approximately 10 days before the start of harvest, or in the 13 or 14 weeks after full flowering (cannot coincide with harvest).
	< 3 insufficient 3–33 below normal 34–63 normal 64–93 above normal > 93 excessive	CQFS (2004)	Collect complete leaves (blade + petiole) from the middle part of the season's branches on different sides of the plants, between 13 and 15 weeks after full flowering. If the sampling coincides with the harvest, bring it forward by one to two weeks, so that the sampling is before the harvest.

(*Continued*)

Table 2.4 (Continued)

Crop	B (mg/kg)	Reference	Diagnostic leaves sampling description
Grape (*Vitis vinifera*)	45–53	Raij et al. (1997)	Sample the newest newly mature leaf, counted from the apex of vine branches.
	45–53 (mature leaf) 30 (after maturity)	SBCS/ NEPAR (2019)	Collect freshly mature leaves from productive and flowering branches.
	< 15 insufficient 15–22 below normal 23–60 normal 61–100 above normal > 100 excessive	CQFS (2004)	Collect the leaf and blade opposite the first bunch of the fruit branch sampled at the beginning of ripening (color change in the berries).
Guava (*Psidium guajava*)	20–25	SBCS/ NEPAR (2019)	Collect the third pair of newly mature leaves with petiole, from the end of the branch, in full bloom, four pairs of leaves per tree, one in each cardinal position.
Plum (*Prunus salicina*)	20–60	SBCS/ NEPAR (2019)	Collect the fourth or fifth leaf from the apex of unshaded branches, collecting four leaves per plant, approximately 10 days before harvest or in the 13th or 14th week after full flowering.
	< 3 insufficient 3–33 below normal 34–63 normal 64–93 above normal > 93 excessive	CQFS (2004)	Collect complete leaves (blade + petiole) from the middle part of the branches emitted in that season, on the different sides of the plants, between 13 and 15 weeks after full flowering, so that it is in a period before harvesting. If the indicated time coincides with the harvest, bring it forward by one to two weeks.

Blackberry (*Rubus* spp.)	30–80	SBCS/NEPAR (2019)	Collect, in spring, the sixth leaf from the branches of the previous year fully expanded and with the pedicel, counted from the apex.
	< 25 insufficient 25–29 below normal 30–80 normal 81–100 above normal > 100 excessive	CQFS (2004)	Collect the sixth leaf with pedicel, fully expanded, counted from the apex. Leaves should be collected from the branches of the previous growing season in spring.
Blueberry (*Vaccinium myrtillus*)	< 20 insufficient 20–30 below normal 31–69 normal 70–200 above normal > 200 excessive	CQFS (2004)	Collect five complete leaves, fully developed, located in the fifth or sixth node, contacted from the end of the young fruiting branches, in the second half of November.
Atemoia (*Annona* sp)	57–73	SBCS/NEPAR (2019)	Collect the third or fourth pair of newly ripe leaves (with petiole) from the end of the branch, at the stage of full flowering, in at least eight leaves per tree, two in each cardinal position.
Strawberry (*Fragaria* x *ananassa*)	40–80	SBCS/NEPAR (2019)	Collect the newly developed third leaf, without petiole, at the beginning of flowering.
	35–100	CQFS (2004)	Collect the newly developed third and fourth leaves (without petiole) at the beginning of flowering.

(*Continued*)

Table 2.4 (Continued)

Crop	B (mg/kg)	Reference	Diagnostic leaves sampling description
Pecan walnut (*Carya illinoinensis*)	20–45	CQFS (2004)	At late summer, collect a pair of leaflets from the leaves located in the middle portion of the branches, in the four quadrants of the plants.
Pear (european) (*Pyrus* sp)	< 20 insufficient 20–29 below normal 30–50 normal 51–140 above normal > 140 excessive	CQFS (2004)	Collect complete and normal leaves from the middle part of the year's shoots, in late summer.
Pear (asian) (*Pyrus* sp)	30–90	CQFS (2004)	Collect complete and normal leaves from the middle part of the year's shoots, in late summer.
Kiwi (*Actinidia deliciosa*)	< 20 insufficient 40–50 normal >100 excessive	CQFS (2004)	Collect the second normal leaf with petiole, located after the fruits, at late summer season.

Table 2.5 Exportation of boron in some fruit crops according to NEPAR/SBCS (2019).

Fruit crop	Exportation (g/t)
Banana (*Musa* sp.)	8.1
Citrus (*Citrus* sp.)	1.9
Guava (*Psidium guajava*)	0.3
Apple (*Malus domestica*)	1.75 (var. Gala); 2.23 (var. Fuji)
Mango (*Mangifera indica*)	0.7 (var. Tommy Atkins); 0.9 (var. Palmer)
Passion fruit (*Passiflora edulis*; P. sp.)	2.05
Plum tree (*Prunus salicina*)	4.5
Atemoia (*Annona* sp.)	24.3

Table 2.6 Recommendation for fertilization with boron for some fruit crops.

Fruit crop	Fertilizer recommendation	Reference
Avocado (*Persea americana*)	Incorporate 20–30 kg of borax/ha, along with liming and preplanting fertilizer. In established orchards, when the leaf content is less than 50 mg/kg, add B to the soil. To spray borax on the soil, on both sides of the crop rows, in the fertilized range, use a dose not exceeding 10 kg of borax/ha.	CQFS (2004)
Apple (*Malus domestica*)	When foliar contents are low, carry out 2–3 biweekly foliar fertilization sprays with borax ($Na_2B_4O_70.10H_2O$) at 0.4% or solubor at 0.2%, starting in mid-spring. Apply in the pink bud stage to favor the fertilization of the flowers. In the Gala cultivar, three applications of boron spaced 30 days apart, the first in mid-spring, can improve the skin color of the fruits at harvest.	
Pear (*Pyrus* sp)	In case of deficiency proven by foliar analysis, apply foliar fertilization spray with two to three foliar fertilization sprays with 0.4% borax or 0.2% solubor from the fall of the petals. Apply boron when the flowers are in the balloon stage, if the objective is to favor effective fertilization and fruiting.	

(Continued)

Table 2.6 (Continued)

Fruit crop	Fertilizer recommendation	Reference
Peach (*Prunus persica*)	If the B content in the soil is less than 0.1 mg/dm^3, 10 kg of boron/ha can be incorporated, together with liming and preplanting fertilization. If the B content is between 0.1 and 0.3 mg/dm^3, apply 7.5 kg B/ha.	
Kiwi (*Actinidia deliciosa*)	The kiwi tree is very sensitive to excess boron, this nutrient should be applied via the leaves only if the deficiency is proven by leaf analysis.	
Citrus (*Citrus* sp)	Incorporate 20–30 kg of borax/ha, along with liming and preplanting fertilizer. In established orchards, when the leaf content is less than 50 mg/kg, add B to the soil. To spray borax on the soil, on both sides of the crop rows, in the fertilized range, use a dose not exceeding 10 kg of borax/ha.	
	If the B content in the soil is below 0.2 mg/dm^3, apply 1 g of B/linear m (within furrow row) at planting with Boric acid, borax or similar.	SBCS/NEPAR (2019)
Papaya (*Carica papaya*)	If the B content in the soil is below 0.2 mg/dm^3, apply between 1 and 2 kg/ha of B.	Raij et al. (1997)

Acknowledgements

We acknowledge the State University of Midwestern of Paraná for supporting this study.

References

Askew, H.O., 1944. A case of combined K and B deficiencies in grapes, N.Z.J. Sci. Technol, 26. pp. 146–152. Nova Zelândia.

Baranwal, D., Tomar, S., Singh, J.P., Maurya, J.K., 2017. Effect of foliar application of zinc and boron on fruit growth, yield and quality of winter season guava (*Psidium guajava* L.). Int. J. Curr. Microbiol. Appl. Sci. 6, 1525–1529.

Bell, R.W., 1997. Diagnosis and prediction of boron deficiency for plant production. Plant. Soil. 193, 149–168.

Boaretto, R.M., Quaggio, J.A., Mattos Jr, D., Muraoka, T., Boaretto, A.E., 2011. Boron uptake and distribution in field grown citrus trees. J. Plant. Nutr. 34, 839–849.

Brdar-Jokanović, M., 2020. Boron toxicity and deficiency in agricultural plants. Int. J. Mol. Sci. 21, 1424.

Broadley, M., Brown, P., Cakmak, I., Rengel, Z., Zhao, F., 2012. Chapter 7 - Function of nutrient: micronutrients. In: Marschner, P. (Ed.), Marschner's Mineral Nutrition of Higher Plants, third (ed.) Academic Press, pp. 191–248.

Brown, P.H., Hu, H., 1996. Phloem mobility of boron is species dependent: evidence for phloem mobility in sorbitol-rich species. Ann. Bot. 77, 497–506.

Chen, L.S., Han, S., Qi, Y.P., Yang, L.T., 2012. Boron stresses and tolerance in citrus. Afr. J. Biotechnol. 11, 5961–5969.

Cook, J.A., Lynn, C.D., Kissler, J.J., 1960. Boron deficiency in California vineyards. Am. Journ.Enol. Vitic. 11 (4), 185–199.

CQFS - Comissão de Química e Fertilidade do Solo - RS/SC. 'Fertilization and liming manual for the states of Rio Grande do Sul and Santa Catarina' = Manual de adubação e calagem para os Estados do Rio Grande do Sul e Santa Catarina. 10ª (ed.) Sociedade Brasileira de Ciência do Solo - Núcleo Regional Sul; 2004. 400p. (in Portuguese).

Cunha, J.M., Freitas, M.S.M., Caetano, L.C.S., Carvalho, A.J.C.D., Peçanha, D.A., Santos, P.C.D., 2019. Fruit quality of pineapple 'Vitória' under macronutrients and boron deficiency. Rev. Bras. Frutic. 41, 1–10.

Dong, X., Wang, M., Ling, N., Shen, Q., Guo, S., 2016. Effects of iron and boron combinations on the suppression of Fusarium wilt in banana. Sci. Rep. 6, 1–11.

Dordas, C., 2008. Role of nutrients in controlling plant diseases in sustainable agriculture: a review. Agron. Sustain. Dev. 28, 33–46.

Dordas, C., Chrispeels, M.J., Brown, P.H., 2000. Permeability and channel-mediated transport of boric acid across membrane vesicles isolated from squash roots. Plant. Physiol. 124, 1349–1361.

Du, W., Pan, Z.Y., Hussain, S.B., Han, Z.X., Peng, S.A., Liu, Y.Z., 2020. Foliar supplied boron can be transported to roots as a boron-sucrose complex via phloem in citrus trees. Front. Plant. Sci. 11, 250.

Elrashidi, M.A., O'Connor, G.A., 1982. Boron Sorpt. desorption soils. Soil. Sci. Soc. Am. J. 46, 27–31.

Epstein E, Bloom AJ. Nutrição mineral das plantas: princípios e perspectivas. 2nd ed. Editora Planta; 2006.

Fang, K., Zhang, W., Xing, Y., Zhang, Q., Yang, L., Cao, Q., et al., 2016. Boron toxicity causes multiple effects on *Malus domestica* pollen tube growth. Front. Plant. Sci. 7, 208.

Fernández, V., Gil-Pelegrín, E., Eichert, T., 2021. Foliar water and solute absorption: an update. Plant. J. 105, 870–883.

Fregoni, M., 1980. Nutrizione e fertilizzazione della vite. Edagricol 418.

Fregoni, M., Scienza, A., 1976. Aspetti della micronutrizione di alcune zone viticole italiana. Vignevini 3, 5–8.

Freitas, M.S.M., Monnerat, P.H., Carvalho, A.J.C.D., Vasconcellos, M.A.D.S., 2011. Deficiency symptoms of macronutrientes and boron in sweet passion fruit plant. Rev. Bras. Frutic. 33, 1329–1341.

Ganie, M., Akhter, F., Bhat, M.A., Malik, A.R., Junaid, J.M., Shah, M.A., et al., 2013. Boron – a critical nutrient element for plant growth and productivity with reference to temperate fruits. Curr. Sci. 104, 76–85.

Goldberg, S., 2014. Chemical modeling of boron adsorption by humic materials using the constant capacitance model. Soil. Sci. 179, 561–567.

Graham, D.R., Webb, M.J., 1991. Micronutrients and disease resistance and tolerance in plants. In: Mortvedt, J.J., Cox, F.R., Shuman, L.M., Welch, R.M. (Eds.),

Micronutrients in agriculture, second (ed.) Soil Science Society of America, Madison, Wisconsin, pp. 329–370.
Gravite, I., Kaufmane, E., Militaru, M., 2017. Influence of boron foliar fertilization on plum pollen viability, germination and fruit set. Acta Hortic. 1175, 67–72.
Gupta, U., 1993. Deficiency, sufficiency, and toxicity levels of boron in crops. Boron and Its Role in Crop Production. CRC Press, pp. 137–145.
Han, S., Chen, L.S., Jiang, H.X., Smith, B.R., Yang, L.T., Xie, C.Y., 2008. Boron deficiency decreases growth and photosynthesis, and increases starch and hexoses in leaves of citrus seedlings. J. Plant. Physiol. 165, 1331–1341.
Havlin, J.L., Tisdale, S.L., Nelson, W.L., Beaton, J.D., 2016. Soil fertility and fertilizers, An Introduction to Nutrient Management, eighth (ed.) Pearson, p. 536.
Huang, J.H., Qi, Y.P., Wen, S.X., Guo, P., Chen, X.M., Chen, L.S., 2016. Illumina microRNA profiles reveal the involvement of miR397a in citrus adaptation to long-term boron toxicity via modulating secondary cell-wall biosynthesis. Sci. Rep. 6, 1–14.
Hu, H., Brown, P.H., Labavitch, J.M., 1996. Species variability in boron requirement is correlated with cell wall pectin. J. Exp. Bot. 47, 227–232.
Huber, Don, Römheld, Volker, Weinmann, Markus, Petra, Marschner, 2012. Chapter 10 - Relationship between Nutrition, Plant Diseases and Pests, Marschner's Mineral Nutrition of Higher Plants (Third Edition), 3rd. Academic Press, pp. 283–298. Available from: https://www.sciencedirect.com/science/article/pii/B9780123849052000108, (accessed 30.03.22).
Kot, F.S., Farran, R., Kochva, M., Shaviv, A., 2012. Boron in humus and inorganic components of Hamra and Grumosol soils irrigated with reclaimed wastewater. Soil. Res. 50, 30–43.
Kuniyuki, H., Terra, M.M., Hiroce, R., Costa, A.S., 1985. Chocolate das bagas e clorose foliar: anomalia da videira causada por deficiência de boro. Bragantia, Camp. 44 (1), 187–196.
Lee, S.H., Kim, W.S., Han, T.H., 2009. Effects of post-harvest foliar boron and calcium applications on subsequent season's pollen germination and pollen tube growth of pear (*Pyrus pyrifolia*). Sci. Hort. 122, 77–82.
Lopes, A.S., 1999. Micronutrientes: Filosofias de aplicação e eficiência agronomica. ANDA. Bol. técnico 8.
Macedo, L.O., Mattos, D., Jacobassi, R., Hippler, F.W.R., Quaggio, J.A., Boaretto, R. M., 2020. Efficiency of foliar application of sparingly soluble sources of boron and zinc in citrus. Sci. Agric. 78.
Marschner, P., 2012. Marschner's Mineral Nutrition of Higher Plants, third (ed.) Academic Press, p. 672.
Matterson KJ. Borate ore discovery, mining and beneficiation. In: Thompson R., Welch A.J.E. (eds.), Suppl to Mellor's Comprehensive Treatise on Inorganic and Theoretical Chemistry, Sect. A3, vol. 5. Longman; 1980.
Mortvedt, J.J., Woodruff, J.R., 1993. Technology and application of boron fertilizers for crops. In: Gupta, U.C. (Ed.), Boron and Its Role in Crop Production. CRC Press, Boca Raton, pp. 156–176.
Nagy, P.T., Kincses, I., Nyéki, J., Soltész, M., Szabó, Z., 2011. Importance of boron in fruit nutrition. Int. J. Hortic. Sci. 17, 39–44.
Nava, G.A., Dalmago, G.A., Bergamaschi, H., Marodin, G.A.B., 2009. Fenologia e produção de pessegueiros "granada" com aplicação de cianamida hidrogenada e boro. Rev. Bras. Frutic. 31, 297–304.
Nogueira, P.V., Silva, D.F., Pio, R., Silva, P.A.D.O., Bisi, R.B., Balbi, R.V., 2015. Germinação de pólen e aplicação de ácido bórico em botões florais de nespereiras. Bragantia 74, 9–15.

Noor, Y., Shah, Z., Tariq, M., 2019. Effect of zinc and boron using different application methods on yield of citrus (sweet orange) in calcareous soils. SJA 35, 1247–1258.
Nyomora, A.M., Brown, P.H., Freeman, M., 1997. Fall foliar-applied boron increases tissue boron concentration and nut set of almond. J. Am. Soc. Hortic. Sci. 122, 405–410.
Nyomora, A., Brown, P.H., Pinney, K., Polito, V.S., 2000. Foliar application of boron to almond trees affects pollen quality. J. Am. Soc. Hortic. Sci. 125, 265–270.
Perica, S., Brown, P.H., Connell, J.H., Nyomora, A.M., Dordas, C., Hu, H., et al., 2001. Foliar boron application improves flower fertility and fruit set of olive. HortScience 36, 714–716.
Qin, G.Y., Zong, Q., Chen, D., HuaTian, S., 2010. Inhibitory effect of boron against *Botrytis cinerea* on table grapes and its possible mechanisms of action. Int. J. Food Microbiol. 138, 145–150.
Raij BV, Cantarella H, Quaggio JA, et al. Technical Bulletin, 100 - Fertilization and liming recommendations for the State of São Paulo' = Boletim Técnico, 100 - Recomendações de adubação e calagem para o Estado de São Paulo. 2nd ed. IAC; 1997. 285 p. (in Portuguese).
Reetz Jr, H.F., 2016. Fertilizers and their efficient use. IFA 110.
Rolshausen, P.E., Gubler, W.D., 2005. Use of boron for the control of Eutypa dieback of grapevines. Plant. Dis. 89, 734–738.
Rosolem, C.A., Almeida, D.S., Cruz, C.V., 2020. Polyol-ester impact on boron foliar absorption and remobilization in cotton and coffee trees. Rev. Bras. Ciênc. Solo 44.
Sá, A.A., Ernani, P.R., Nava, G., Amarante, C.V.T.D., Pereira, A.J., 2014. Influência de formas de aplicação de boro na qualidade e no rendimento de maçãs (*Malus domestica*). Rev. Bras. Frutic. 36, 487–494.
SBCS/NEPAR—Sociedade Brasileira de Ciência do Solo/ Núcleo do estado do Paraná, Fertilization and liming manual for the state of Paraná. = Manual de adubação e calagem para o estado do Paraná. 2ed. SBCS/NEPAR; 2019. 289 p. (in Portuguese).
Shorrocks, V.M., 1997. The occurrence and correction of boron deficiency. Plant. Soil. 193, 121–148.
Siebeneichler, S., Monnerat, P., Silva, J., 2008. Deficiência de boro na cultura do abacaxi "Pérola.". Acta Amaz. 38, 651–656.
Sims, J.T., Johnson, G.V., 1991. Micronutrient soil tests. In: Mortvedt, J.J., Cox, F.R., Shuman, L.M., Walch, R.M. (Eds.), Micronutrients in Agriculture. Soil Science Society of America, pp. 427–477.
Sotiropoulos, T.E., Molassiotis, A., Almaliotis, D., Mouhtaridou, G., Dimassi, K., Therios, I., et al., 2006. Growth, nutritional status, chlorophyll content, and antioxidant responses of the apple rootstock MM 111 shoots sultured under hygh boron concentrations in vitro. J. Plant. Nutr. 29, 575–583.
Souza, F.B.M., Pio, R., Coelho, V.A.T., Rodas, C.L., Silva, I.P.D., 2015. Visual symptoms of micronutrients, boron and iron deficiency and mineral composition of blackberry tree. Pesq. Agropec. Trop. 45, 241–248.
Taiz, L., Zeiger, E., Møller, I.M., Murphy, A., 2017. Fisiologia e desenvolvimento vegetal, Artmed, sixth (ed.)p. 888.
Tarabih, M.E., El-Metwelly, M.A., 2014. Effect of jojoba oil and boric acid as postharvest treatments on the shelf life of Washington navel orange fruits. Int. J. Agric. Res. 9, 1–16.
Terra MM. Nutrição, calagem e adubação. In: Pommer, C.V. (ed.), Uva: Tecnologia de produção, pós-colheita, Mercado. Porto Alegre: Cinco Continentes; 2003. p. 405–475.
Terra, M.M., Pires, E.J.P., Botelho, R.V., Tecchio, M.A., Tucci, M.L.S.A., 2007. Toxicity from boron in 'Italia'(*Vitis vinifera* l.) Grapevines. Ambiência 3, 435–440.

Thurzó, S., Szabó, Z., Nyéki, J., Nagy, P.T., Silva, A.P., Gonçalves, B., 2010. Effect of boron and calcium sprays on photosynthetic pigments, total phenols and flavonoid content of sweet cherry (*Prunus avium* L.). Acta Hortic. 868, 457–461.

Torun, A.A., Bozgeyik, S., Duymuş, E., 2021. The effect of boron application in increasing doses on shoot boron concentration and uptake of other nutrients of pistachio (*Pistacia vera* L.). TURJAF 9, 855–862.

Wang, N., Yang, C., Pan, Z., Liu, Y., Peng, S.A., 2015. Boron deficiency in woody plants: various responses and tolerance mechanisms. Front. Plant. Sci. 6, 916.

Wells, M.L., Conner, P.J., Funderburk, J.F., Price, J.G., 2008. Effects of foliar-applied boron on fruit retention, fruit quality, and tissue boron concentration of pecan. HortScience. 43, 696–699.

Will, S., Eichert, T., Fernández, V., Müller, T., Römheld, V., 2012. Boron foliar fertilization of soybean and lychee: effects of side of application and formulation adjuvants. J. Plant. Nutr. Soil. Sci. 175, 180–188.

Winkler, A.J., Cook, J.A., Kliewer, W.M., Lider, L.A., Cerruti, L., 1974. General Viticulture, 2nd ed. University of California Press, pp. 1–710.

Wojcik, P., 2005. Response of primocane-fruiting "Polana" red raspberry to boron fertilization. J. Plant. Nutr. 28, 1821–1832.

Wojcik, P., Wojcik, M., 2003. Effect of boron fertilization on 'conference' pear tree vigor, nutrition, and fruit yield and storability. Plant. Soil. 256, 413.

Wojcik, P., Wojcik, M., Klamkowski, K., 2008. Response of apple trees to boron fertilization under conditions of low soil boron availability. Sci. Hort. 1, 58–64.

Woods, W.G., 1994. An introduction to boron: history, sources, uses, and chemistry. Environ. Health Perspect. 102, 5–11.

Wu, X., Riaz, M., Yan, L., Jiang, C., 2020. Distribution and mobility of foliar-applied boron (10B) in citrange rootstock under different boron conditions. J. Plant. Growth Regul. 39, 575–582.

Yadav, A., Kanwar, J., Dubey, R., Upadhyay, M., 2020. Effect of foliar application of urea, zinc sulphate and borax on flowering, fruiting and yield of acid lime (*Citrus aurantifolia* Swingle) vari. Kagzi lime under Malwa Plateau conditions. J. Pharmacogn. Phytochem. 9, 483–485.

Zambon, C.R., Silva, L.F.D.O.D., Pio, R., Figueiredo, M.A.D., Silva, K.N., 2014. Estabelecimento de meio de cultura e quantificação da germinação de grãos de pólen de cultivares de marmeleiros. Rev. Bras. Frutic. 36, 400–407.

Zhou, G., Li, B., Chen, J., Yao, F., Guan, G., Liu, G., et al., 2020. Physiological and nutritional responses of 'HB' Pummelo (*Citrus grandis* (L.) osbeck 'Hirado Buntan') to the combined effects of low pH levels and boron deficiency. HortScience. 55, 1–8.

CHAPTER 3

Boron deficiency and toxicity symptoms in plants

Monireh Mones Sardrodi[1], Zahra Dehghanian[2], Khashayar Habibi[3], Arash Hemati[4], Behnam Asgari Lajayer[4], Janhvi Pandey[5] and Tess Astatkie[6]

[1]Department of Plant Breeding and Biotechnology, Faculty of Agriculture, University of Tabriz, Tabriz, Iran
[2]Department of Biotechnology, Faculty of Agriculture, Azarbaijan Shahid Madani University, Tabriz, Iran
[3]Department of Biotechnology, College of Agriculture, Isfahan University of Technology, Isfahan, Iran
[4]Department of Soil Science, Faculty of Agriculture, University of Tabriz, Tabriz, Iran
[5]Division of Agronomy and Soil Science, CSIR-Central Institute of Medicinal and Aromatic Plants, Lucknow, India
[6]Faculty of Agriculture, Dalhousie University, Truro, NS, Canada

3.1 Introduction

Boron belongs to the metalloid subgroup III and possesses properties that are halfway among the metals and nonmetals (Marschner, 2011; Khadem Moghadam et al., 2020). Unlike its deficiency in the natural mediums, boron is extensively dispersed in the hydrosphere and lithosphere, with concentrations ranging from 5 to 10 mg/kg in rocks (Shorrocks, 1997), 3–30 g/kg in rivers, and 4.5 mg/L in the oceans (Power and Woods, 1997; Lemarchand et al., 2000).

Boron is a critical element for plants, and its accessibility in the soil and irrigation water is a significant factor governing plant production (Tanaka and Fujiwara, 2008). In the soil, dissolved boron is found commonly as boric acid ($B(OH)_3$) which could be washed off under extreme rainfall conditions (Shorrocks, 1997; Yan et al., 2006). Plants that grow in such rainfall prone environments naturally suffer from boron deficiency, which is exhibited in symptoms such as inhibited growth. Contrary to that, under low rainfall conditions, a higher percentage of boron in the soil might also pose harm to the plant's development (Reid, 2007a). This occurs frequently in dry or semiarid places with increased boron groundwater concentration and its deposition in the topsoil. As a result of extreme evaporation of groundwater levels, agricultural production is impacted severely (Tanaka and Fujiwara, 2008).

Boron in Plants and Agriculture
DOI: https://doi.org/10.1016/B978-0-323-90857-3.00010-2

Borate and boric acid could make compounds with an extended range of molecules that have two groups of hydroxyl arranged in a cis-form configuration. One of boron's principal activities in complex plants has been attributed to the ability to create borate esters with rhamnogalacturonan II (RG-II) residues (Kobayashi et al., 1996). It contributes significantly to control cell wall pores (Fleischer et al., 1999) and tensile strength, which is critical for the functioning and structure of the cell wall (O'Neill et al., 2004; Ryden et al., 2003). Boron deficiency has been linked to the excessively inflated cell walls and the structure of impaired RG-II dimer. Furthermore, in mur1−1 and mur1−2 (mutants of *Arabidopsis thaliana*) with a lower quantity of this complex, the availability of RGII-borate compound for the normal plant development was demonstrated (Matoh and Ochiai, 2005). Under limited boron supply, Noguchi et al. (2003) reported a low cross-linking degree in the *Arabidopsis* bor1−1 mutant plant's cell wall, when compared to wild-types, which can be attributed to the decrement in shoot boron concentration.

Boron is required for the meristematic tissue and floral organ cell division/elongation, pollen tube germination/elongation, male fertility of the flowers, and the production of seed/fruit. Furthermore, the seeds with boron deficiency produce aberrant seedlings (Marschner, 2011). The most important physiological impact of boron deficiency at the cellular level is the reduction in root elongation by lowering cell expansion and division in the growth region of root tips. As root growth stops and root cap disintegrates, the mortality of root tip increases, which subsequently leads to an increase in the shoot to root length ratio. This may increase the vulnerability of plants and their responses to nutrition and water stress leading to the formation of vessels in boron deficient roots. The early effects of boron deficiency on the initiation of vessels and their differentiation are still unknown. Although low boron levels stifle leaf growth and reduce photosynthesis indirectly, the exact magnitude of reduction and the role of boron in photosynthesis are still unknown. The lower concentration of boron has a significant effect on sexual reproduction in field plants, and large grain yield losses might occur without visible signs during the vegetative growth. Low boron impairs microsporogenesis and pollen tube formation in flowers, which affects male fertility. Impaired embryogenesis leads to the abortion of seed or incomplete/damaged production of embryos, as well as deformed fruits. These effects are considered postfertilization impacts (Dell and Huang, 1997).

3.2 Symptoms of boron deficiency

Boron-deficient soils can be found all over the world. Lower boron concentration is the most general micronutrient insufficiency in numerous agronomic locations, ranging from tropical to temperate. Deficiency is particularly common in the humid regions' coarse-textured (sandy) soils, where substantial boron leaching losses (e.g., podsols, podzols) lead to soil's low boron concentration to the critical phases. According to various studies, boron shortage is frequent in Nepal, Philippines, India, and Thailand. Some of the most common boron deficiency symptoms in plants are top sickness of tobacco, corky core or internal cork, drought spot of apple, turnip water core, citrus fruit that is hard to eat, Alfa alfa yellows, grape hen and chicken, sugar beet heart rot, celery stem crack, cauliflower and broccoli hollow stems, swade brown heart, and tip burn of Chinese cabbage. Symptoms of boron deficiency are commonly seen in the terminal shoots as tiny, malformed leaves on the critically compacted stems. The development of apical meristems is halted, and the apex of the shoot necrotizes, resulting in various axillary branches which also necrotize shortly. Due to this, boron-deficient plants seem apparently bushy. The stem of boron deficient plants become brittle and display longitudinal cortical breaks. Moreover, such plants' petioles and stems develop longitudinal and transverse cracks, or corky ridges (Thompson et al., 1976; Anderson et al., 1994; Goldberg, 1997).

Sunflower is a vulnerable crop to boron deficiency, displaying boron deficient signs on the reproductive organs, leaves, and stems (Wren et al., 1983; Saiki et al., 1993). Younger leaves develop bronze color and morph into rigid, distorted, and necrotic forms. Moreover, the stems get corky, the seeds turn to bad shapes, and the capitulum is distorted (Wren et al., 1983). An experiment conducted by Deverel and Millard (1988) revealed that the growth of the roots of boron-deficient sunflower plants halt in less than 6 h after boron is removed from the growth medium.

Boron deficiency spawns a poor root system in pea plants and is the main reason behind bud abortion on the top and leaf hardening (Dyer and Caprara, 1997; Howe, 1998). Boron-deficient onion leaves turn thick, brittle, and blue-green in hue, with hard and rough surface scales and necrosis of the inner scales. The outer leaves of the onion head appear thick and brittle under a mild shortage. Cauliflower plants display some of these symptoms when they become boron deficient. Leaf petiole and midribs develop little blisters and fissures, and the outer leaves become thick and brittle (Gerike et al., 1976). A dense layer of epicuticular wax coats the leaves of boron-deficient cabbage and cauliflower leaves, giving them a bluish appearance. Young boron deficient lettuce leaves' margins become burnt and ragged, appearing "hooked" and deformed.

Many researchers have reported that the fastest reaction to boron depletion or shortage in the developed plants is the deterrence or root growth cessation in the main and lateral roots (Bergmann et al., 1995). It has been reported that root tip elongation ceases within hours of removing boron from the nutritional solution (Schöberl and Huber, 1988). Boron deficiency inhibits root extension in the plants, but if boron is re-provided before the impairment of the meristematic zone of root tips, the repression can be reversed (Howe, 1998). It was reported that after 3 h, the root tip elongation of boron-deficient squash plants (*Cucurbita pepo*) decelerated, significantly hindered after 6 h, and entirely stopped after 24 h of nutrient solution delivery with inadequate boron concentration. Meanwhile, the boron replenishment after 12 h, normalized the root elongation process within 12–18 h. Likewise, within 3 h of removing boron from the nutrient solution, elongation of the root in tomato plants (*Lycopersicum esculentum*) was also inhibited (Deverel and Millard, 1988).

Boron also promotes nodulation in the roots of dhaincha plants (*Sesbania cannabina*), according to Singh (1972). Discoloration and curling of young terminal leaves in the root crops including carrot (*Daucus carota*), sugar beet (*Beta vulgaris*), turnip, and radish is an early indication of boron deficiency (Maier and Knight, 1991). As the shortage aggravates, wrinkles and cracks appear on the root surface, and water-soaked patches appear on the softer tissue of the core, resulting in tissue disintegration. Shelp et al. (1987) reported external roughness and brown necrosis in the roots of rutabaga with severe boron deficit. It has been reported that boron-deficient sunflower plants produce a lot of lateral roots and adventitious roots with swollen and necrotic tips (Hamilton, 1995). Moreover, boron deficiency also impairs flowering and fruit production in various plant species. Abnormal flower production in boron-deficient cotton (Wallen et al., 1957) and *Zinnia elegans* plants (Hamilton and Buhl, 1990) were reported. Boron shortage causes internode shortening in most fruit trees. The terminal shoots may die, and the terminal buds stay seminal in severe deficiency. Meanwhile, fruits of such plants develop a rough surface, fractures, and are frequently deformed, while stems exhibit bark breaking.

In boron-deficient peaches, premature fruit falling has been documented (Taylor et al., 1985). The development of corky patches followed by fruit drop has been reported in boron-deficient papaya and grape plants (Birge et al., 1984; Black et al., 1993). The emergence of white translucent lines in the center of the leaf lamina thereby reducing leaf area, failure of young emerging leaves to unroll, development of short and thick internodes, and cessation of apical growth are some of the impacts of boron deficit in maize plants (*Zea mays*). Reduced growth and leaf area, chlorosis, thickening and curling of leaves, terminal chlorosis, preexuding leaves' downward cupping, internode shortening, and developing tips necrosis of stem are all boron deficient signs in Periwinkle plants (Bingham, 1982).

3.3 The deficiency of boron

It has been reported that boron deficiency has diverse impacts on very different mechanisms in vascular plants including, sugar translocation, oxidase activity of IAA, root elongation, pollen tube development, nucleic acid synthesis, and the metabolism of carbohydrates (Blevins and Lukaszewski, 1998; Goldbach and Wimmer, 2007). Other reports also mentioned similar impacts in the ion fluxes across membranes, plasmalemma-bound enzymes, membrane potential (Blaser-Grill et al., 1989; Goldbach et al., 2001), and cytoskeletal proteins (Yu et al., 2003). Boron can also play a role in the polyamines and phenolics aggregation (Camacho-Cristóbal et al., 2004, 2005; Camacho-Cristóbal and González-Fontes, 2007), as well as in nitrogen metabolism (Camacho-Cristóbal and González-Fontes, 1999).

3.3.1 Membranes, cytoskeleton, and cell wall

Boron insufficiency is well recognized to provoke the production of aberrant cell walls with altered physical characteristics in plant cells (Fleischer et al., 1999; Ryden et al., 2003). Boron participation in the wall of the cell, cross-linking of RG-II and pectin synthesis causes these effects. Boron shortage inhibits aggregation inside the parts of the endocytic direction which ends in enhanced development of pectin in the cell wall (Yu et al., 2002). Furthermore, after 24 h of boron deficiency, the expression of numerous cell wall modification enzymes is downregulated, which could impact cell wall softening and result in the elongation of cells. Same deficiency can alter and decrease the transcript levels of several xyloglucan endotransglycosylase/hydrolases (Cosgrove, 1999), which might be attributed to its influence over the xyloglucan cross-linked microfibrillar network re-arrangement leading to an altercation in tensile properties of the cell walls (Ryden et al., 2003).

Some studies have suggested that boron might play a role in the cytoskeleton composition and associated activities (Yu et al., 2001; Bassil et al., 2004). It has been reported that in *Arabidopsis* roots, boron deficiency elevates actin and tubulin protein concentrations and alters polymerization patterns of cytoskeletal in maize root apices cells. Under boron deficiency, this surge in cytoskeletal proteins is thought to represent an adaptive reaction to help with the cells' physical strengthening in the root environment (Yu et al., 2003).

Boron seems to be necessary for the formation of membrane and their function, particularly the plasma membrane (Cakmak and Römheld, 1997; Goldbach et al., 2001; Brown et al., 2002). In the roots of *Helianthus annuus* (Ferrol and Donaire, 1992) and *Daucus carota* (Blaser-Grill et al., 1989) plants, the deficiency of boron impacted the potential of the membrane and lowered the proton-pumping ATPase activity. Furthermore, boron insufficiency has been reported to alter the susceptibility of plasma membrane ion perforation (Cakmak et al., 1995; Wang et al., 1999). Regarding the immediate boron deficiency impacts, the fundamental processes that are affected and the impacts on the plasma membrane shape and function remain unclear (Blevins and Lukaszewski, 1998; Brown et al., 2002; Goldbach and Wimmer, 2007). As a result, various molecules of the membrane with hydroxylated ligands, including glycolipids and glycoproteins, have been identified as proper sites for a potential boron performance inside the membrane (Goldbach and Wimmer, 2007). According to a study by Redondo-Nieto et al. (2007), at the minimum, three B-binding membrane glycoproteins were not found in boron-deficient pea nodes, which suggests that boron and membrane glycoproteins are involved in the membrane activities related to general cell proliferation in boron-deficient plant tissues. Furthermore, arabinogalactan proteins (AGPs), which are connected to the cell membrane by a glycosyl-phosphatidylinositol anchor, have been proposed as possible boron-binding constituents (Goldbach and Wimmer, 2007). In pollen tubes that are facing deficiency or subjected to Yariv reagent, a component that cross-links plasma membrane in association with AGPs, resembling the modification has been reported in the pectin of the cell wall. Furthermore, it has recently been discovered that a lack of boron induces a significant reduction in various AGP gene expressions in the roots of *Arabidopsis* (Camacho-Cristóbal and González-Fontes, 2007). Boron may function in cellular membranes not only by fixation of the membrane molecules with cis–diol groups but also by modulating the expression of the genes involved in membrane composition and function (Bolaños et al., 2004).

3.3.2 Nitrate uptake and fixation

Boron is required for N_2 fixation in the cyanobacterium *Anabaena* PCC 7119 heterocyst and the actinomycetes vesicles of the genus *Frankia*, according to several investigations (Bolaños et al., 2002). Two types of microbes require boron for the envelope's stability, which protects nitrogenase from oxygen inactivation under an N_2-fixing environment. Furthermore, it has been reported that under boron insufficiency, legumes produce nodules with diminished ability to fix N_2 (Bolaños et al., 1994; Yamagishi and Yamamoto, 1994). This may be related to boron's potential participation in the cell surface interaction of Rhizobium-legume (Bolaños et al., 1996). Boron is required for the nodule-specific plant-derived glycoproteins aiming (Bolaños et al., 2001), which act as messengers for bacteroid development toward an N_2 fixer (Bolaños et al., 2004). Furthermore, the nodules' cell walls with boron deficiency possess low amounts of hydroxyproline-/proline rich proteins such as ENOD2, which lead to an enhanced oxygen penetration toward the nodules and, as a result, nitrogenase inactivation (Bonilla et al., 2004).

Several studies have reported that boron could assist in nitrogen absorption. In plants with boron deficiency, the small amount of nitrate reductase (NR) activity and elevated nitrate buildup has already been observed (Kastori and Petrović, 1989; Ramón et al., 1989; Shen et al., 1993). The mentioned impacts could be ascribed to boron's participation in the NR protein's de novo production or the nitrate absorption simplification (Ruiz et al., 1998). In comparison to the control plants, tobacco plants exposed to an extreme boron deficiency (boron-deprived for 6 weeks) had a considerable drop in the activity of leaf NR, as well as potassium, calcium, magnesium, and, most importantly, the concentrations of nitrate (Camacho-Cristóbal and González-Fontes, 1999). Time-limited boron deficiency has been demonstrated to cause a decrease in nitrate content of the root and the leaf; however, this is not impacting the activity of NR and the other macronutrients concentrations including potassium, calcium, magnesium, and phosphate (Camacho-Cristóbal and González-Fontes, 2007). This lower nitrate amount has been attributed to boron-deficient plants' lower net nitrate absorption ratio, which was likely caused by a dip in the plasma membrane H + -ATPase (PMA2) transcript amounts. Following the boron-deficient treatment, it was found that asparagine synthetase promotes ammonium assimilation in the roots of tobacco (Camacho-Cristóbal and González-Fontes, 2007).

3.3.3 Oxidative stress and secondary metabolism

Boron appears to be one of the nutrients involved in phenolic compound accumulation and metabolism in plants (vascular). It is recognized that boron shortage leads to phenolic buildup via stimulating the enzyme phenylalanine-ammonium lyase (Cakmak and Römheld, 1997; Camacho-Cristóbal et al., 2002). According to various studies, boron deprivation causes quantitative and qualitative changes in the plant's phenolic pool (Camacho-Cristóbal et al., 2002; Camacho-Cristóbal et al., 2004; Karioti et al., 2006). As a result of boron shortage, two polyamine-phenolic conjugates aggregate, which have not been seen in boron-sufficient conditions.

Reports suggest that boron deficiency also enhances polyphenol oxidase activity (PPO) (Pfeffer et al., 1999; Camacho-Cristóbal et al., 2002), the enzyme that catalyzes the phenolic components' oxidation to quinones. It has been suggested that accumulating phenolics and their oxidation products might be the cause for loss of membrane integrity under boron deficiency, but its supplementation to leaves (deficient) does not restore plasma membrane integrity while complexing phenols or suppressing PPO action (Pfeffer et al., 1998; Ruiz et al., 1998; Cara et al., 2002).

Several studies have revealed that boron deficiency impacts the cycle of glutathione/ascorbate, which functions as a critical reaction in the detoxification process of oxygen toxic species from the cells. Both glutathione and ascorbate amounts have been shown to reduce in leaves and roots under boron deficiency environments (Lukaszewski and Blevins, 1996; Cakmak and Römheld, 1997). In BY-2 cells of tobacco, an elevation in glutathione S-transferase and glucosyltransferase expression has recently been reported, which might be a resistant pathway toward oxidative damage under boron deficiency (Kobayashi et al., 2004). Moreover, it increased glutathione amounts in aluminum-stressed maize and sunflower plants (Ruiz et al., 2006; Corrales et al., 2008), which supports the idea that enough boron activates antioxidant resistance in plants with aluminum stress.

3.4 The boron toxicity

The toxicity of boron is a global concern that severely affects crop yield in the zones (related to agriculture) of West Asia, North Africa, and Australia that are exposed to saline soils, poor rainfall, alkaline, and limited leaching. Furthermore, irrigation with water containing a higher concentration of boron can lead to boron-rich soils (Nable et al., 1997).

Boron poisoning has a variety of impacts in vascular plants, including modified metabolism, slower root cell division, reduction in leaf chlorophyll content and photosynthetic rates, and lower lignin and suberin amounts, to name a few. As a result, plants challenged with high boron concentration have stunted root and shoot development (Nable et al., 1990).

Boron aggregation in plants follow a passage from the base of the leaf to the tip, resulting in conventional toxicity symptoms including chlorosis (marginal/tip), and necrosis on older leaves (Marschner, 2011; Roessner et al., 2006). Even though the physiological reason for the toxicity of boron is unknown, three key explanations have been hypothesized based on the understanding of boron chemistry (its capability to bind cis-configured molecules having two hydroxyl groups): (1) metabolic disturbance by attaching to the ribose moieties of molecules including nicotinamide adenine dinucleotide (reduced form), adenosine triphosphate, nicotinamide adenine dinucleotide phosphate (reduced form), (2) binding to ribose, either as a free sugar or inside RNA, and (3) disrupting cell division and development (Reid et al., 2004).

It has not been proven that leaf toxicity (which is merged by osmotic stress) is caused by boron buildup. Although a low internal boron dose reduces the rate of plant growth, this inhibition is not caused by the impacts of boron on energy supply or the protein synthesis inhibition. The toxicity to older organs is related to the cumulative slowing down of several cellular mechanisms, which is exacerbated by photo-oxidative tension in the receiving sunlight (Reid et al., 2004).

3.4.1 Boron application for abiotic stress relief

B's toxicity and inadequacy are compounded by its chemical composition that impacts an extensive range of biological actions in plants, including assisting the plant in dealing with both abiotic and biotic stressors. The distribution of B in plants is shown in Fig. 3.1. In nature, stress does not generally come in isolation and many of them act synergistically.

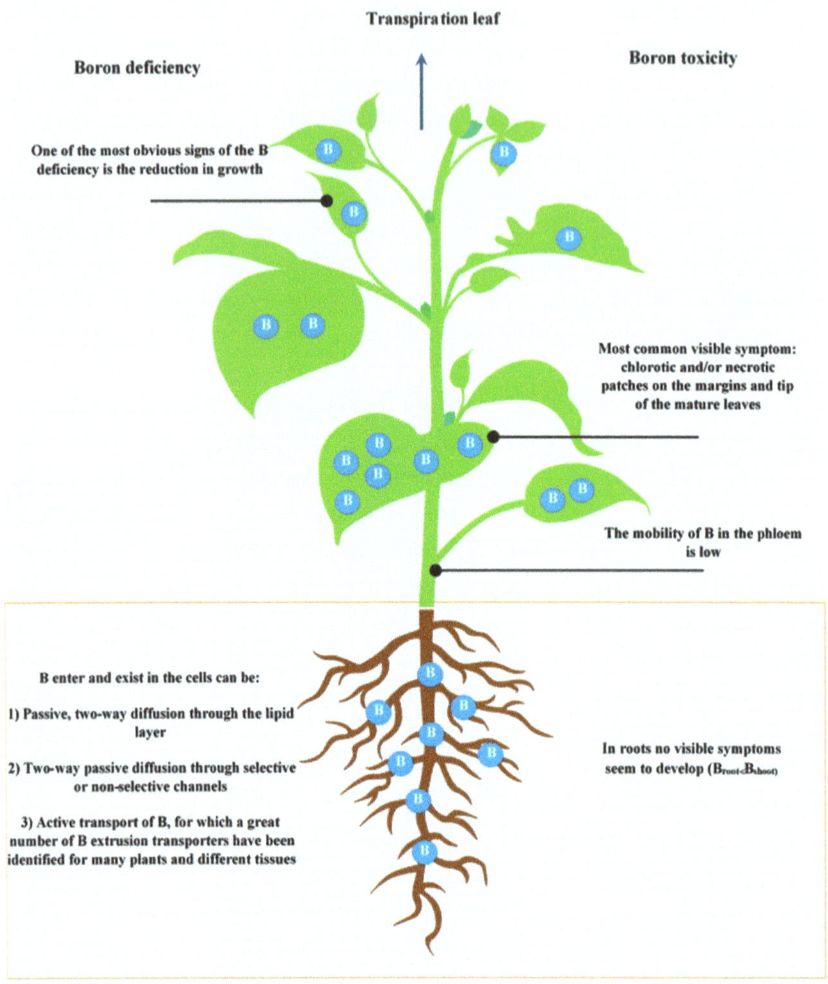

Figure 3.1 Boron distribution in plants.

Boron usage has been proven to help plants cope with the negative salt stress impacts on development, yield, and nutrient absorption. Boron is known for its involvement in the transport, carbohydrate metabolism, along with alterations under salinity tension. This relationship is particularly important due to its connections to processes like photosynthesis and respiration. Furthermore, boron is involved in the wall of cell production, structure, membrane, and composition. Its administration reduces the harmful salinity impacts, by maintaining cell viability that maintains the flexibility of the wall and helps in the re-establishment of K^+ levels.

B supplementation has been shown to restore nutritional balance and enhance the nitrogen-fixing legume plants' salt tolerance (Marschner, 2011). Drought-stressed crops frequently exhibit boron shortage because its availability is reduced along with a decrease in water delivery to the roots. Furthermore, Marschner (2011) reported that plant mineral nutrient condition acts as a vital factor for enhancing the drought tolerance of the plant.

Boron nutrition improves photosynthetic efficiency, synthesis of hormones, metabolism of lipid, flower retention, pollen development, and the germination of seeds, which help plants to tolerate drought stress. Pollen generation, seed/grain production, and seed germination are all aspects of pollination. Furthermore, plants with a high B-nutrient content have shown greater drought resilience by developing more root hairs and mycorrhizae, which enhances water intake from the soil rhizosphere. The nutritional condition of the boron, zinc, and manganese have two effects on plant drought sensitivity. Zn, B, and Mn are engaged in the ROS detoxification and thus have a preservative function in avoiding photo-oxidative harm in chloroplasts catalyzed by ROS. Second, these micronutrients might help with drought tolerance by shielding membranes from oxidative damage. The resistance process toward water deficiency is linked to high total ascorbate and glutathione, to control hydrogen peroxide overproduction, and alleviate the unfavorable repercussions of electrolyte leakage in the plasma membrane. Boron usage in drought-stricken areas has been demonstrated to improve people's nutritional health (Table 3.1) (Arbones and Rufat, 2005; Wróbel, 2009). If the theory that adding more B enhances Al toxicity in most crops grown in acidic soils is correct, B could be a viable candidate for increasing soil acidity and, as a result, Al toxicity is relatively less expensive than current soil-acidity amendment methods (calcium carbonate, liming).

3.4.2 Tolerance to boron toxicity

Boron-tolerant cultivars have lower boron concentrations in the leaf organs than nontolerant cultivars (Nable et al., 1997), which is most likely due to decreased boron absorption into the roots and shoots. In this view, the proper ability of the landrace cv Sahara to efflux boron has been

Table 3.1 Boron application for abiotic stress relief.

Stress	Crop	B application	Level of stress	Cause	Reference
Salinity	Varieties of tomato (*Solanum lycopersicum*) "Poncho negro."	20 milligrams per liter (boric acid). Solution for nutrient	NaCl concentrations of 75 and 150 mM	Reduce Na^+ transport to the leaves while increasing photosynthetic rates and osmo-regulators (proline and soluble carbohydrates) consumption.	Díaz et al. (2011)
	Badami pistachio (*Pistacia vera*)	5.0 milligrams per kilogram of soil1 (boric acid). Water used for irrigation	800, 1600, 2400, and 3200 mg/kg soil, respectively. Water used for irrigation.	Improve CO_2 assimilation and growth	Karimi and Tavallali (2017)
Drought	Wheat cv Ismena (*Triticum aestivum*)	7 cma^3 plant − 1 (0.3% H_3BO_3). Tillering phase foliar application	Field capacity is between 40 and 60 percent.	Grain and straw yields should be improved.	Wróbel (2009)
	apricots (*Prunus dulcis*)	7 L $tree^{-1}$ 832 ppm B Solubor (20.8 percent B). Application on the leaves	Dryland in the Mediterranean area	increase output	Arbones and Rufat (2005)
Toxicity of aluminum	*Trifolium vesiculosum* (arrowleaf clover)	0.25 kg of B per hectare (boric acid)	Aluminum saturation of 46.5 percent	Root and shoot development, as well as root spread in the soil profile, should be increased.	Favaretto et al. (2007)

attributed to its boron tolerance, and two theories for this active efflux of boron have been proposed, namely anion (borate) exchange and an anion channel (Hayes and Reid, 2004). Although BOR1 transports B from roots to shoots when B supply is low when B toxic level is high, BOR1 gets degraded via endocytosis (Takano et al., 2005), and fails to result in better plant growth (Miwa et al., 2006). These findings indicate that BOR1 does not have a function in the resistance of boron toxicity. Nonetheless, a transgenic *A. thaliana* line has been developed demonstrating that BOR4 (another efflux-type borate transporter) is not reduced in the posttranslational phase like BOR1 (Miwa et al., 2007). BOR4-GFP aggregation and boron resistance are considerably associated, and BOR4-GFP overproduction promotes plant development under high boron concentrations by boron efflux. Furthermore, the fluorescence of GFP originated from BOR4-GFP was reported to be visible in the epidermal cell's plasma membrane in the *A. thaliana*'s elongation region of root (Miwa et al., 2007).

Bot1 (a BOR1 ortholog) has also been recognized as a responsible gene for boron toxicity resistance in the barley landrace Sahara 3771. Bot1 transcript amounts in the leaf and roots are around 160-fold and 18-fold more in Sahara than in Clipper (Sutton et al., 2007). Moreover, in comparison with Clipper Bot1 or Arabidopsis BOR1, the protein of Sahara Bot1 has a better capability to confer boron resistance in the yeast. Bot1 transcript amounts indicate a decreased net entrance of boron toward the barley root and a larger boron removal from leaf guttation by hydathodes (Sutton et al., 2007). As a result, the Sahara cultivar's capability to hold lower shoot B concentrations is at least attributable to an active boron efflux mechanism from the root, that is, Bot1 encodes a functioning boron efflux transporter responsible for boron resistance. It has also been hypothesized that the BOR2 gene encodes a transporter (efflux-type borate) that is important in wheat and barley resistance to the toxicity of boron (Reid, 2007b).

Boron has also been shown to suppress a step of the pre-mRNA splicing reaction (in vitro) so that the toxicity of boron is predominantly caused by RNA splicing disruption. Multiple boron-resistant genes from *Arabidopsis* and *Lupin*, which encode the factors of transcription or ribosomal proteins have been shown to impart tolerance to increased boron concentration in yeast. These proteins may protect splicing sites due to boron attack, implying the existence of a mechanism for conferring boron resistance more than the ability to efflux boron from cells (Nozawa et al., 2006; Reid, 2007b).

3.5 Gene expression and boron

Several studies suggested that the deficiency of boron, impacts the gene expression involved in nitrogen metabolism (Redondo-Nieto et al., 2001; Camacho-Cristóbal and González-Fontes, 2007), oxidative stress (Kobayashi et al., 2004), boron absorption (Takano et al., 2006; Kasajima and Fujiwara, 2007), and the formation of the cell wall (Kasajima and Fujiwara, 2007). However, no concrete evidence exists to clarify how the boron insufficiency message is transmitted to the nuclei. After the cellular redox imbalance caused by boron shortage is postulated, a fast message transmitted from the wall of the cell to the cytoplasm could participate in the induction of the genes. According to Goldbach and Wimmer (2007), changes in boron concentrations might trigger the signals that extend toward the cytoplasm by the plasma membrane-cytoskeleton continuum of the cell wall, possibly including AGPs. The fact that boron deficiency affects the cytoskeletal protein's polymerization pattern (tubulin and actin) assemblies, support this concept (Yu et al., 2001, 2003). In brefeldin A-induced compartments, this leads to the endocytic route blockage in the internalization of boron cross-linked RG-II pectins (Yu et al., 2002). Finally, González-Fontes et al. (2008) hypothesized that boron serves as a biological signal that interacts with transcription agents, at the very moment when plants (vascular) are exposed to the deficiency of boron, and numerous gene expressions related in many physiological mechanisms are swiftly impacted. It is possible that boron is interfering with the mediated signaling (Bolaños et al., 2004). Furthermore, high boron concentration has been shown to stimulate numerous gene expressions in *Arabidopsis* rosette leaves and the roots (Kasajima and Fujiwara, 2007). For instance, the gene (At1g03770) has been discovered as a transcription factor of the zinc finger family that regulates the downstream gene expression which participates in the plant physiological resistant ways toward boron toxicity (Kasajima and Fujiwara, 2007). This proves that boron has a proposed position as a cellular signal proper of interacting with transcription agents (González-Fontes et al., 2008).

3.6 Conclusion and future prospect

Boron is required for vascular plant development, and proper boron nutrition is critical for agricultural production. This finding emphasizes the function of boron's necessity in plants and the molecular processes underlying plant responses to the deficiency of boron and its toxicity to develop crop resistance to boron stressors. The pectin assembly crosslinking and cell wall RG-II has been proven as one of the boron's main functions (Kobayashi et al., 1996). Nonetheless, Bolaños et al. (2004) proposed that boron's primary purpose in the bio-systems is to stabilize molecules with cis–diol groups, regardless of their function. As a result, it has been proposed that boron concentration in the membranes play its role by forming complexes with glycoproteins (Goldbach and Wimmer, 2007). Recently, boron's potential action as a signal (cellular) in the interaction with transcription agents has been hypothesized (González-Fontes et al., 2008). The boron-binding ligand's identification besides their activities in plant physiology should be considered in future research.

References

Anderson, D.L., Kitto, M.E., McCarthy, L., Zoller, W.H., 1994. Sources and atmospheric distribution of particulate and gas-phase boron. Atmos. Environ. 28 (8), 1401–1410.

Arbones, A., Rufat, J., 2005. Foliar application of boron to almond trees under dry conditions. Viticultura Enología Profesional. Extraordin. (España).

Bassil, E., Hu, H., Brown, P.H., 2004. Use of phenylboronic acids to investigate boron function in plants. Possible role of boron in transvacuolar cytoplasmic strands and cell-to-wall adhesion. Plant. Physiol. 136 (2), 3383–3395.

Bergmann, W., Bruchlos, P., Marks, G., 1995. Zum toxischen Grenzwert für Bor/The toxic limiting value of boron. Tenside Surfactants Detergents 32 (3), 229–237.

Bingham, F., 1982. The Boron Concentration of Wild Trout Streams in California. Report to the Procter & Gamble Company, Ivorydale Technical Center, Cincinnati, OH, p. 45.

Birge, W., Black, J., Westermann, A., Short, T., Taylor, S., Parekh, M., 1984. Toxicity of Boron to Embryonic and Larval Stages of Rainbow Trout (*Salmo Gairdneri*) Exposed in Reconstituted and Natural Waters. Completion Report prepared for the Procter & Gamble Company, Cincinnati, USA.

Black, J.A., Barnum, J.B., Birge, W.J., 1993. An integrated assessment of the biological effects of boron to the rainbow trout. Chemosphere 26 (7), 1383–1413.

Blaser-Grill, J., Knoppik, D., Amberger, A., Goldbach, H., 1989. Influence of boron on the membrane potential in *Elodea densa* and *Helianthus annuus* roots and H^+ extrusion of suspension cultured *Daucus carota* cells. Plant. Physiol. 90 (1), 280–284.

Blevins, D.G., Lukaszewski, K.M., 1998. Boron in plant structure and function. Annu. Rev. plant. Biol. 49 (1), 481–500.

Bolaños, L., Brewin, N.J., Bonilla, I., 1996. Effects of boron on Rhizobium-legume cell-surface interactions and nodule development. Plant. Physiol. 110 (4), 1249–1256.

Bolaños, L., Cebrián, A., Redondo-Nieto, M., Rivilla, R., Bonilla, I., 2001. Lectin-like glycoprotein PsNLEC-1 is not correctly glycosylated and targeted in boron-deficient pea nodules. Mol. Plant-Microbe Interact. 14 (5), 663–670.

Bolaños, L., Esteban, E., de Lorenzo, C., Fernandez-Pascual, M., de Felipe, M.R., Garate, A., et al., 1994. Essentiality of boron for symbiotic dinitrogen fixation in pea (*Pisum sativum*) rhizobium nodules. Plant. Physiol. 104 (1), 85–90.

Bolaños, L., Lukaszewski, K., Bonilla, I., Blevins, D., 2004. Why boron? Plant. Physiol. Biochem. 42 (11), 907–912.
Bolaños, L., Redondo-Nieto, M., Bonilla, I., Wall, L.G., 2002. Boron requirement in the *Discaria trinervis* (Rhamnaceae) and Frankia symbiotic relationship. Its essentiality for Frankia BCU110501 growth and nitrogen fixation. Physiologia Plant. 115 (4), 563–570.
Bonilla, I., El-Hamdaoui, A., Bolaños, L., 2004. Boron and calcium increase *Pisum sativum* seed germination and seedling development under salt stress. Plant. Soil. 267 (1), 97–107.
Brown, P.H., Bellaloui, N., Wimmer, M., Bassil, E.S., Ruiz, J., Hu, H., et al., 2002. Boron in plant biology. Plant. Biol. 4 (2), 205–223.
Cakmak, I., Kurz, H., Marschner, H., 1995. Short-term effects of boron, germanium and high light intensity on membrane permeability in boron deficient leaves of sunflower. Physiologia Plant. 95 (1), 11–18.
Cakmak, I., Römheld, V., 1997. Boron deficiency-induced impairments of cellular functions in plants. Plant. Soil. 193 (1), 71–83.
Camacho-Cristóbal, J.J., Anzellotti, D., González-Fontes, A., 2002. Changes in phenolic metabolism of tobacco plants during short-term boron deficiency. Plant. Physiol. Biochem. 40 (12), 997–1002.
Camacho-Cristóbal, J.J., González-Fontes, A., 1999. Boron deficiency causes a drastic decrease in nitrate content and nitrate reductase activity, and increases the content of carbohydrates in leaves from tobacco plants. Planta 209 (4), 528–536.
Camacho-Cristóbal, J.J., González-Fontes, A., 2007. Boron deficiency decreases plasmalemma H + -ATPase expression and nitrate uptake, and promotes ammonium assimilation into asparagine in tobacco roots. Planta 226 (2), 443–451.
Camacho-Cristóbal, J.J., Lunar, L., Lafont, F., Baumert, A., González-Fontes, A., 2004. Boron deficiency causes accumulation of chlorogenic acid and caffeoyl polyamine conjugates in tobacco leaves. J. Plant. Physiol. 161 (7), 879–881.
Camacho-Cristóbal, J.J., Maldonado, J.M., González-Fontes, A., 2005. Boron deficiency increases putrescine levels in tobacco plants. J. Plant. Physiol. 162 (8), 921–928.
Cara, F.A., Sánchez, E., Ruiz, J.M., Romero, L., 2002. Is phenol oxidation responsible for the short-term effects of boron deficiency on plasma-membrane permeability and function in squash roots? Plant. Physiol. Biochem. 40 (10), 853–858.
Corrales, I., Poschenrieder, C., Barceló, J., 2008. Boron-induced amelioration of aluminum toxicity in a monocot and a dicot species. J. Plant. Physiol. 165 (5), 504–513.
Cosgrove, D.J., 1999. Enzymes and other agents that enhance cell wall extensibility. Annu. Rev. Plant. Biol. 50 (1), 391–417.
Dell, B., Huang, L., 1997. Physiological response of plants to low boron. Plant. Soil. 193 (1), 103–120.
Deverel, S.J., Millard, S.P., 1988. Distribution and mobility of selenium and other trace elements in shallow groundwater of the western San Joaquin Valley, California. Environ. Sci. & Technol. 22 (6), 697–702.
Dyer, S.D., Caprara, R.J., 1997. A method for evaluating consumer product ingredient contributions to surface and drinking water: boron as a test case. Environ. Toxicol. Chem. 16 (10), 2070–2081.
Díaz, M., Bastías, E., Pacheco, P., Tapia, L., Martínez-Ballesta, M., Carvajal, M., 2011. Characterization of the physiological response of the highly-tolerant tomato cv.'poncho negro'to salinity and excess boron. J. Plant. Nutr. 34 (9), 1254–1267.
Favaretto, N., Motta, A.C.V., Barcik, C., Lustosa, S.B.C., Comin, J.J., 2007. Shoot and root responses of Trifolium vesiculosum to boron fertilization in an acidic Brazilian soil. Braz. Arch. Biol. Technol. 50, 597–604.

Ferrol, N., Donaire, J.P., 1992. Effect of boron on plasma membrane proton extrusion and redox activity in sunflower cells. Plant. Sci. 86 (1), 41–47.
Fleischer, A., O'Neill, M.A., Ehwald, R., 1999. The pore size of non-graminaceous plant cell walls is rapidly decreased by borate ester cross-linking of the pectic polysaccharide rhamnogalacturonan II. Plant. Physiol. 121 (3), 829–838.
Gerike, P., Fischer, W., Holtmann, W., 1976. Der Einfluß von Bor auf die aerobe biologische Abwasserreinigung. Tenside Surfact Detergents 13 (5), 249–252.
Goldbach, H.E., Wimmer, M.A., 2007. Boron in plants and animals: is there a role beyond cell-wall structure? J. Plant. Nutr. Soil. Sci. 170 (1), 39–48.
Goldbach, H.E., Yu, Q., Wingender, R., Schulz, M., Wimmer, M., Findeklee, P., et al., 2001. Rapid response reactions of roots to boron deprivation. J. Plant. Nutr. Soil. Sci. 164 (2), 173–181.
Goldberg, S., 1997. Reactions of boron with soils. Plant. Soil. 193 (1), 35–48.
González-Fontes, A., Rexach, J., Navarro-Gochicoa, M.T., Herrera-Rodríguez, M.B., Beato, V.M., Maldonado, J.M., et al., 2008. Is boron involved solely in structural roles in vascular plants? Plant. Signal. & Behav. 3 (1), 24–26.
Hamilton, S.J., 1995. Hazard assessment of inorganics to three endangered fish in the Green River, Utah. Ecotoxicol. Environ. Saf. 30 (2), 134–142.
Hamilton, S.J., Buhl, K.J., 1990. Acute toxicity of boron, molybdenum, and selenium to fry of chinook salmon and coho salmon. Arch. Environ. Contamin Toxicol. 19 (3), 366–373.
Hayes, J.E., Reid, R.J., 2004. Boron tolerance in barley is mediated by efflux of boron from the roots. Plant. Physiol. 136 (2), 3376–3382.
Howe, P.D., 1998. A review of boron effects in the environment. Biol. Trace Elem. Res. 66 (1), 153–166.
Karimi, S., Tavallali, V., 2017. Interactive effects of soil salinity and boron on growth, mineral composition and CO2 assimilation of pistachio seedlings. Acta Physiologiae Plant. 39 (11), 1–10.
Karioti, A., Chatzopoulou, A., Bilia, A.R., Liakopoulos, G., Stavrianakou, S., Skaltsa, H., 2006. Novel secoiridoid glucosides in *Olea europaea* leaves suffering from boron deficiency. Biosci, Biotechnol, Biochem. 70 (8), 1898–1903.
Kasajima, I., Fujiwara, T., 2007. Identification of novel *Arabidopsis thaliana* genes which are induced by high levels of boron. Plant. Biotechnol. 24 (3), 355–360.
Kastori, R., Petrović, N., 1989. Effect of boron on nitrate reductase activity in young sunflower plants. J. Plant. Nutr. 12 (5), 621–632.
Khadem Moghadam, K., Asgari Lajayer, B., Ghorbanpour, M., 2020. Boron tolerance in plants: physiological roles and transport mechanisms. In: Deshmukh, R., Tripathi, D. K., Guerriero, G. (Eds.), Metalloids in Plants: Advances and Future Prospects. John Wiley & Sons Ltd, pp. 301–314.
Kobayashi, M., Matoh, T., Azuma, J., 1996. Two chains of rhamnogalacturonan II are cross-linked by borate-diol ester bonds in higher plant cell walls. Plant. Physiol. 110 (3), 1017–1020.
Kobayashi, M., Mutoh, T., Matoh, T., 2004. Boron nutrition of cultured tobacco BY-2 cells. IV. Genes induced under low boron supply. J. Exp. Botany 55 (401), 1441–1443.
Lemarchand, D., Gaillardet, J., Lewin, E., Allegre, C., 2000. The influence of rivers on marine boron isotopes and implications for reconstructing past ocean pH. Nature 408 (6815), 951–954.
Lukaszewski, K.M., Blevins, D.G., 1996. Root growth inhibition in boron-deficient or aluminum-stressed squash may be a result of impaired ascorbate metabolism. Plant. Physiol. 112 (3), 1135–1140.

Maier, K.J., Knight, A.W., 1991. The toxicity of waterborne boron to *Daphnia magna* and *Chironomus decorus* and the effects of water hardness and sulfate on boron toxicity. Arch. Environ. Contamin Toxicol. 20 (2), 282–287.
Marschner, H., 2011. Mineral Nutrition of Higher Plants. Academic press.
Matoh, T., Ochiai, K., 2005. Distribution and partitioning of newly taken-up boron in sunflower. Plant. Soil. 278 (1), 351–360.
Miwa, K., Takano, J., Fujiwara, T., 2006. Improvement of seed yields under boron-limiting conditions through overexpression of BOR1, a boron transporter for xylem loading, in *Arabidopsis thaliana*. Plant. J. 46 (6), 1084–1091.
Miwa, K., Takano, J., Omori, H., Seki, M., Shinozaki, K., Fujiwara, T., 2007. Plants tolerant of high boron levels. Science 318 (5855), 1417. -1417.
Nable, R.O., Bañuelos, G.S., Paull, J.G., 1997. Boron toxicity. Plant. Soil. 193 (1), 181–198.
Nable, R.O., Cartwright, B., Lance, R.C., 1990. Genotypic differences in boron accumulation in barley: relative susceptibilities to boron deficiency and toxicity. Genetic Aspects of Plant Mineral Nutrition. Springer, pp. 243–251.
Noguchi, K., Ishii, T., Matsunaga, T., Kakegawa, K., Hayashi, H., Fujiwara, T., 2003. Biochemical properties of the cell wall in the Arabidopsis mutant bor1-1 in relation to boron nutrition. J. Plant. Nutr. Soil. Sci. 166 (2), 175–178.
Nozawa, A., Miwa, K., Kobayashi, M., Fujiwara, T., 2006. Isolation of *Arabidopsis thaliana* cDNAs that confer yeast boric acid tolerance. Biosci, Biotechnol, Biochem. 70 (7), 1724–1730.
O'Neill, M.A., Ishii, T., Albersheim, P., Darvill, A.G., 2004. Rhamnogalacturonan II: structure and function of a borate cross-linked cell wall pectic polysaccharide. Annu. Rev. Plant. Biol. 55, 109–139.
Pfeffer, H., Dannel, F., Römheld, V., 1998. Are there connections between phenol metabolism, ascorbate metabolism and membrane integrity in leaves of boron-deficient sunflower plants? Physiologia Plant. 104 (3), 479–485.
Pfeffer, H., Dannel, F., Römheld, V., 1999. Are there two mechanisms for boron uptake in sunflower? J. Plant. Physiol. 155 (1), 34–40.
Power, P.P., Woods, W.G., 1997. The chemistry of boron and its speciation in plants. Plant. Soil. 193 (1), 1–13.
Ramón, A.M., Ruiz, R.O.C., Gárate, A., 1989. In vitro stabilization and distribution of nitrate reductase in tomato plants. Incidence of boron deficiency. J. Plant. Physiol. 135 (1), 126–128.
Redondo-Nieto, M., Rivilla, R., El-Hamdaoui, A., Bonilla, I., Bolaños, L., 2001. Research note: boron deficiency affects early infection events in the pea-Rhizobium symbiotic interaction. Funct. Plant. Biol. 28 (8), 819–823.
Redondo-Nieto, M., Pulido, L., Reguera, M., Bonilla, I., Bolanos, L., 2007. Developmentally regulated membrane glycoproteins sharing antigenicity with rhamnogalacturonan II are not detected in nodulated boron deficient *Pisum sativum*. Plant, Cell Environ. 30 (11), 1436–1443.
Reid, R.J., Hayes, J.E., Post, A., Stangoulis, J.C.R., Graham, R.D., 2004. A critical analysis of the causes of boron toxicity in plants. Plant. Cell Environ. 27 (11), 1405–1414.
Reid, R., 2007a. Update on boron toxicity and tolerance in plants. In: Xu, F., Goldbach, H.E., Brown, P.H., Bell, R.W., Fujiwara, T., Hunt, C.D., et al.,Advances in Plant and Animal Boron Nutrition: Proceedings of the 3rd International Symposium on All Aspects of Plant and animal Boron Nutrition. Springer, Dordrecht, pp. 83–90.
Reid, R., 2007b. Identification of boron transporter genes likely to be responsible for tolerance to boron toxicity in wheat and barley. Plant. Cell Physiol. 48 (12), 1673–1678.

Roessner, U., Patterson, J.H., Forbes, M.G., Fincher, G.B., Langridge, P., Bacic, A., 2006. An investigation of boron toxicity in barley using metabolomics. Plant. Physiol. 142 (3), 1087–1101.

Ruiz, J., Baghour, M., Bretones, G., Belakbir, A., Romero, L., 1998. Nitrogen metabolism in tobacco plants (*Nicotiana tabacum* L.): role of boron as a possible regulatory factor. Int. J. Plant. Sci. 159 (1), 121–126.

Ruiz, J., Rivero, R., Romero, L., 2006. Boron increases synthesis of glutathione in sunflower plants subjected to aluminum stress. Plant. Soil. 279 (1), 25–30.

Ryden, P., Sugimoto-Shirasu, K., Smith, A.C., Findlay, K., Reiter, W.D., McCann, M. C., 2003. Tensile properties of Arabidopsis cell walls depend on both a xyloglucan cross-linked microfibrillar network and rhamnogalacturonan II-borate complexes. Plant. Physiol. 132 (2), 1033–1040.

Saiki, M.K., Jennings, M.R., Brumbaugh, W.G., 1993. Boron, molybdenum, and selenium in aquatic food chains from the lower San Joaquin River and its tributaries, California. Arch. Environ. Contamin Toxicol. 24 (3), 307–319.

Schöberl, P., Huber, L., 1988. Ökologisch relevante Daten von nichttensidischen Inhaltsstoffen in Wasch-und Reinigungsmitteln. Tenside Surfact Detergents 25 (2), 99–107.

Shelp, B.J., Shattuck, V.I., Proctor, J.T.A., 1987. Boron nutrition and mobility, and its relation to the elemental composition of greenhouse grown root crops. II Radish. Commun. Soil. Sci. Plant. Anal. 18 (2), 203–219.

Shen, Z., Liang, Y., Shen, K., 1993. Effect of boron on the nitrate reductase activity in oilseed rape plants. J. Plant. Nutr. 16 (7), 1229–1239.

Shorrocks, V.M., 1997. The occurrence and correction of boron deficiency. Plant. Soil. 193 (1), 121–148.

Singh, R.G., 1972. Effect of phosphate and boron on growth, nodulation and seed yield of dhaincha (*Sesbania cannabina*. Retz. Pers). Indian. J. Agric. Sci. 42, 139–144.

Sutton, T., Baumann, U., Hayes, J., Collins, N.C., Shi, B.J., Schnurbusch, T., et al., 2007. Boron-toxicity tolerance in barley arising from efflux transporter amplification. Science 318 (5855), 1446–1449.

Takano, J., Miwa, K., Yuan, L., von Wirén, N., Fujiwara, T., 2005. Endocytosis and degradation of BOR1, a boron transporter of Arabidopsis thaliana, regulated by boron availability. Proc. Natl Acad. Sci. 102 (34), 12276–12281.

Takano, J., Wada, M., Ludewig, U., Schaaf, G., Von Wirén, N., Fujiwara, T., 2006. The Arabidopsis major intrinsic protein NIP5; 1 is essential for efficient boron uptake and plant development under boron limitation. Plant. Cell 18 (6), 1498–1509.

Tanaka, M., Fujiwara, T., 2008. Physiological roles and transport mechanisms of boron: perspectives from plants. Pflügers Archiv-Eur J. Physiol. 456 (4), 671–677.

Taylor, D., Maddock, B., Mance, G., 1985. The acute toxicity of nine 'grey list'metals (arsenic, boron, chromium, copper, lead, nickel, tin, vanadium and zinc) to two marine fish species: dab (*Limanda limanda*) and grey mullet (*Chelon labrosus*). Aquat. Toxicol. 7 (3), 135–144.

Thompson, J., Davis, J., Drew, R., 1976. Toxicity, uptake and survey studies of boron in the marine environment. Water Res. 10 (10), 869–875.

Wallen, I., Greer, W., Lasater, R., 1957. Toxicity to *Gambusia Affinis* of certain pure chemicals in turbid waters. Sew. Ind. Wastes 29 (6), 695–711.

Wang, Z., Tang, Y., Zhang, F., Wang, H., 1999. Effect of boron and low temperature on membrane integrity of cucumber leaves. J. Plant. Nutr. 22 (3), 543–550.

Wren, C., MacCrimmon, H., Loescher, B., 1983. Examination of bioaccumulation and biomagnification of metals in a precambrian shield lake. Water, Air, Soil. Pollut. 19 (3), 277–291.

Wróbel, S., 2009. Response of spring wheat to foliar fertilization with boron under reduced boron availability. J. Elementol 14 (2), 395–404.

Yamagishi, M., Yamamoto, Y., 1994. Effects of boron on nodule development and symbiotic nitrogen fixation in soybean plants. Soil. Sci. Plant. Nutr. 40 (2), 265–274.

Yan, X., Wu, P., Ling, H., Xu, G., Xu, F., Zhang, Q., 2006. Plant nutriomics in China: an overview. Ann. Botany 98 (3), 473–482.

Yu, Q., Baluška, F., Jasper, F., Menzel, D., Goldbach, H.E., 2003. Short-term boron deprivation enhances levels of cytoskeletal proteins in maize, but not zucchini, root apices. Physiologia Plant. 117 (2), 270–278.

Yu, Q., Hlavacka, A., Matoh, T., Volkmann, D., Menzel, D., Goldbach, H.E., et al., 2002. Short-term boron deprivation inhibits endocytosis of cell wall pectins in meristematic cells of maize and wheat root apices. Plant. Physiol. 130 (1), 415–421.

Yu, Q., Wingender, R., Schulz, M., Baluška, F., Goldbach, H., 2001. Short-term boron deprivation induces increased levels of cytoskeletal proteins in Arabidopsis roots. Plant. Biol. 3 (4), 335–340.

CHAPTER 4

Molecular regulatory mechanisms in plants that underlie phenotypic adaptations to low boron levels

Michaela S. Matthes
Institute for Crop Science and Resource Conservation, Crop Functional Genomics, University of Bonn, Bonn, Germany

4.1 Introduction

The essential micronutrient boron (Warington, 1923) cannot be produced by plants themselves and has to be taken up from the soil. Therefore, deficiencies of this nutrient in plants occur when soil boron levels are low, which prevail in many parts of the world (Shorrocks, 1997). Plant boron deficiencies lead to striking defects in plant growth and development. Typical phenotypic defects are an inhibition of root growth including a reduction in lateral root number, defects in leaf expansion, with small and brittle leaves, and most pronounced defects in reproductive development, including flowering and seed setting (as reviewed in Matthes et al., 2020). The defects in reproductive development consequently lead to drastic yield declines in many major crops (Dell and Huang, 1997 and references therein). In addition, symptoms like heart browning and heart rotting have been associated with boron deficiency (Sotta et al., 2019).

The first defects under boron deficiency have been observed in plant meristems (Warington, 1923; Sommer and Sorokin, 1928; Chatterjee et al., 2014; Durbak et al., 2014). Meristems are groups of undifferentiated stem cells that are responsible for all postembryonic growth, both above (shoot apical meristem) and below (root apical meristem) ground. Meristem development is halted in boron deficient conditions, yet the molecular mechanisms of how boron deficiency leads to meristem defects or the reasons why particularly meristems are sensitive to boron deficiency are just beginning to be understood.

Boron in Plants and Agriculture
DOI: https://doi.org/10.1016/B978-0-323-90857-3.00016-3

The best characterized functional role of boron is associated with its cross-linking capabilities of the pectic subunits, rhamnogalacturonan-II (RG-II), forming boron-RG-II dimers in the primary cell wall (Kobayashi et al., 1996; Matoh et al., 1996; O'Neill et al., 2001). This boron-RG-II crosslinking contributes to the stability of the primary cell wall. In boron deficient conditions, less boron enters the cell, leading to less boron-RG-II dimers in the cell wall and consequently to a weakening of the primary cell wall. Indeed, tolerance to low boron conditions in certain oilseed rape genotypes could be associated with the higher mechanical strength of their cell walls due to variations in pectin concentrations (Zhou et al., 2017). Yet, the defects of the cell walls in boron deficient conditions cannot explain all phenotypic defects seen under boron deficiency (Brown et al., 2002; Matthes et al., 2020), suggesting additional molecular regulatory mechanisms targeted by boron. Particularly the identification and characterization of boron transporters in various plant species progressed considerably. Although it was long believed that boron is only passively diffusing into cells, it is now established that passive boron transport prevails under adequate soil boron levels (Dordas and Brown, 2000), while facilitated and active transport of boron is needed, when soil boron levels are low (Noguchi et al., 1997; Dordas et al., 2000; Takano et al., 2006). Boron import into the plant root is facilitated by channel proteins of the MIP family and transported within the plant by exporters of the BOR family (reviewed in Yoshinari and Takano, 2017). In Arabidopsis, NIP5;1 is the main boron importer protein, which transports boron from the soil into the root (Takano et al., 2006). HIGH BORON REQUIRING1 (BOR1) is involved in the export of boron out of the cell. Both of these transporter families are polarly localized within cells, with BOR1 displaying polar localization towards the inner/stele side in various root cell types, the hypocotyl, and cotyledons of Arabidopsis (Takano et al., 2010; Yoshinari et al., 2019) and NIP5;1 showing polar localization towards the soil side in Arabidopsis root epidermal and root cap cells (Takano et al., 2010). The polar localization of BOR1 and NIP5;1 leads to a directional, radial transport of boron from the soil into the stele, from where boron is loaded into the xylem and transported through the transpiration stream to the shoot. Through mathematical modeling and experimental evidence it was shown in the Arabidopsis root, that the region around the quiescent center has a higher boron concentration of soluble boron compared to other regions of the root

(Shimotohno et al., 2015). More detailed laser ablation (LA)-inductively coupled plasma (ICP)-mass spectrometry (MS) analysis showed that boron concentration is highest near the tip and lower in the more proximal region of the meristem zone (Shimotohno et al., 2015). Research in the last 20 years has identified additional boron transporters in Arabidopsis (Tanaka et al., 2008; Li et al., 2011; Miwa et al., 2013; Miwa et al., 2014; Routray et al., 2018), but also orthologs of *AtNIP5;1* and *AtBORs* in various other plant species including maize (Chatterjee et al., 2014; Durbak et al., 2014; Leonard et al., 2014; Chatterjee et al., 2017), rapeseed (Zhang et al., 2017), barley (Sutton et al., 2007), and rice (Nakagawa et al., 2007; Hanaoka et al., 2014; Shao et al., 2018). Similar to cell wall mutants, phenotypes of boron transporter mutants match only partially the phenotypes of normal plants growing under boron deficiency (as reviewed in Matthes et al., 2020), again suggesting additional molecular mechanisms targeted by boron levels.

Boron interaction partners were identified not only in the cell wall but also in the plasma membrane (Wimmer et al., 2009; Voxeur and Fry, 2014) and various interactions of boron with phytohormones were suggested (Abreu et al., 2014; Eggert and von Wirén, 2017; Gómez-Soto et al., 2019; Huang et al., 2021; Li et al., 2015; Martín-Rejano et al., 2011; Matthes et al., 2022; Matthes and Torres-Ruiz, 2016; Poza-Viejo et al., 2018; Wang et al., 2006; Zhang et al., 2020). Boron deficiency can lead to altered metabolite abundances (Alves et al., 2011; Camacho-Cristóbal and González-Fontes, 2007; Dong et al., 2016; Wilder et al., 2022) and to changes in transcript levels (Camacho-Cristóbal et al., 2011; Peng et al., 2012), micro-RNA abundances (Lu et al., 2014; Lu et al., 2015) and phosphorylation intensities (Chen and Hoehenwarter, 2019), that affect genes and proteins of various cellular processes. The identification of the first boron dependent transcription factors (Kasajima et al., 2010; Feng et al., 2020) further suggests a direct involvement of boron in signaling processes as previously postulated (González-Fontes et al., 2013), that could explain why many diverse cellular processes are affected in boron deficient conditions.

Most of our current knowledge about the molecular mechanisms underlying root and shoot meristem development in plants comes from the model organism Arabidopsis and the crop maize, representing dicotyledonous and monocotyledonous plants, the two groups of flowering plants. Root meristem development in Arabidopsis is under transcriptional control by specific transcription factors, mainly the WUSCHEL-LIKE HOMEOBOX5 (WOX5) transcription factor, members of the PLETHORA (PLT) transcription factor family, the GRAS transcription factor SCARECROW, its main interaction partner SHORTROOT, and the PLANT-SPECIFIC TEOSINTE-BRANCHED CYCLOIDEA PCNA transcription factors (as reviewed for example in Motte et al., 2019). Arabidopsis, but also maize root development additionally result from the coordinated action of multiple plant hormones, including and specifically auxin (as reviewed in Brumos et al., 2018; Hochholdinger et al., 2018). In shoot meristems, genetic factors that control stem cell number, fate, and function center around the CLAVATA (CLV)-WUSCHEL (WUS) pathway (as reviewed in Wu et al., 2018) and the action of transcription factors (TFs), like the class I KNOTTED-LIKE homeobox transcription factors, exemplified by KNOTTED1 (KN1) in maize (Vollbrecht et al., 1991) and SHOOTMERISTEMLESS (STM) in Arabidopsis (Long et al., 1996). While *CLV-WUS* pathway genes form a negative feedback loop to regulate the formation and maintenance of shoot meristems (as reviewed in Kitagawa and Jackson, 2019), *kn1* and *STM* are required for shoot meristem maintenance to prevent cell differentiation. (Vollbrecht et al., 1991; Jackson et al., 1994; Long et al., 1996; Kerstetter et al., 1997). In addition, phytohormones like auxin and cytokinin contribute to shoot meristem function, where cytokinin promotes meristem maintenance and auxin regulates organ initiation and differentiation (as reviewed in Lee et al., 2019). How boron levels integrate into these genetic pathways to shape plant architecture above or below ground is not known. It seems apparent though that due to the involvement of different players in either meristem type, the targets of varying boron levels are different in the shoot and the root.

Due to their importance for meristem development and their observed alterations in boron deficient conditions, particularly transcription factors and their target genes and phytohormone pathways offer avenues to understand why particularly plant meristems are sensitive to boron deficiency. In this chapter, the current knowledge of the molecular regulatory mechanisms underlying boron deficiency-induced phenotypic defects in plants is compiled, with a special emphasis on interactions of boron with phytohormones, boron-dependent transcription factors, and transcript level changes, as they appear in close relation to observed root and shoot apical meristem defects.

4.2 Cessation of plant growth in boron deficient conditions: cell elongation or cell division

Boron deficiency defects in plants are characterized through cessation of growth, resulting in root, leaf, flower, and meristem defects in many plant species (as reviewed in Matthes et al., 2020). Plants grown without boron develop short and thick roots with a stunted appearance (Warington, 1923; Sommer, 1927; Sommer and Sorokin, 1928; Martín-Rejano et al., 2011; Li et al., 2016; Poza-Viejo et al., 2018; Housh et al., 2020). Above ground, vegetative plant growth can appear normal (Warington, 1923; Dell and Huang, 1997; Lordkaew et al., 2011), plants can develop leathery leaves with or without yellow dots and plant development can be slower up until flowering (Warington, 1923). During reproductive development, plants can be smaller, stunted and dark green in comparison to control plants (Warington, 1923) and can reach from withering and falling off of flower buds (Warington, 1923) to plants not reaching flowering (Sommer, 1927). These phenotypes originate in plant meristems, where the first boron deficiency defects are observed (Sommer and Sorokin, 1928; Chatterjee et al., 2014; Durbak et al., 2014). In the root, boron deficiency leads to an absence of a recognizable meristem (Sommer and Sorokin, 1928) and progressive shoot meristem defects, as shown in maize reproductive meristems, cause the observed flower defects in maize boron transporter mutants (Chatterjee et al., 2014; Durbak et al., 2014). Boron deficiency defects are characterized through shorter or smaller organs, which can cellularly occur through cell elongation, cell division defects, or a combination of both. During the cessation of root elongation in boron deficient conditions, both impairment of cell enlargement and reduced division in the meristematic zone were reported (Dell and Huang, 1997 and references therein), yet which process is affected first, and therefore primary, is debated. Early research stated that the inhibition of root elongation in squash results from the cessation of mitosis and cell division rather than cell elongation (Cohen and Albert, 1974). Yet, cell division defects upon boron deficiency were later proclaimed secondary (Dell and Huang, 1997; Martín-Rejano et al., 2011), since they were only observed under prolonged boron deficiency conditions (Cohen and Albert, 1974) and since the cell division marker *CyclinB1* was unaltered after four days in low boron conditions in the primary root meristem of Arabidopsis (Martín-Rejano et al., 2011). These results were challenged by studies pointing out the importance of cell division for boron-deficiency induced growth defects in the Arabidopsis root

(Abreu et al., 2014; Poza-Viejo et al., 2018; Zhang et al., 2020). The root meristem was more sensitive to boron deficiency than the elongation of the differentiated hypocotyl cells (Abreu et al., 2014) and the inhibition of cell proliferation in the root meristem coincided in time with cell elongation defects (Poza-Viejo et al., 2018), therefore showing that cell division defects are early events contributing to root growth arrest under boron deficiency (Poza-Viejo et al., 2018). The different conclusions and outcomes of these studies can likely be attributed to the different boron conditions applied (0.4 μM boron and 10 μM boron in Martín-Rejano et al., 2011 in comparison to 0.03 μM boron and 30 μM boron in Poza-Viejo et al., 2018). However, in combination these findings suggest that both cell elongation inhibition and reduced cell division act in concert under boron deficiency causing root growth defects. In shoot meristems, the roles of cell elongation and cell division defects in boron deficient conditions remain to be assessed.

4.3 Interactions of boron with phytohormones

Phytohormones play crucial roles during all aspects of plant development in control, but also in stress conditions. Studies showing interactions between boron levels and particularly auxin (Jarvis et al., 1983; Wang et al., 2006; Martín-Rejano et al., 2011; Camacho-Cristóbal et al., 2015; Li et al., 2015; Matthes and Torres-Ruiz, 2016), cytokinin (Li et al., 2001; Wang et al., 2006; Abreu et al., 2014; Eggert and von Wirén, 2017; Poza-Viejo et al., 2018), ethylene (Martín-Rejano et al., 2011; Camacho-Cristóbal et al., 2015), abscisic acid (Eggert and von Wirén, 2017; Gómez-Soto et al., 2019), jasmonic acid (Peng et al., 2012; Huang et al., 2021), and brassinosteroids (Zhang et al., 2020) have been reported. Interactions of boron with these phytohormones are reported to occur on the levels of biosynthesis, signaling, transport, or catabolism (Fig. 4.1).

4.3.1 Boron–auxin interactions

Reports about potential boron–auxin interactions go back to the observation that boron deficiency symptoms in cotton resembled auxin deficiency symptoms and that auxin could at least to some extent replace boron (Eaton, 1940). Since then, boron-auxin interactions have been proposed on various levels, including biosynthesis, signaling, transport, and catabolism (Fig. 4.1A). While the mechanisms behind such interactions are not resolved, boron–auxin catabolism interactions were shown on the level of auxin oxidases (as reviewed in Blevins and Lukaszewski, 1998).

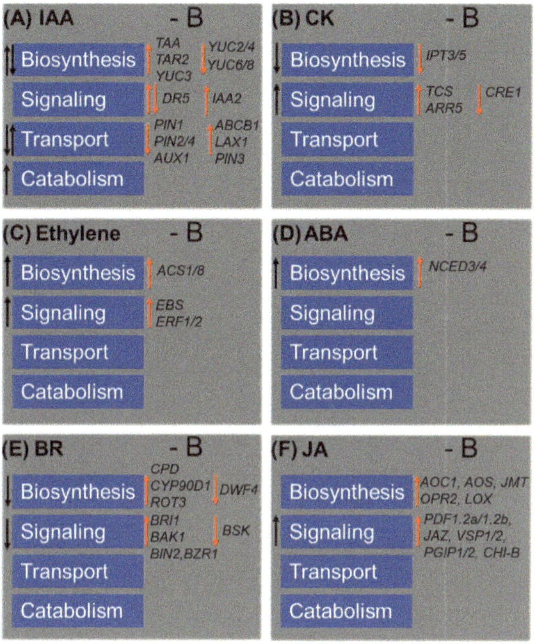

Figure 4.1 Overview of genes and steps in phytohormone pathways that are affected in boron deficient conditions. (A) Auxin, (B) cytokinin, (C) ethylene, (D) abscisic acid, (E) brassinosteroids, (F) jasmonic acid. For all pathways biosynthesis, signaling, transport, and catabolism are indicated. Black arrows indicate whether the specific step in the phytohormone pathway is up- or downregulated. Red arrows indicate, which genes in the specific step are up- or downregulated. Abbreviations: ABA = abscisic acid, *ABCB = ATP BINDING CASSETTE SUBFAMILY B*, *ACS = 1-AMINOCYCLOPROPANE-1-CARBOXYLIC ACID SYNTHASE*, *AOC = ALLENE OXIDE CYCLASE*, *AOS = ALLENE OXIDE SYNTHASE*, *ARR = ARABIDOPSIS RESPONSE REGULATOR*, *AUX = AUXIN*, -B = boron deficient conditions, *BAK = BRI1-ASSOCIATED RECEPTOR KINASE*, *BIN = BRASSINOSTEROID-INSENSITIVE*, BR = brassinosteroids, *BRI = BRASSINOSTEROID INSENSITIVE*, *BSK = BRASSINOSTEROID SIGNALING KINASE*, *BZR = BRASSINAZOLE-RESISTANT*, *CHI-B = BASIC CHITINASE*, *CRE = CYTOKININ RESPONSE*, CK = cytokinin, *CPD = CONSTITUTIVE PHOTOMORPHOGENESIS AND DWARFISM*, *Dr5* = auxin responsive promoter, *DWF = DWARF*, *ERF = ETHYLENE RESPONSE FACTOR*, IAA = auxin, *IAA2 = INDOLE ACETIC ACID 2*, *IPT = ISOPENTENYLTRANSFERASE*, JA = jasmonic acid, *JAZ = JASMONATE ZINC FINGER*, *JMT = JASMONIC ACID CARBOXYL METHYLTRANSFERASE*, *LAX = LIKE-AUX1*, *LOX = LIPOXYGENASE*, *NCED = 9-CIS-EPOXYCAROTENOID DIOXYGENASE*, *OPR = 12-OXOPHYTODIENOIC ACID REDUCTASE*, PDF = *PLANT DEFENSIN*, PGIP = *POLYGALACTURONASE-INHIBITING PROTEIN*, *PIN = PINFORMED*, *ROT = ROTUNDIFOLIA*, *TAA = TRYPTOPHAN AMINOTRANSFERASE*, *TAR2 = TRYPTOPHAN AMINOTRANSFERASE-RELATED PROTEIN 2*, TCS = Two Component signaling Sensor, *VSP = VEGETATIVE STORAGE PROTEIN*, *YUC = YUCCA*.

Through the enhancement of auxin oxidases, it was postulated, that boron could control effective auxin concentrations in the root, therefore regulating the development and growth of root primordia (Jarvis et al., 1983). Boron levels also affected the expression of auxin biosynthesis genes in trifoliate oranges (Li et al., 2016). *TRYPTOPHANE AMINO TRANSFERASE OF ARABIDOPSIS1*, *TRYPTOPHANE AMINO TRANSFERASE RELATED2*, and *YUCCA3* (*YUC3*) were significantly induced under boron deficiency in roots, while *YUC8*, *YUC2*, *YUC4*, and *YUC6* were slightly lower compared to normal boron conditions (Li et al., 2016). Notably, recent studies about auxin level measurements in boron deficient tissue report both increased (Eggert and von Wirén, 2017) and decreased levels of auxin (Li et al., 2016; Zhou et al., 2016). While these discrepancies are likely due to different experimental setups, they exemplify the need for reporting phytohormone levels together with boron levels *in planta*.

Boron–auxin interactions on the level of auxin transport were reported in several species including pea (Li et al., 2001; Wang et al., 2006), trifoliate orange (Li et al., 2016), and Arabidopsis (Camacho-Cristóbal et al., 2015; Li et al., 2015; Matthes and Torres-Ruiz, 2016). Boron deficiency reduced auxin export in pea (Li et al., 2001; Wang et al., 2006) and this effect was postulated to be direct (Wang et al., 2006). Boron deficiency was shown to reduce the abundance and polar localization of the auxin efflux transporter PINFORMED1 (PIN1) in both the Arabidopsis root meristem (Li et al., 2015) and the Arabidopsis embryo (Matthes and Torres-Ruiz, 2016), suggesting reduced auxin transport in the Arabidopsis root and embryo. The assessment of auxin response monitored through the auxin-inducible promoters DR5 (Ulmasov et al., 1997) or IAA2, led to an increase of the DR5 signal and the IAA2 signal in the Arabidopsis root (Martín-Rejano et al., 2011; Camacho-Cristóbal et al., 2015; Li et al., 2015) and a decrease in the Arabidopsis embryo (Matthes and Torres-Ruiz, 2016). These results indicated an increase in rootward but also shootward auxin transport in Arabidopsis seedlings (Martín-Rejano et al., 2011; Camacho-Cristóbal et al., 2015; Li et al., 2015) and a decrease in rootward auxin transport in the Arabidopsis embryo (Matthes and Torres-Ruiz, 2016). In addition, low boron conditions resulted in a greater reduction in root meristem size of the Arabidopsis *pin2* mutant compared to the wild-type control (Li et al., 2015). *PIN2* and *AUX1* expression (both involved in shootward auxin transport) were reduced in Arabidopsis roots (Camacho-Cristóbal et al., 2015) and Arabidopsis *pin2* and *aux1* mutants were less sensitive to boron deprivation with respect to root cell elongation. It was therefore concluded that shootward auxin transport through PIN2 and AUX1 is involved in the inhibition of root cell elongation under boron deprivation (Camacho-Cristóbal et al., 2015). These studies additionally highlighted the differential roles of PIN2 in the root meristem and root elongation zone during the adaptation of the root to boron deficiency.

In trifoliate oranges, auxin transport gene expression was shown to be affected by boron levels in the root (Li et al., 2016). Expression of the transporter genes *ATP BINDING CASSETTE SUBFAMILY B MEMBER1* (*ABCB1*), *LIKE-AUXIN1* (*LAX1*), and *PIN3*, which are involved in shootward auxin transport, was significantly higher, and expression of the transporter genes *AUXIN1* (*AUX1*), *PIN1*, and *PIN4*, which are involved in rootward auxin transport, were significantly lower in boron deficient conditions (Li et al., 2016). The authors, therefore, concluded that shootward auxin transport increases and root auxin transport decrease in boron deficient conditions in trifoliate oranges (Li et al., 2016), which appears to be in contrast to Arabidopsis (Camacho-Cristóbal et al., 2015).

4.3.2 Boron–cytokinin interactions

Targeted analysis of potential interactions of boron with the cytokinin pathway (Fig. 4.1B) arose because cytokinin levels were considerably reduced in boron deficient pea plants (Li et al., 2001; Wang et al., 2006). Cytokinin levels are also positively correlated with shoot boron concentrations in *Brassica napus* (Eggert and von Wirén, 2017), which led the authors to conclude that cytokinins are a major boron-dependent growth signal. Interestingly, correlations between boron and cytokinin levels appear to be species and/or tissue-specific, as cytokinin levels are elevated in the vasculature of plantain grown in low boron conditions (Pommerrenig et al., 2019). Potential boron–cytokinin interactions were also identified in Arabidopsis (Peng et al., 2012; Abreu et al., 2014; Poza-Viejo et al., 2018) and maize (Matthes et al., 2022).

Expression of the cytokinin biosynthesis genes *ISOPENTENYLTRANSFERASE3/5* (*IPT3/5*) and expression of the cytokinin receptor gene *CYTOKININ RESPONSE1* (*CRE1*), which mediate cytokinin signaling, were downregulated in boron deficient conditions in Arabidopsis (Peng et al., 2012; Abreu et al., 2014). Cytokinins played a regulatory role for meristem development in boron deficient conditions (Poza-Viejo et al., 2018). Cytokinin treatment caused a significant increase in root meristem size due to increased mitotic activity of the Arabidopsis root apical meristem in transfer experiments from control conditions to severe boron deficiency conditions (Poza-Viejo et al., 2018). This showed that cytokinins control the inhibition of cell proliferation during root growth under boron deficiency. An early plant response in boron deficient conditions was an increase in cytokinin signaling, and through mutant analyses, it could be shown that boron deficiency defects are alleviated, when cytokinin signaling is suppressed (Poza-Viejo et al., 2018). Type B-ARABIDOPSIS RESPONSE REGULATORs, which are cytokinin responsive transcription factors, might have a specific role in the boron stress response, potentially related to their regulatory role in auxin, but also ethylene signaling (Poza-Viejo et al., 2018).

4.3.3 Boron–ethylene interactions

It was suggested that the ethylene pathway contributes to the inhibition of root cell elongation in boron deficient conditions in Arabidopsis (Fig. 4.1C) (Martín-Rejano et al., 2011; Camacho-Cristóbal et al., 2015). These conclusions were based on the observations that the ethylene reporter line, EBS::GUS, and the ethylene biosynthesis marker line, ACS11::GUS showed enhanced signals in low boron conditions in the Arabidopsis primary root, suggesting an accumulation of ethylene in low boron and boron deficient conditions (Martín-Rejano et al., 2011; Camacho-Cristóbal et al., 2015). It is interesting to note that primary root inhibition in boron deficient conditions did not depend on the central regulator of ethylene signaling, ETHYLENE-INSENSITIVE2 (EIN2), while the ethylene-dependent induction of root hairs in boron deficient conditions did work through EIN2 (Martín-Rejano et al., 2011), highlighting the importance of tissue and process-specific studies. Further, ethylene was proposed to act upstream of auxin signaling during root elongation inhibition in Arabidopsis upon boron deficiency (Camacho-Cristóbal et al., 2015). This ethylene-auxin signaling pathway was suggested to be triggered by boron deficiency-induced defects in cell wall integrity (Camacho-Cristóbal et al., 2015).

4.3.4 Boron–abscisic acid (ABA) interactions

A potential interaction between ABA (Fig. 4.1D) and boron has so far only been associated with adaptations to boron toxic conditions (Macho-Rivero et al., 2017). Yet, different phytohormones can directly affect boron transport in Arabidopsis through binding in the promoter of the boron transporter gene *NIP5;1* (Gómez-Soto et al., 2019). Most notably, ABA treatment induced *NIP5;1* expression in boron deficient conditions and led to increased boron concentrations in wild type and *nip5;1* mutant seedlings grown in boron deficient conditions. ABA treatment further caused an improvement of root growth in *nip5;1* mutant seedlings, when transferred to boron deficiency conditions (Gómez-Soto et al., 2019). Increases in ABA levels in *Brassica napus* shoot tissue when grown in boron deficient conditions were reported (Eggert and von Wirén, 2017) and an increase in the ABA biosynthesis genes *9-CIS-EPOXYCAROTENOID DIOXYGENASE 3/4 (NCED3/4)* in boron deficient conditions was observed in the Arabidopsis root (Peng et al., 2012).

4.3.5 Boron–brassinosteroid interactions

A new avenue was recently opened up, by showing a potential interaction of boron deficiency with the brassinosteroid pathway (Fig. 4.1E) in the Arabidopsis root (Zhang et al., 2020). By comparing transcriptomic data sets of boron deficiency response genes and brassinosteroid regulated genes, the authors showed a high percentage of coregulation, from which about 60% showed an opposite regulation in response to brassinosteroids compared to boron (Zhang et al., 2020). Root growth was inhibited in the Arabidopsis brassinosteroid receptor mutant, *brassinosteroid insensitive 1* (*bri1*), in boron deficient conditions, and could not be fully restored upon boron fertilization (Zhang et al., 2020), which led the authors to propose that boron deficiency downregulates brassinosteroid signaling to inhibit root growth. Physiologically active brassinosteroid levels were reduced in Arabidopsis roots (Zhang et al., 2020) and in *Brassica napus* shoots (Eggert and von Wirén, 2017), while brassinosteroid levels accumulated in plantain mesophyll (Pommerrenig et al., 2019). In Arabidopsis, the decrease in brassinosteroid biosynthesis in boron deficient conditions correlated with a downregulation of brassinosteroid biosynthesis genes (Zhang et al., 2020).

4.3.6 Boron–jasmonic acid interactions

Potential interactions between boron and the jasmonic acid pathway (Fig. 4.1F) were uncovered by untargeted transcriptomics studies (Peng et al., 2012; Zhou et al., 2016) that detected many jasmonic acid-related genes to be differentially regulated by boron deficiency in Arabidopsis seedlings (Peng et al., 2012) and *Brassica napus* seedlings (Zhou et al., 2016). Jasmonic acid inhibited root growth by restricting cell division in the meristem zone and cell elongation in the elongation zone (Huang et al., 2021). Enhanced jasmonic acid signaling was an early event involved in the inhibition of primary root growth under boron deficiency. Interestingly, the expression of the boron transporter gene *NIP5;1* was not upregulated under boron deficiency in the *jasmonate-resistant 1−1* (*jar1−1*) mutant, which suggests that jasmonic acid is involved in the transcriptional regulation of *NIP5;1* under boron deficiency (Huang et al., 2021).

4.4 Boron deficiency and transcript level changes

Targeted and untargeted transcriptomics studies in various plant species identified diverse pathways that are altered during boron deficiency (Kobayashi et al., 2004; Kasajima and Fujiwara, 2007; Peng et al., 2012; Redondo-Nieto et al., 2012; Leonard et al., 2014; Lu et al., 2014; Zhou et al., 2016; Zhang et al., 2020). Such approaches can identify novel molecular players involved in the boron deficiency-induced response of plants. Pathways and molecular players that are differentially regulated in boron deficient conditions include cell wall biosynthesis genes, transmembrane transporter related genes, transcription factors, oxidative damage, maintenance and structure of membranes, phytohormone pathways, and lower nitrate assimilation (as reviewed in Camacho-Cristóbal et al., 2011). It is not resolved, why so many different molecular pathways are altered in boron deficient conditions. Yet, potential interactions of boron with or a boron-dependent regulation of transcription factors offer avenues for boron targeting several unrelated processes. Several transcription factors were differentially regulated in boron deficient conditions (as reviewed in González-Fontes et al., 2013) and two boron dependent WRKY transcription factors were identified; WRKY6 in Arabidopsis (Kasajima et al., 2010) and WRKY47 in *Brassica napus* (Feng et al., 2020). The WRKY6 transcript was induced by boron deficiency close to the root tip and the *wrky6−3* mutant showed a stochastic reduction in root length in boron deficient conditions. The reduction in root length was caused by cell division defects, rather than cell elongation defects (Kasajima et al., 2010). Only a small fraction of genes affected by boron deficiency are regulated by WRKY6, indicating a specific rather than a universal role of WRKY6 in response to boron deficiency in Arabidopsis. While the role of WRKY6 appeared unrelated to regulating boron accumulation (Kasajima et al., 2010), WRKY47 in *Brassica napus* was suggested to directly target boron import, as WRKY47 could bind in the promoter of the boron transporter gene *BnaNIP5;1* in a yeast one-hybrid assay and *wrky47* mutants showed reduced expression of *BnaNIP5;1* genes in leaves (Feng et al., 2020). Although these WRKY transcription factors were not characterized as global players in boron deficiency responses in plants, their identification highlights the possibility that such master regulators might exist.

4.5 Conclusions and outlook

Next to the functional role of boron in crosslinking RG-II (Kobayashi et al., 1996; Matoh et al., 1996; O'Neill et al., 2001) and therefore stabilizing the cell wall, additional regulatory mechanisms, that underly the phenotypic adaptations of plants to low boron conditions, are beginning to be identified. In particular interesting are those that can be directly connected with meristem defects, like alterations of phytohormone pathways and boron dependent transcription factors. Currently, studies that aim at understanding boron deficiency on a molecular level, focus on the root meristems. While shoot meristems are functionally and molecularly very different from the root meristem (as reviewed in Wachsman et al., 2015; Slovak et al., 2016; Somssich and Je, 2016), they also respond to boron deficiency with a cessation of growth. Molecular studies of boron deficiency in shoot meristems are therefore highly needed.

Although cell wall dependency or independency of such additional regulatory mechanisms is challenging to prove, evidence increases that some of the observed effects, for example, the effects of boron on auxin transport, might be direct and independent of the cell wall (Wang et al., 2006). While particularly boron—phytohormone interactions were mainly studied in a targeted way on a whole organ level in the past, additional untargeted approaches in defined cellular structures, ideally combined with targeted boron analytics might help to unravel the molecular mechanisms that underly specifically the observed meristem defects under boron deficiency. A recently published global ribosome-sequencing analysis in Arabidopsis identified amongst others the meristem development regulators CLAVATA3/ESR (CLE)-RELATED PROTEIN20, CLE3, SCARECROW, and a GRAS transcription factor as potentially boron regulated factors (Sotta et al., 2021). In addition, defects in the meristem maintenance regulator gene *kn1* were recently identified as early responses in boron deficient maize shoot meristems (Matthes et al., 2022). These studies could suggest a previously unexplored and exciting direct role of boron during meristem development in plants. With advanced technologies like single-cell RNA-sequencing and LA ICP-MS, combined with the availability of genome sequences of many models and crop plants, urgently needed organ and cell-type-specific boron studies seem within reach.

References

Abreu, I., Poza, L., Bonilla, I., Bolaños, L., 2014. Boron deficiency results in early repression of a cytokinin receptor gene and abnormal cell differentiation in the apical root meristem of *Arabidopsis thaliana*. Plant. Physiol. Biochem. 77, 117–121.

Alves, M., Chicau, P., Matias, H., Passarinho, J., Pinheiro, C., Ricardo, C.P., 2011. Metabolic analysis revealed altered amino acid profiles.pdf. Physiol. Plant. 142, 224–232.

Blevins, D., Lukaszewski, K., 1998. Boron in plant structure and function. Annu. Rev. Plant. Physiol. Plant Mol. Biol. 49, 481–500.

Brown, P.H., Bellaloui, N., Wimmer, M.A., Bassil, E.S., Ruiz, J., Hu, H., et al., 2002. Boron in plant biology. Plant. Biol. 4, 205–223.

Brumos, J., Robles, L.M., Yun, J., Vu, T.C., Jackson, S., Alonso, J.M., et al., 2018. Local auxin biosynthesis is a key regulator of plant development. Dev. Cell 47, 306–318. e5.

Camacho-Cristóbal, J.J., González-Fontes, A., 2007. Boron deficiency decreases plasmalemma H + -ATPase expression and nitrate uptake, and promotes ammonium assimilation into asparagine in tobacco roots. Planta 226, 443–451.

Camacho-Cristóbal, J.J., Martín-Rejano, E.M., Herrera-Rodríguez, M.B., Navarro-Gochicoa, M.T., Rexach, J., González-Fontes, A., 2015. Boron deficiency inhibits root cell elongation via an ethylene/auxin/ROS-dependent pathway in Arabidopsis seedlings. J. Exp. Bot. 66, 3831–3840.

Camacho-Cristóbal, J.J., Rexach, J., Herrera-Rodríguez, M.B., Navarro-Gochicoa, M.T., González-Fontes, A., 2011. Boron deficiency and transcript level changes. Plant. Sci. 181, 85–89.

Chatterjee, M., Liu, Q., Menello, C., Galli, M., Gallavotti, A., 2017. The combined action of duplicated boron transporters is required for maize growth in boron-deficient conditions. Genetics 206, 2041–2051.

Chatterjee, M., Tabi, Z., Galli, M., Malcomber, S., Buck, A., Muszynski, M., et al., 2014. The boron efflux transporter ROTTEN EAR Is required for maize inflorescence development and fertility. Plant. Cell 26, 2962–2977.

Chen, Y., Hoehenwarter, W., 2019. Rapid and reproducible phosphopeptide enrichment by tandem metal oxide affinity chromatography: application to boron deficiency induced phosphoproteomics. Plant. J. 98, 370–384.

Cohen, M.S., Albert, L.S., 1974. Autoradiographic examination of meristems of intact boron-deficient squash roots treated with tritiated thymidine. Plant. Physiol. 54, 766–768.

Dell, B., Huang, L., 1997. Physiological response of plants to low boron. Plant. Soil. 193, 103–120.

Dong, X., Liu, G., Wu, X., Lu, X., Yan, L., Muhammad, R., et al., 2016. Different metabolite profile and metabolic pathway with leaves and roots in response to boron deficiency at the initial stage of citrus rootstock growth. Plant. Physiol. Biochem. 108, 121–131.

Dordas, C., Brown, P., 2000. Membrane biology permeability of boric acid across lipid bilayers and factors affecting it. J. Membr. Biol. 105, 95–105.

Dordas, C., Chrispeels, M.J., Brown, P.H., 2000. Permeability and channel-mediated transport of boric acid across membrane vesicles isolated from squash roots 1. Plant. Physiol. 124, 1349–1361.

Durbak, A.R., Phillips, K.A., Pike, S., O'Neill, M.A., Mares, J., Gallavotti, A., et al., 2014. Transport of boron by the *tassel-less1* aquaporin is critical for vegetative and reproductive development in maize. Plant. Cell 26, 2978–2995.

Eaton, F.M., 1940. Interrelations in the effects of boron and indoleacetic acid on plant growth. Bot. Gaz. 101, 700–705.

Eggert, K., von Wirén, N., 2017. Response of the plant hormone network to boron deficiency. N. Phytol. 216, 868–881.

Feng, Y., Cui, R., Wang, S., He, M., Hua, Y., Shi, L., et al., 2020. Transcription factor BnaA9.WRKY47 contributes to the adaptation of *Brassica napus* to low boron stress by up-regulating the boric acid channel gene BnaA3.NIP5;1. Plant. Biotechnol. J. 18, 1241–1254.

González-Fontes, A., Rexach, J., Quiles-Pando, C., Herrera-Rodríguez, M.B., Camacho-Cristóbal, J.J., Navarro-Gochicoa, M.T., 2013. Transcription factors as potential participants in the signal transduction pathway of boron deficiency. Plant. Signal. Behav. Available from: https://doi.org/10.4161/psb.26114.

Gómez-Soto, D., Galván, S., Rosales, E., Bienert, P., Abreu, I., Bonilla, I., et al., 2019. Insights into the role of phytohormones regulating pAtNIP5; 1 activity and boron transport in *Arabidopsis thaliana*. Plant. Sci. 287, 110198.

Hanaoka, H., Uraguchi, S., Takano, J., Tanaka, M., Fujiwara, T., 2014. OsNIP3;1, a rice boric acid channel, regulates boron distribution and is essential for growth under boron-deficient conditions. Plant. J. 78, 890–902.

Hochholdinger, F., Yu, P., Marcon, C., 2018. Genetic control of root system development in maize. Trends Plant. Sci. 23, 79–88.

Housh, A.B., Matthes, M.S., Gerheart, A., Wilder, S.L., Kil, K., Schueller, M., et al., 2020. Assessment of a 18 F-phenylboronic acid radiotracer for imaging boron in maize. Int. J. Mol. Sci. Available from: https://doi.org/10.3390/ijms21030979.

Huang, Y., Wang, S., Shi, L., Xu, F., 2021. JASMONATE RESISTANT 1 negatively regulates root growth under boron deficiency in Arabidopsis. J. Exp. Bot. 72, 3108–3121.

Jackson, D., Veit, B., Hake, S., 1994. Expression of maize KNOTTED1 related homeobox genes in the shoot apical meristem predicts patterns of morphogenesis in the vegetative shoot. Development 120, 405–413.

Jarvis, B., Ali, A., Shaheed, A., 1983. Auxin and boron in relation to the rooting response and ageing of mung bean cuttings. N. Phytol. 95, 509–518.

Kasajima, I., Fujiwara, T., 2007. Identification of novel *Arabidopsis thaliana* genes which are induced by high levels of boron. Plant. Biotechnol. 24, 355–360.

Kasajima, I., Ide, Y., Yokota Hirai, M., Fujiwara, T., 2010. WRKY6 is involved in the response to boron deficiency in *Arabidopsis thaliana*. Physiol. Plant. 139, 80–92.

Kerstetter, R.A., Laudencia-Chingcuanco, O., Smith, L.G., Hake, S., 1997. Loss-of-function mutations in the maize homeobox gene, knotted1, are defective in shoot meristem maintenance. Development 124, 3045–3054.

Kitagawa, M., Jackson, D., 2019. Control of meristem size. Annu. Rev. Plant. Biol. 70, 269–291.

Kobayashi, M., Matoh, T., Azuma, J., 1996. Two chains of rhamnogalacturonan II are cross-linked by borate-diol ester bonds in higher plant cell walls. Plant. Physiol. 110, 1017–1020.

Kobayashi, M., Mutoh, T., Matoh, T., 2004. Boron nutrition of cultured tobacco BY-2 cells. IV. Genes induced under low boron supply. J. Exp. Bot. 55, 1441–1443.

Lee, Z.H., Hirakawa, T., Yamaguchi, N., Ito, T., 2019. The roles of plant hormones and their interactions with regulatory genes in determining meristem activity. Int. J. Mol. Sci. Available from: https://doi.org/10.3390/ijms20164065.

Leonard, A., Holloway, B., Guo, M., Rupe, M., Yu, G., Beatty, M., et al., 2014. Tassel-less1 encodes a boron channel protein required for inflorescence development in maize. Plant. Cell Physiol. 55, 1044–1054.

Li, T., Choi, W., Wallace, I.S., Baudry, J., Roberts, D.M., 2011. *Arabidopsis thaliana* NIP7;1: an anther-specific boric acid transporter of the aquaporin superfamily regulated by an unusual tyrosine in helix 2 of the transport pore. Biochemistry 50, 6633–6641.

Li, K., Kamiya, T., Fujiwara, T., 2015. Differential roles of PIN1 and PIN2 in root meristem maintenance under Low-B conditions in arabidopsis thaliana. Plant. Cell Physiol. 56, 1205–1214.

Li, Q., Liu, Y., Pan, Z., Xie, S., Peng, S.A., 2016. Boron deficiency alters root growth and development and interacts with auxin metabolism by influencing the expression of auxin synthesis and transport genes. Biotechnol. Biotechnol Equip. 30, 661–668.

Li, C., Pfeffer, H., Dannel, F., Römheld, V., Bangerth, F., 2001. Effects of boron starvation on boron compartmentation, and possibly hormone-mediated elongation growth and apical dominance of pea (*Pisum sativum*) plants. Physiol. Plant. 111, 212–219.

Long, J.A., Moan, E.I., Medford, J.I., Barton, M.K., 1996. A member of the KNOTTED class of homeodomain proteins encoded by the STM gene of Arabidopsis. Nature 379, 66–69.

Lordkaew, S., Dell, B., Jamjod, S., Rerkasem, B., 2011. Boron deficiency in maize. Plant. Soil. 342, 207–220.

Lu, Y.Bin, Qi, Y.P., Yang, L.T., Guo, P., Li, Y., Chen, L.S., 2015. Boron-deficiency-responsive microRNAs and their targets in *Citrus sinensis* leaves. BMC Plant. Biol. 15, 1–15.

Lu, Y.Bin, Yang, L.T., Qi, Y.P., Li, Y., Li, Z., Chen, Y.Bin, et al., 2014. Identification of boron-deficiency-responsive microRNAs in *Citrus sinensis* roots by illumina sequencing. BMC Plant. Biol. 14, 1–16.

Macho-Rivero, M.Á., Camacho-Cristóbal, J.J., Herrera-Rodríguez, M.B., Müller, M., Munné-Bosch, S., González-Fontes, A., 2017. Abscisic acid and transpiration rate are involved in the response to boron toxicity in Arabidopsis plants. Physiol. Plant. 160, 21–32.

Martín-Rejano, E.M., Camacho-Cristóbal, J.J., Herrera-Rodríguez, M.B., Rexach, J., Navarro-Gochicoa, M.T., González-Fontes, A., 2011. Auxin and ethylene are involved in the responses of root system architecture to low boron supply in Arabidopsis seedlings. Physiol. Plant. 142, 170–178.

Matoh, T., Kawaguchi, S., Kobayashi, M., 1996. Ubiquity of a borate-rhamnogalacturonan II complex in the cell walls of higher plants. Plant. Cell Physiol. 37, 636–640.

Matthes, M.S., Darnell, Z., Best, N.B., Guthrie, K., Robil, J.M., Amstutz, J., Durbak, A., McSteen, P., 2022. Defects in meristem maintenance, cell division, and cytokinin signaling are early responses in the boron deficient maize mutant *tassel-less1*. Physiol. Plantarum. Available from: https://doi.org/10.1111/ppl.13670.

Matthes, M.S., Robil, J.M., Mcsteen, P., 2020. From element to development: the power of the essential micronutrient boron to shape morphological processes in plants. J. Exp. Bot. 71, 1681–1693.

Matthes, M., Torres-Ruiz, R.A., 2016. Boronic acid treatment phenocopies monopteros by affecting PIN1 membrane stability and polar auxin transport in *Arabidopsis thaliana* embryos. Development 143, 4053–4062.

Miwa, K., Aibara, I., Fujiwara, T., Miwa, K., Aibara, I., Fujiwara, T., 2014. Arabidopsis thaliana BOR4 is upregulated under high boron conditions and confers tolerance to high boron. Soil. Sci. Plant. Nutr. 60, 349–355.

Miwa, K., Wakuta, S., Takada, S., Ide, K., Takano, J., Naito, S., et al., 2013. Roles of BOR2, a boron exporter, in cross linking of rhamnogalacturonan II and root elongation under boron limitation in Arabidopsis. Plant. Physiol. 163, 1699–1709.

Motte, H., Vanneste, S., Beeckman, T., 2019. Molecular and environmental regulation of root development. Annu. Rev. Plant. Biol. 70, 465–488.

Nakagawa, Y., Hanaoka, H., Kobayashi, M., Miyoshi, K., Miwa, K., Fujiwara, T., 2007. Cell-type specificity of the expression of Os BOR1, a rice efflux boron transporter gene, is regulated in response to boron availability for efficient boron uptake and xylem loading. Plant. Cell 19, 2624–2635.

Noguchi, K., Yasumori, M., Imai, T., Naito, S., Matsunaga, T., Oda, H., 1997. Bor1-1, an *Arabidopsis thaliana* mutant that requires a high level of boron. Plant. Physiol. 901–906.

O'Neill, M.A., Eberhard, S., Albersheim, P., Darvill, A.G., 2001. Requirement of borate cross-linking of cell wall rhamnogalacturonan II for Arabidopsis growth. Science 294, 846–849.

Peng, L., Shi, L., Cai, H., Xu, F., Zeng, C., 2012. Transcriptional profiling reveals adaptive responses to boron deficiency stress in Arabidopsis. Z. fur Naturforsch - Sect. C. J. Biosci. 67, 510–524.

Pommerrenig, B., Eggert, K., Bienert, G.P., 2019.) Boron deficiency effects on sugar, ionome, and phytohormone profiles of vascular and non-vascular leaf tissues of common plantain (*Plantago major* L.). Int. J. Mol. Sci. Available from: https://doi.org/10.3390/ijms20163882.

Poza-Viejo, L., Abreu, I., González-García, M.P., Allauca, P., Bonilla, I., Bolanos, L., et al., 2018. Boron deficiency inhibits root growth by controlling meristem activity under cytokinin regulation. Plant. Sci. 270, 176–189.

Redondo-Nieto, M., Maunoury, N., Mergaert, P., Kondorosi, E., Bonilla, I., Bolaños, L., 2012. Boron and calcium induce major changes in gene expression during legume nodule organogenesis. Does boron have a role in signalling? N. Phytol. 195, 14–19.

Routray, P., Li, T., Yamasaki, A., Yoshinari, A., Takano, J., Choi, G., 2018. Nodulin intrinsic protein 7;1 is a tapetal boric acid channel involved in pollen cell wall formation. Plant. Physiol. 178, 1269–1283.

Shao, J.F., Yamaji, N., Liu, W., Yokosho, K., Shen, F., Ma, J.F., 2018. Preferential distribution of boron to developing tissues is mediated by the intrinsic protein OsNIP3. Plant. Physiol. 176, 1739–1750.

Shimotohno, A., Sotta, N., Sato, T., De Ruvo, M., Marée, A.F.M., Grieneisen, V.A., et al., 2015. Mathematical modeling and experimental validation of the spatial distribution of boron in the root of arabidopsis thaliana identify high boron accumulation in the tip and predict a distinct root tip uptake function. Plant. Cell Physiol. 56, 620–630.

Shorrocks, V.M., 1997. The occurrence and correction of boron deficiency. Plant. Soil. 193, 121–148.

Slovak, R., Ogura, T., Satbhai, S.B., Ristova, D., Busch, W., 2016. Genetic control of root growth: From genes to networks. Ann. Bot. 117, 9–24.

Sommer, A.L., 1927. The search for elements essential in only small amounts for plant growth. Science (80), 482–484.

Sommer, A.L., Sorokin, H., 1928. Effects of the absence of boron and of some other essential elements on the cell and tissue structure of the root tips of *Pisum sativum*. Plant. Physiol. 3, 237–260.

Somssich, M., Je, B.Il, 2016. CLAVATA-WUSCHEL signaling in the shoot meristem. Development 3238–3248.
Sotta, N., Bian, B., Peng, D., Hongkham, P., Kamiya, T., Niikura, S., et al., 2019. Local boron concentrations in tuberous roots of Japanese radish (*Raphanus sativus* L.) negatively correlate with distribution of brown heart. Plant. Physiol. Biochem. 136, 58–66.
Sotta, N., Chiba, Y., Miwa, K., Takamatsu, S., Tanaka, M., Yamashita, Y., et al., 2021. Global analysis of boron-induced ribosome stalling reveals its effects on translation termination and unique regulation by AUG-stops in Arabidopsis shoots. Plant. J. 106, 1455–1467.
Sutton, T., Baumann, U., Hayes, J., Collins, N.C., Shi, B.J., Schnurbusch, T., et al., 2007. Boron-toxicity tolerance in barley arising from efflux transporter amplification. Science 318 (80), 1446–1449.
Takano, J., Tanaka, M., Toyoda, A., Miwa, K., Kasai, K., Fuji, K., et al., 2010. Polar localization and degradation of Arabidopsis boron transporters through distinct trafficking pathways. Proc. Natl Acad. Sci. USA 107, 5220–5225.
Takano, J., Wada, M., Ludewig, U., Schaaf, G., von Wiren, N., Fujiwara, T., 2006. The Arabidopsis major intrinsic protein NIP5;1 is essential for efficient boron uptake and plant development under boron limitation. Plant. Cell 18, 1498–1509.
Tanaka, M., Wallace, I.S., Takano, J., Roberts, D.M., Fujiwara, T., 2008. NIP6;1 is a boric acid channel for preferential transport of boron to growing shoot tissues in Arabidopsis. Plant. Cell 20, 2860–2875.
Ulmasov, T., Murfett, J., Hagen, G., Guilfoyle, T.J., 1997. Aux/IAA proteins repress expression of reporter genes containing natural and highly active synthetic auxin response elements. Plant. Cell 9, 1963.
Vollbrecht, E., Veit, B., Sinha, N., Hake, S., 1991. The developmental gene Knotted-1 is a member of a maize homeobox gene family. Nature 350, 241–243.
Voxeur, A., Fry, S.C., 2014. Glycosylinositol phosphorylceramides from Rosa cell cultures are boron-bridged in the plasma membrane and form complexes with rhamnogalacturonan II. Plant. J. 79, 139–149.
Wachsman, G., Sparks, E.E., Benfey, P.N., 2015. Genes and networks regulating root anatomy and architecture. N. Phytol. 208, 26–38.
Wang, G., Römheld, V., Li, C., Bangerth, F., 2006. Involvement of auxin and CKs in boron deficiency induced changes in apical dominance of pea plants (*Pisum sativum* L.). J. Plant. Physiol. 163, 591–600.
Warington, K., 1923. The effect of boric acid and borax on the broad bean and certain other plants. Ann. Bot. 37, 629–672.
Wilder, S.L., Scott, S., Waller, S., Powell, A., Benoit, M., Guthrie, J.M., Schueller, M.J., Awale, P., McSteen, P., Matthes, M.S., Ferrieri, R.A., 2022. Carbon-11 radiotracing reveals physiological and metabolic responses of maize grown under different regimes of boron treatment. Plants 11 (3), 241.
Wimmer, M.A., Lochnit, G., Bassil, E., Muehling, K.H., Goldbach, H.E., 2009. Membrane-associated, boron-interacting proteins isolated by boronate affinity chromatography. Plant. Cell Physiol. 50, 1292–1304.
Wu, Q., Xu, F., Jackson, D., 2018. All together now, a magical mystery tour of the maize shoot meristem. Curr. Opin. Plant. Biol. 45, 26–35.
Yoshinari, A., Hosokawa, T., Amano, T., Beier, M.P., Kunied, T., Shimada, T., et al., 2019. Polar localization of the borate exporter bor1 requires AP2-dependent endocytosis. Plant. Physiol. 179, 1569–1580.
Yoshinari, A., Takano, J., 2017. Insights into the mechanisms underlying boron homeostasis in plants. Front. Plant. Sci. 8, 1–8.

Zhang, Q., Chen, H., He, M., Zhao, Z., Cai, H., Ding, G., et al., 2017. The boron transporter BnaC4.BOR1;1c is critical for inflorescence development and fertility under boron limitation in *Brassica napus*. Plant, Cell Env. 40, 1819–1833.
Zhang, C., He, M., Wang, S., Chu, L., Wang, C., Yang, N., et al., 2020. Boron deficiency-induced root growth inhibition is mediated by brassinosteroid signalling regulation in Arabidopsis. Plant. J. Available from: https://doi.org/10.1111/tpj.15311.
Zhou, T., Hua, Y., Huang, Y., Ding, G., Shi, L., Xu, F., 2016. Physiological and transcriptional analyses reveal differential phytohormone responses to boron deficiency in brassica napus genotypes. Front. Plant. Sci. 7, 1–16.
Zhou, T., Hua, Y., Zhang, B., Zhang, X., Zhou, Y., Shi, L., et al., 2017. Low-boron tolerance strategies involving pectin-mediated cell wall mechanical properties in *Brassica napus*. Plant. Cell Physiol. 58, 1991–2005.

CHAPTER 5

From outside to inside: mechanisms modulating plant responses to boron stress

Greice Leal Pereira[1], João Antonio Siqueira[1], Paula da Fonseca-Pereira[1], Rebeca P. Omena-Garcia[1], Vitor L. Nascimento[2], Adriano Nunes-Nesi[1] and Wagner L. Araújo[1]
[1]Departamento de Biologia Vegetal, Universidade Federal de Viçosa, Viçosa, Minas Gerais, Brazil
[2]Setor de Fisiologia Vegetal - Departamento de Biologia, Universidade Federal de Lavras, Lavras, Minas Gerais, Brazil

5.1 Introduction

The micronutrient boron (B) is known to be essential for completing the life cycle of a wide spectrum of organisms in all domains, being considered a prebiotic element and central to the origin of life on the Earth (Nielsen and Eckhert, 2020; Pereira et al., 2021; Scorei, 2012). About a century ago it was reported that B is necessary for the adequate growth and development of plants wherein its essentiality was first described in 1923 by Warington, showing that the plant growth was reduced in the absence of B, but it was rescued after its replenishment (Warington, 1923). B plays important role in plants, including its involvement in cell wall formation and functionality (Riaz et al., 2018). Briefly, (i) B forms B-ramnogalacturonan-II (B-RG-II) complexes that stabilize the pectin network (O'Neill et al., 2001, 2004; Voxeur et al., 2011; Chormova et al., 2014a,b), (ii) B acts in the maintenance of the structural and functional integrity of the plasma membrane (González-Fontes et al., 2013; Matthes and Torres-Ruiz, 2017), (iii) B is important to the transport and metabolism of sugars, the metabolism of indolacetic acid (Moghadam et al., 2020), the metabolism of RNA (pyrimidine biosynthesis), homeostasis of phenolic, and nitrogen compounds and ascorbate production (Shireen et al., 2018; Camacho-Cristóbal et al., 2008), and (iv) moreover, this element plays a vital role in pollen germination, pollen tube growth, and many other processes during the reproductive period of plants (Wang et al., 2003; Iwai et al., 2006).

Boron in Plants and Agriculture
DOI: https://doi.org/10.1016/B978-0-323-90857-3.00003-5

In the soil solution, B predominates in the form of boric acid (H_3BO_3), an uncharged molecule (Hu and Brown, 1997), which strongly influences its availability. Thus, two extremes, deficiency and toxicity, are usually observed in cultivable areas that drastically affect agricultural production (Brdar-Jokanović, 2020; Takano et al., 2008). B deficiency is a general problem and is commonly associated with highly leached sandy soils with low pH (acidic conditions) and low organic matter content (Shorrocks, 1997; Yan et al., 2006). On the other hand, toxicity occurs predominantly in regions characterized by alkaline and saline soils, with low precipitation and little leaching (arid and semiarid regions). Notably, high concentrations of B can occur naturally in soil and groundwater or be added to the soil by fertilizers and irrigation water, making B levels highly toxic to plant growth (Tanaka and Fujiwara, 2008; Camacho-Cristóbal et al., 2008; Smith et al., 2013; Archana and Verma, 2017).

One conspicuous feature of B is that it presents a relatively narrow window between deficient and toxic concentrations. Therefore, the inadequate supply of B may promote biochemical, physiological, anatomical, and molecular changes. Due to this amplitude in the B levels present in the soil solution, plants developed different adaptive and well-orchestrated and finely adjusted mechanisms in response to the availability of B (Ozhuner et al., 2013; Pereira et al., 2021). It is worth mentioning that these responses mainly depend on the species and its developmental stage.

The primary mechanism of tolerance is associated with the entry of B into the roots. Among the mechanisms used by plants, morphoanatomical modifications of the root system are crucially important to mediate the availability of B for plant uptake (Choi et al., 2007). In addition, a deeper understanding of the mechanisms associated with the regulation of B uptake, transport, and distribution has been greatly facilitated by the identification and characterization of a series of transporters involved in this process (Takano et al., 2008; Miwa and Fujiwara, 2011; Reid. 2014; Yoshinari and Takano, 2017). Several B transporters have been described recently showing that plants are able not only to perceive the internal and external conditions of B but also to regulate their transport according to their availability, by the expression of transporters, maintaining the homeostasis of B in tissues (Miwa and Fujiwara, 2011; Miwa et al., 2013). Members of the major intrinsic proteins and BOR transporters have been shown to mediate B uptake in the roots, its translocation from root to shoot and its redistribution for the shoot (Pommerrening et al., 2015; Reid, 2014). In summary, the diversity of B transporters highlights the strict regulation of B homeostasis.

Once B has been absorbed by the root cells, B is transported to the shoot of the plants through the xylem vessels, through the transpiration flow. Subsequently, B can be redistributed via the phloem to organs according to the nutritional demand, but its mobility seems to rely on the presence of molecules that can form complexes and vary considerably between species (Huang et al., 2008; Stangoulis et al., 2010; Liu et al. 2012; Reid, 2014; Mosa et al., 2016; Oikonomou et al., 2019). Therefore, plants need to optimize the mechanisms of distribution and redistribution of B in conditions of deficiency and toxicity.

Under adverse conditions of B availability, there is extensive reprogramming of the central and specialized metabolism of the plants (Pereira et al., 2021; Reid and Fitzpatrick, 2009; Roessner et al., 2006), causing in many cases oxidative damage. However, plants have developed an efficient system to control the production of ROS, through enzymatic and nonenzymatic antioxidant mechanisms. Several genes related to those mechanisms have been found to be regulated by B levels, suggesting that plants may sense, respond and maintain B homeostasis, directly or indirectly (Kobayashi et al., 2004; Takano et al., 2006; Camacho-Cristóbal et al., 2008; Ozhuner et al., 2013; Abreu et al., 2014).

B is especially important in the reproductive phase of plants (Wongmo et al., 2004). During the transition from the vegetative to the reproductive phase plants need to acquire enough B while protecting themselves from its toxic effects. This is related to the fact that meristems are extremely sensitive to both B deficiency and excess (Camacho-Cristóbal et al., 2008). Accordingly, it is extremely important to clarify how and to which extent the components that regulate this adaptive mechanism are associated with B homeostasis.

Lastly, both B deficiency and toxicity remains as important questions regarding the challenges associated with the mechanisms by which plants modulate the various levels of responses to tolerate this adverse condition. Detailed knowledge of the role played by participants in these mechanisms, as well as root changes, transporters and channels of B, physiological, biochemical, metabolic changes and their interactions, will be essential to advance our understanding of B function in plants. It would aid during the selection of plants resistant to deficiency or toxicity to B in the soils. In the next topics of this chapter, the main plant modifications to cope with adverse B conditions are discussed and at the end, we summarize the major findings and challenges in understanding these adaptive mechanisms to B stress during plant growth and development.

5.2 Architectural adaptation of the root system to boron availability

The essential nutrients for the growth and development of plants are present in the soil solution and, in response to the fluctuations of their levels, plants have developed a variety of adaptive mechanisms. Among them, the plastic development of the root system is crucially important, ensuring the exploration and acquisition of resources (Hell and Hillebrand 2001; López-Búcio et al., 2003; Lambers et al., 2006; Martín-Rejano et al., 2011; Gruber et al., 2013; Potocka and Szymanowska-Pułka, 2018). Thus, a well-developed root system is central for nutrient uptake processes (Forde and Lorenzo, 2001; Brundrett et al., 2017). Not surprisingly, changes in the morphology and anatomy of the root may mediate the adaptation of plants and therefore their ability to grow in soils where nutrient availability is extreme. Among these changes, an increase in the surface area for uptake and root length, as well, in the cell division of root meristems, and the lateral root (LR) and/or root hairs formation are usually observed (Bell and Sultan, 1999; Sorgonà et al., 2005; Desnos, 2008; Pang et al., 2010; Rao et al., 2016).

B played an important role during the evolutionary development of the root system (Lewis, 1980). In fact, this micronutrient was considered necessary for the root development, being a key factor for the origin of the root system of the first vascular plants (Lewis, 1980; Kutschera and Niklas, 2017). It has been shown that by restricting B the architecture of the root system is modified, resulting in a rapid and severe inhibition of primary root (PR) elongation, coupled with an increased root hair elongation and LR growth, as well as with changes in root polarity and morphogenesis (Kocabek et al., 2009; Martín-Rejano et al., 2011; Mei et al., 2011; Oiwa et al., 2013; Li et al., 2015; Kobayashi et al., 2017; Wu et al., 2017; Riaz et al., 2018).

Deficiency symptoms are explained in part because B plays an important structural role in cell wall architecture through the formation of the covalent complex with ramnogalacturonan-II (RG-II), that is, B-RG-II, which is omnipresent in sporophyte and gametophyte generations of vascular land plants (O'Neill et al., 2004; Chormova et al., 2014a,b). B-RG-II crosslinking is necessary to maintain both the mechanical and the physical properties of the cell wall (Ryden et al., 2003; O'Neill et al., 2004; Goldbach and Wimmer 2007). Therefore, B deficiency promotes destabilization of cell wall integrity, causing the undesired inhibition of mechanical strength and cell wall elasticity (Chormova and Fry, 2016; Goldbach et al., 2001; Goldbach and Wimmer, 2007; Liu et al., 2014). Thus, these changes are most likely related to the rapid inhibition of root growth induced by B deficiency. Furthermore, the expression of several cell wall modifying enzymes is downregulated under B deficiency (Camacho-Cristóbal et al., 2008), which most likely alter the cell wall loosening required for elongation.

The hormones auxin and ethylene are involved in root growth responses to B deficiency (Camacho-Cristóbal et al., 2015; Martín-Rejano et al., 2011; Li et al., 2015, 2016; Eggert and Wirén, 2017). Indeed, root system growth, development and, therefore, architecture are closely related to the synthesis, distribution, and transport of auxin (Blilou et al., 2005; Tanaka et al., 2006). Thus, the inhibition of root elongation is caused by the accumulation of excessive levels of auxin in the meristematic regions of B-deficient roots (Dugger 1983). Therefore, PR inhibition under B deficiency may be caused, at least in part, by changes in the distribution of this hormone. Briefly, auxin synthesis takes place at the root tip and it is then transported to the elongation zone, where the auxin levels remain below a growth inhibitory threshold. However, B deficiency seems to increase auxin synthesis in the meristematic region and its transport to the elongation zone, where it would accumulate and induce a local response, inhibiting root cell elongation. In addition to auxin, ethylene also affects root growth by inhibiting the rapid expansion of root meristem cells and inducing the formation and elongation of root hairs (Le et al., 2001; Swarup et al., 2007; Stepanova et al., 2007). The main changes at the root level that contribute to the tolerance to B deficiency are possibly associated with the allocation of biomass into the root system, higher root: shoot ratio and amounts of LR (Steingrobe et al., 2001; Li et al., 2016). These changes can be plant-specific or adaptive responses to B deficiency (Dong et al., 2016). Therefore, improving B acquisition by changing root morphology is seemingly important to reduce the impacts of B deficiency (Rerkasem and Jamjod, 2004).

The primary symptom of B toxicity observed in different crops is the alteration of the architecture of the root system, through its inhibition (Reid et al., 2004; Choi et al., 2006; Cervilla et al., 2009; Huang et al., 2019). Moreover, B toxicity is associated with abnormal cell division at the root meristem (Liu et al., 2000), and hypodermis formation along with progressive suberin deposition in the cortical root cell wall (Ghanati et al., 2005). The root tip is the main target of B toxicity, as growth inhibition occurs when B excess is applied to the root tip, but not when this excess is applied to mature sections (Reid et al., 2004). Two fundamental developmental processes sustain root growth: (1) mitotic cell division in the root apical meristem (RAM) and (2) cell elongation after mitosis (Scheres et al., 2002). Excess B affects RAM by reducing or inhibiting cell division activity (Liu et al., 2000; Sakamoto et al., 2011, 2018; Aquea et al., 2012) and the inhibition of cell division is attributed to DNA damage caused by excess B in RAM, leading to severe defects in RAM morphology (Sakamoto et al., 2011, 2018). In this context, B toxicity is associated with reduced mitotic index in the root meristem, and this is associated with an increase in chromosomal fragments, chromosomal viscosity, and micronucleus development (Liu et al., 2000; Konuk et al., 2007). These observations provide insight into how B toxicity impedes root growth by increasing the incidence of abnormal mitosis which, in turn, affects cell division and ultimately decreases the rate of root elongation.

B is required for the regulation of lignin biosynthesis and suberin precursors (Ghanati et al., 2002), thus, the inhibition of root growth by B excess occurs due to the formation of hypodermis and the progressive deposition of suberin in the cortical cell walls, leading to the structural rigidity of those tissues (Ghanati et al., 2005). Accordingly, increased cell wall lignification may have a negative effect on root elongation, reducing cell wall extensibility (Schopfer, 2001). However, lignification may be a mechanism of adaptation to stress. In environments enriched with copper (Cu) or cadmium (Cd), lignification is considered part of the defense reaction that reduces the entry of metal into the plant tissue, by reducing membrane permeability (Ederli, et al., 2004; Kováčik and Klejdus, 2008). In this sense, the relationship between ligno-suberization of roots and reduced uptake of B may denote a high genotypic variation in the responses to excess B (Ghanati et al., 2005).

There is a wide range of genotypic variations in response to B stress in different crop. Thus, variations in root elongation have been used as an indicator of tolerance to B stress (Jefferies et al., 2000; Hayes and Reid, 2004; Torun et al., 2006; Choi et al., 2006; Ochiai et al., 2008; Hayes et al., 2013; Kaur et al., 2006; Pallotta et al., 2014; Guo et al., 2014). Plants tolerant to B toxicity present adaptability to a wide range of concentrations of B in the soil solution, without an evident decrease in root growth (Ozturk et al., 2010). All these observations demonstrate the significance of the root system in the adaptive responses following B stress.

It is important to mention that B tolerance is usually associated with root morphological and anatomical changes, for instance, increased branching and development of thinner roots, which is associated with an enhanced elongation rate, allowing higher soil exploration (Choi et al., 2006; Cervilla et al., 2009; Princi et al., 2013). Understanding the morphoanatomical mechanisms of the root system associated with the adaptation to extreme levels of B in the soil solution is of crucial importance. We further posit that to enhance agricultural production, allowing plants to grow normally in environments with B deficiency or toxicity, the selection through genetic variation of these aforementioned root traits related to the tolerance to B stress becomes necessary.

5.3 Regulation of boron uptake: a central process for homeostasis

In the soil solution, B predominates in the form of H_3BO_3, a neutral molecule, that readily permeates cell membranes (Dordas et al., 2000). It has been extensively accepted that the uptake of B in the roots was exclusively a passive process (Shelp et al., 1995). However, depending on the B availability, the transport of this element can occur through three different mechanisms: (1) passive transport across the plasma membrane by the diffusion process, (2) transport facilitated by NIPs (Nodulin 26-like Intrinsic Proteins), and (3) energy-dependent transport (against a concentration gradient) (Marschner, 2012).

Under these circumstances, plants tightly regulate the abundance of transporters to deal with the optimal narrow range of B (Yoshinari and Takano, 2017). Although significant advances have been made in understanding the regulation of B transport and the mechanisms related to the maintenance of B levels in plants, the physiological impact of the regulation of these transporters on the distribution and dynamics of B at the cellular, tissue and individual levels are still incomplete (Fig. 5.1). The search for a deeper understanding of the mechanisms associated with the regulation of B uptake has been greatly facilitated by the identification and characterization of a series of transporters involved in this process. During the last two decades, using molecular and phylogenetic analyzes in Arabidopsis (*Arabidopsis thaliana*), several B transporters have been described showing that plants are able not only to perceive the internal and external conditions of B, but also to regulate their transport according to their availability (Miwa and Fujiwara, 2011; Miwa et al., 2013; Onuh and Miwa, 2021).

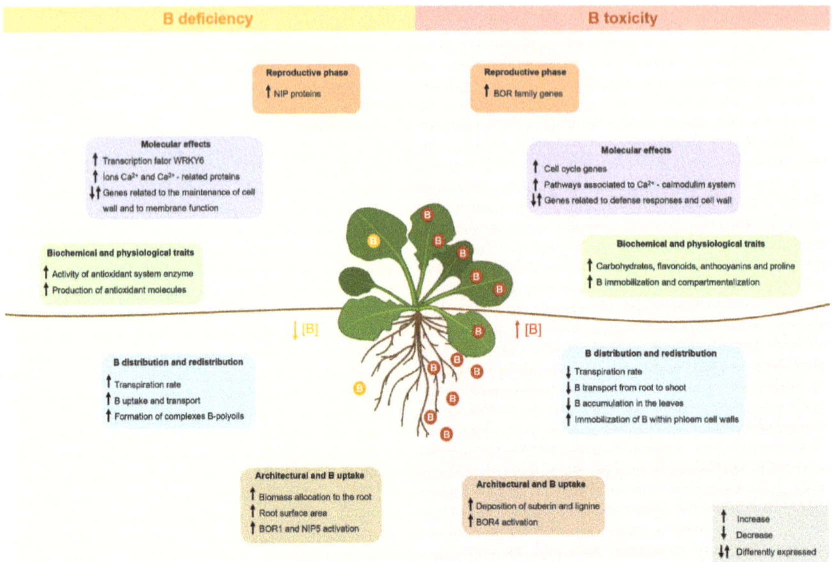

Figure 5.1 Representative scheme showing the main effects of boron on plants. The figure displays the positive and negatives impacts of B on plants in general.

Although these results were obtained in a model plant, several genes homologous to B transporters have been further characterized in a wide range of other plant species (Miwa and Fujiwara, 2010; Ozyigit et al., 2020; Onuh and Miwa, 2021).

The first B transporter identified in Arabidopsis was BOR1 (Takano et al., 2002). The BOR1 transporter plays a fundamental role in loading the xylem against a concentration gradient. It is usually located at the plasma membrane of the root cells, predominantly in the cells of the pericarp around the xylem, and is of great importance in the transport of B from the roots to the shoot of plants, specifically under low availability of B (Takano et al., 2002, 2010). Interestingly, the expression and activity of BOR1 are regulated by B levels. Briefly, under high levels of B in the soil solution, BOR1 is transported to the vacuole where it undergoes ubiquitination in a lysine residue (K590) in the C-terminal tail, followed by degradation through multivesicular bodies (Takano et al., 2005, 2010; Kasai et al., 2011). In addition, another mechanism for controlling BOR1 levels has been recently demonstrated, where high B levels have reduced the translational efficiency of this transporter, and this downregulation is mediated by several upstream to the 5′- UTR open reading frames (uORFs) (Aibara et al., 2018). Briefly, two uORFs closest to the BOR1 start codon play a role in its B-dependent regulation, and those uORFs affected the rate of translational reset, as it shortened the distance between the last uORF and the main one (Aibara et al., 2018). These responses are apparently important to avoid the accumulation of B at toxic levels in plant tissues when B levels are relatively high in the soil.

NIP5;1 is a H_3BO_3 channel located in the root epidermal cells with external polarity, facing the soil (Takano et al., 2010). This membrane protein was characterized as the main aquaporin associated with the uptake of B under limiting conditions of this nutrient (Takano et al., 2006). Notably, NIP5;1, the main member of the intrinsic protein family, is an important facilitator of the passive flow of water and small uncharged molecules (e.g., glycerol, urea, and H_3BO_3). The abundance of NIP5;1 protein is controlled by the B-dependent mRNA degradation (Tanaka et al., 2011, 2016). Briefly, at the mRNA level, the expression of NIP5;1 in the roots is positively regulated by B deficiency (Takano et al., 2006); moreover, under high B conditions, the expression of NIP5;1 is negatively regulated by posttranscriptional regulation (Tanaka et al., 2011). During translation, the ribosomes are stopped in their 5′-UTR region, which increases the destabilization of the NIP5;1 mRNA, which probably increases mRNA cleavage in the 5 ′ portion of the stagnant ribosome (Tanaka et al., 2016). Thus, NIP5;1 mRNA is present at lower levels under sufficient B conditions than under low availability of B. Briefly, both BOR1 and NIP5;1 transporters are often present in the same cells, but located on opposite sides, allowing the directional B flow (Yoshinari et al., 2019). This location pattern is consistent with its role under B-limiting conditions, which allows the coordinated functioning of NIP5;1 for inflow and BOR1 for efflux maintaining an efficient radial transport of B to the vascular tissues (Onuh and Miwa, 2021).

A mathematical model of temporal changes in cells was conducted to determine the physiological significance of these mechanisms of rapid regulation of NIP5;1 and BOR1. Remarkably, it demonstrates that this B-dependent protein regulation allows plants to prevent an unstable flow of B within the root cells (Sotta et al., 2017). This model revealed that rapid regulation is necessary for stable B flow, both under constant and fluctuating B conditions. In addition, another mathematical model was used to examine the spatial distribution of B in the root cells based on the location of NIP5;1 and BOR, revealing a higher accumulation of B in the quiescent center at the root tip (Shimotohno et al., 2015). In response to B excess, many plant species can reduce B accumulation, excluding it mainly in the roots, to maintain the overall plant homeostasis of the plant (Hamurcu et al., 2016). Under B excess, there is an induction of the B-efflux BOR4 transporter located in the plasma membrane (mainly in the endoderm), which has been reported as a homologous BOR1 gene (Miwa et al., 2007, 2014). Unlike BOR1, BOR4 is located on the distal side of the root epidermis and accumulates stably under high B conditions (Miwa et al., 2007), allowing BOR4 to exclude B for the soil solution. Plants with an increased expression and activity of this transporter usually exhibit greater tolerance to B conditions. It further allowed an increasing plant growth when exposed to high levels of B, by contrast, the functional lack of this transporter is associated with reductions in shoot growth and symptoms of toxicity under high B levels (Miwa et al., 2007, 2013; Kajikawa et al., 2011; Miwa and Fujiwara, 2011; Lv et al., 2017). Taken together, these B transporters and their regulatory mechanisms are central hubs for growth and nutrient homeostasis in plants.

5.4 The role of boron distribution and redistribution in plant adaptation

Once B is absorbed by the root cells mediated by the NIP5;1 and BOR1 transporters, it is translocated to the shoot until the xylem, through transpiration flow that is driven by the water potential gradient (Shelp et al., 1995; Marschner, 2012). There is a direct relationship between transpiration rates and B accumulation in the shoot (Dannel et al., 2002; Ben-Gal and Shani, 2002). This relationship is associated with mechanisms that limit the supply of B from the roots to the leaves. It seems reasonable to assume that the distribution of B, via the xylem, may provide better adaptation to deficiency or toxicity (Fig. 5.1).

The maintenance of transpiration under conditions of low B availability is vital to the proper distribution of this element between roots and shoot, preventing the general B deficiency within the plant. B distribution is related to the efficiency of water use, probably due to its transport into the xylem being associated with aquaporins (Wimmer and Eichert, 2013). Accordingly, ten aquaporin related genes were detected in plant species that are recognized to be efficient in the uptake and distribution of B; however, only two of those genes have been upregulated in B-inefficient species, under conditions of low availability of B (Zhou et al., 2015). In plants, the aquaporin family functions as membrane channels and can facilitate the absorption as well as the long-distance transport (via xylem) of water and other substances with low molecular weight (Ishibashi et al., 2011; Maurel et al., 2015). These aquaporins can influence water relations and the availability of B for plants. Therefore, under conditions of B deficiency, a greater capacity to uptake both water and B and transport them through the xylem to the shoot is likely related to the functions of aquaporin. Consequently, the transport through transpiration flow is a significant source of B, even under soil deficient B conditions (Matthes et al., 2018).

Under toxic conditions, plants accumulate more B in the shoot than in the roots (Mosa et al., 2016). Consequently, symptoms that manifest as necrosis on the leaf margins are usually observed (Camacho-Cristóbal et al., 2008). This response occurs because B is passively transported through the transpiration flow accumulating on the leaf margins. (Brown and Shelp 1997; Marschner, 2012). It has been recently reported that the B toxicity culminated with a rapid decrease in transpiration rates in Arabidopsis, which could be associated with a mechanism that limits the transport of excessive B from the root to the shoot and, consequently, its accumulation in the leaves (Sutton et al., 2007; Macho-Rivero et al., 2017). In tomato (*Solanum lycopersicum*), transpiration rate decreased as the B content increased in the irrigation water (Ben-Gal and Shani, 2002). The reduced water flow to the shoot, under conditions of B toxicity, is not only related to the reduction in the transpiration rate, but also to the negative regulation of several genes that encode aquaporins, which could result in a suppressed uptake and/or distribution of water and B in the roots and leaves (Macho-Rivero et al., 2017; Macho-Rivero et al., 2018). Previous studies have shown that B toxicity is associated with a decreased expression of several plasma membrane intrinsic proteins (PIPs) genes in Arabidopsis (Aquea et al., 2012). Since PIP family play a key role in the permeability of B, this reduced expression of PIP may result in lower uptake and/or water distribution in plants.

Abscisic acid (ABA) plays a key role in reducing stomatal conductance and, therefore, in the transpiration rate, which contributes to a reduction in the distribution of water from the root to the shoot of the plants, affecting the responses to B toxicity (Macho-Rivero et al., 2018). NCED3 (9-*cis*-epoxycarotenoid dioxygenase) gene expression and the shoo ABA content increases rapidly under conditions B toxicity. NCED3 knockout plants are characterized by a higher transpiration rate and stomatal conductance, showing also a high sensitivity to B toxicity as revealed by the higher accumulation of B in the shoot than wild-type plants (Iuchi et al., 2001; Frey et al., 2012; Macho-Rivero et al., 2018). Taken together, these results highlight the importance of the role of ABA signaling in the adaptive response of plants to B toxicity. Finally, all this evidence suggests that the changes caused by the deficiency and toxicity of B in the overall plant water balance worked as a mechanism to prevent the deficiency or excessive accumulation of B in plant tissues.

The efficiency to tolerate B excess is associated with the combined effect of the distribution and redistribution processes. By contrast to the situation observed with the xylem, the efficiency of B mobilization in long-distance transport (redistribution), via the phloem, depends mainly on the extent of the transfer of the xylem to the phloem and from species to species. It is important to mention that the redistribution of nutrients within plants is an important characteristic that largely affects both the growth and survival of young organs and tissues (Fernández et al., 2013). The direction of transport in the phloem is mainly determined by nutritional demands, occurring from the source to the sink (Marschner, 2012). The main loading sites for nutrients in the phloem (source) are located on the stem and leaves which provide nutrients for tissue and organs in high demand (sink), such as apical meristem, fruits, and flowers, allowing the cycle of nutrients in the plant (Marschner, 2012). In summary, the redistribution of nutrients (B included) is important due to the distribution via xylem being mainly directed to the places of greatest transpiration, which are often not the places of greatest demand.

Knowledge concerning the mobility of nutrients through phloem is important in the diagnosis and correction of deficiency and toxicity (Brown and Shelp, 1997), and the phloem mobility of B remains controversial being rather variable between plant species. For a long time, it was generally accepted that B was an immobile nutrient and therefore tended to accumulate in old leaves with high rates of transpiration (Stangoulis et al., 2001). This apparent immobility was associated with the inability of B to enter the phloem and the absence of long-distance flow in the phloem due to the return into the xylem because of its high membrane permeability. Following this hypothesis, since B reenter the xylem, it is simply distributed over the transpiration flow (Oertli, 1993; Reid et al., 2004). Therefore, immobile elements present in mature tissues cannot be redistributed to meet the demand of developing tissues and, therefore, must be constantly supplied via soil. However, B can be transported to adjacent tissues or organs through the phloem, via a complex with some metabolites (Brown and Shelp, 1997; Reid, 2014). These metabolites are related to photosynthetic assimilates, such as sucrose (Stangoulis et al., 2010; Du et al., 2020) and sugar alcohols (polyols) mannitol, dulcitol, and sorbitol (Bellaloui et al., 2003; Lehto et al., 2004; Liakopoulos et al., 2009; Will et al., 2011; Liu et al., 2012). It is reasonable to assume therefore that plants subjected to stress by B are characterized by altered carbohydrate metabolism. In fact, sucrose binds to B, resulting in the effective accumulation of this element in the phloem. As sucrose is discharged, B is released and diffuses into the sink cells (Stangoulis et al., 2010).

The mobility mechanism in species that redistributes B is associated with the presence of stable polyol-B-polyol complexes (Hu et al., 1997, Papadakis et al., 2018; Oikonomou et al., 2019; Tsiantas et al., 2019; Landi et al., 2019). The mobility of B in the phloem ensures that this nutrient can be redistributed through the plant, a fact that is remarkable under B deficient supply (Liakopoulos et al., 2009). Previous studies have shown that when B supply was interrupted, B-efficient cultivars had a greater capacity to redistribute B when compared to B-inefficient cultivars (Stangoulis et al., 2001; Liu et al., 2012; Wang et al., 2014). Therefore, the availability of B in tissues and organs in high demand under deficiency is related to the redistribution process. It has been shown that the introgression of the gene for sorbitol synthesis increased the mobility of B within the plant and consequently improved tolerance to deficiency (Brown et al., 1999, Bellaloui et al., 2003). In summary, plants with the ability to form complexes with polyols can better tolerate environments with low levels of B than plants that do not form these complexes, given that this nutrient can be supplied through redistribution to adjacent organs and tissues. Plants with restricted mobility of B in the phloem require a continuous supply to meet the demand of the different tissues and organs, where B deficiency would quickly induce growth inhibition (Hu et al., 1997).

To fully understand the mechanisms related to the distribution and redistribution of B is of crucial significance since different cultivars employ distinct strategies when submitted to high levels of B (Reid, 2010). Plant species that produce polyols under B excess accumulate this micronutrient in the stems, and not in the margins of old leaves, where B normally accumulates in species that do not redistribute it (El-Motaium et al., 1994). Thus, a completely different pattern of B accumulation and, consequently, in the development of toxicity symptoms, is usually observed in plants where B is considered mobile in the phloem (Papadakis et al., 2003). It seems that a mechanism of adaptation to the excess of B exists through the binding and immobilization of this element within the cell walls, especially in the newly formed nonphotosynthetic phloem tissues (Oikonomou et al., 2019).

In addition to the amount of B immobilized in the phloem cells, the increase in sorbitol (parallel to the marked reduction in sucrose content) represents an additional attempt to capture B ions in the phloem, probably mitigating its toxic effect (Oikonomou et al., 2019). In addition, the accumulation of sorbitol observed in young leaves may represent another form of the B complex that preserves other key polyols and polyol-based compounds (i.e., ATP, NADH). Further studies are clearly required to fully elucidate the range of mechanisms used by plants to cope with toxic B conditions.

5.5 Biochemical and physiological changes regulating stress tolerance by boron

The main purpose of this topic is to highlight the physiological and biochemical responses of plants to deal with both deficiency and toxicity caused by B. To understand the main metabolic plant responses, we will initially point them in each isolated stress condition and then with both stresses simultaneously. It is important to highlight that such responses are described in different species, from model to crop plants (annual and perennial). Next, we will attempt to unify these responses searching for a single pattern, with the most common metabolic mechanism in plants (Fig. 5.1).

First, to understand B deficiency is important to consider that each experiment is quite different from each others. Thus, whilst some experiments are focused on the short-term (hours or days) others are on the long-term (full life cycle of the plants) responses. In addition, another usual difference relies on the exact concentration used by each research group, with great variation between different experiments and different plants, ranging from 0 to 30 μM (close to optimal concentration for most plants) or even between 0 and 7.5 kg per hectare; to summarize, B deficiency stress is considered any values below the optimum for the specific plant. Most of these experiments are performed in pot cultivation (with nutrient solution) and the most common source of B in these experiments is H_3BO_3. That being said, reasonable comparisons between different experiments must be done with caution.

Under B deficiency, it was shown that *n* different organs of white lupine (*Lupinus albus*), there is a minor change in the levels of organic acids and carbohydrates; however, there are major changes in amino acid levels, especially an increase in those related to stress signaling and protection from oxidative stress (Alves et al., 2011). Tobacco plants (*Nicotiana tabacum*) submitted to B deficiency displayed a major alteration in N metabolism enzymes, with marked alteration in the levels of amino acids; a reduction in carbohydrate content and an increase in levels of organic acids and phenolic compounds were also observed (Beato et al., 2011; Camacho-Cristóbal et al., 2002). In rapeseed (*Brassica napus*), B deficiency led to reduced growth and yield associated with lower photosynthetic rates and N-fixing enzymes activity, coupled with an increase in antioxidant compounds with alteration in specialized metabolism (Yuai et al., 1993). Different cotton (*Gossypium hirsutum*) genotypes share the same response pattern to B deficiency characterized by reductions in gas exchange and accumulation of carbohydrates in the leaves due to lack of photoassimilates exported to sink organs (Bogiani et al., 2013; Zhao and Oosterhuis, 2003). It most likely explains the reduction in the number of flowers because of the high number of abortions. In white mulberry (*Morus alba*) B deficiency is usually associated with oxidative damage with a significant increase in both enzyme activity and the production of antioxidant compounds (Tewari et al., 2010).

Tolerance for B excess is highly variable for each species, with control treatments (optimal conditions) in the range of a few tens of μM of B for most plants studied, and the values used in toxic experiments ranging between 10 and 500 times the control values, depending on the plant species and condition studied. H_3BO_3 is commonly used as a source of B and the experiments analyzed here vary from weeks to months, being part of the life cycle or the complete life cycle of the plants. In Arabidopsis, B toxicity leads to a reduction in the content of photosynthetic pigments and accumulation of carbohydrates, in addition to an increase in antioxidant compounds such as flavonoids, anthocyanins and proline (Kayıhan et al., 2016). In tomato plants subjected to B excess, there is an activation of the antioxidant defense machinery, with emphasis on the role of ascorbate as a major player in this process (Cervilla et al., 2007). In carrots (*Daucus carota*) the application of salicylic acid can reduce B toxicity via alterations, in the levels of amino acids, such as proline (Eraslan et al., 2007). In leaves and roots of wheat (*Triticum aestivum*) B toxicity culminated with changes in photosynthetic pigments and glycine betaine content. Furthermore, there is a considerable impact on the content of carbohydrates and amino acids, with minor changes in the content of total proteins, despite a huge variation in gene expression (Kayıhan et al., 2017;

Metwally et al., 2017). B toxicity further reduced growth and yield in both sensitive and tolerant wheat cultivars.

In pistachio (*Pistacia vera*) excess of B did not impact plant growth or proline production, but there is a remarkable reduction in C isotope discrimination (Hokmabadi et al., 2006). In *Citrus macrophylla*, toxicity by B did not promote major phenotypic changes that most likely occurred due to the regulation at various levels: (i) the inhibition of the expression of B transporters associated with an activation of cellular efflux of this element was observed, (ii) in addition, there is a compartmentalization of B in the vacuole and deposition of B in cell walls, reducing cytoplasm damage, and (iii) there is an activation of the antioxidant system as revealed by the proline accumulation (Martinez-Cuenca et al., 2015). In the comparison between *Citrus grandis* (B-sensitive species) and *Citrus sinensis* (B-tolerant species) in conditions of B toxicity, the differences were mostly due to the maintenance of the energy flux and photosynthetic metabolism accompanied by increased antioxidant responses in tolerant plants, while sensitive plants did not show efficient mechanisms to control these pathways (Sang et al., 2015, 2017). In poplar species (*Populus russkii* and *Populus bolleca*) there is a high tolerance to B toxicity as revealed by activation of the antioxidant protection machinery associated with the allocation of this element in vacuoles, coupled to the alteration in the expression of B transporters, thus avoiding damage in these plants (Wu et al. 2015; Ou et al. 2019). In eucalyptus (*Eucalyptus camaldulensis*) C isotope discrimination, associated with the maintenance of transpiration rates is likely a good indicator of tolerance to B toxicity (Poss et al. 2000). Collectively, it demonstrated that plant responses to B toxicity are associated with a range of complementary responses.

A quite common type of experiment is with the application of both B deficiency and toxicity stress. Deficiency values are accepted as anyone below optimal (control treatments) and can be 10 to 50-times lesser than in the control or even no presence of B. The toxicity values reached up to 100-times more than the control value. It is noteworthy that the experiments consider treatments for between days and months, depending on the variety of plants studied. In brown mustard (*Brassica juncea*), both in leaves and roots, the responses to both B deficiency and toxicity are remarkably similar, with a reduction in the content of photosynthetic pigments and accumulation of ROS and antioxidant compounds; also, there is an accumulation of proline under these stressful conditions with changes in the osmotic potential of these tissues (Pandey and Archana, 2013). In highbush blueberry (*Vaccinium corymbosum*), B stress generates oxidative damage, without a clear response of protection with impacts on growth and assimilation of other nutrients, but the level of this impact is dependent on cultivars (Meriño-Gergichevich et al., 2017). In different *Citrus* spp. used as rootstocks the responses, but both in B deficiency and toxicity there is a reduction in gas exchange that is related to anatomical changes in these plants (Mesquita et al., 2016). In *C. grandis* subjected to both B deficiency and toxicity, the gas exchange responses, chlorophyll a fluorescence, carbohydrate production, pigment, ascorbate, and glutathione contents are similar at the two B stress levels, but on a different scale; wherein plants deficient in B have less oxidative damage than plants treated with excess B; moreover, the reduction in photosynthetic parameters of plants deficient in B is related to the accumulation of starch, whereas in plants in toxicity by B it is related to oxidative damage (Han et al., 2009).

B stress (both deficiency and toxicity) directly affects the ascorbate-glutathione cycle, the synthesis of phenolic compounds, N assimilation and photosynthesis rate, mainly altering the electron transport chain (Brown et al., 2002; Landi et al., 2012; Rosolem and Bogiani, 2011). The influence of B stress on photosynthesis is likely related to the reduction in leaf growth, but also the content of photosynthetic pigments and the enzyme activity since there is a strong relationship between B and N metabolism (Dell and Huang, 1997; Chen et al., 2012). B deficiency and toxicity lead to oxidative damage, however, plants develop an efficient system to control the production of ROS through enzymatic and nonenzymatic antioxidant mechanisms. In summary, plants with high antioxidant defenses under B stress are likely to be more tolerant (Cervilla et al., 2007; Beato et al., 2011; Landi et al., 2012; Kayıhan et al., 2016).

B plays an important role in the C metabolism, particularly in the synthesis and transport of carbohydrates. Under B excess, there is an accumulation of monosaccharides and a reduction in sucrose levels. Carbohydrate biosynthesis and degradation affect cellular osmotic potential, which in turn suggest that B-stress-mediated carbohydrate accumulation is part of an adaptive response to B deficiency. Furthermore, amino acid metabolism plays a crucial role in B tolerance, and the branched-chain amino acids isoleucine and leucine act as alternative energy sources and induce cells to produce energy under B stress. Meanwhile, free amino acids are important regulatory and signaling molecules. Regarding proteins, there is a reduction and an increase in protein content under conditions of deficiency and toxicity, respectively (Brown et al., 2002; Cervilla et al., 2007;

Han et al., 2009; Alves et al., 2011; Sang et al., 2015). As for the specialized metabolism, there are quantitative alterations of phenolic compounds production in response to B stress; and the accumulation of phenolic compounds suggests a possible detoxifying mechanism through the sequestration, immobilization, and compartmentalization of excessive B (Camacho-Cristóbal et al., 2002; Lewis, 1980).

B stress responses in gas exchange and chlorophyll fluorescence are generally aimed at reducing the efficiency of these processes; however, a clear (or direct) mechanism for the role of B in photosynthesis remains to be elucidated. It is possible that the role of B is even indirect, affecting membrane stability and consequently electron transport, simultaneously affecting the antioxidant system and ultimately inducing a reduction in photosynthetic activity. Although different studies present rather distinct approaches with great variation in the B values used, it seems that a similar physiological and biochemical response pattern were observed. Therefore, it can be concluded that differences between plant species with respect to B accumulation and tolerance may be due to differences in metabolism and differences in their tissue accumulation patterns.

5.6 Molecular effects underlying the effects of boron in plants

Harmful effects caused by deficiency or excess of B in plants is well studied for years and several differential physiological alterations have been shown to be caused by inadequate B availability, as discussed before. However, the molecular mechanisms underlying these effects remain much less well understood. Based on transcript analyses of the relevant genes to the role and the response to B in plants, several genes related to the B transport mechanism and its fine adjustment in maintaining cell homeostasis are differentially regulated upon B treatment. As transporters will be briefly covered before in this chapter, here we will focus on additional important genes whose transcripts contribute to the response to either deficiency and/or toxicity caused by unbalanced levels of B.

The first transcription factor shown to be affected by B deficiency is WRKY6, which has the potential to mitigate the negative effects promoted by B deficiency (Kasajima et al., 2010). Specifically, Arabidopsis mutants of *WRKY6* showed retarded root growth compared with the WT under B deficiency, but not under normal conditions (Kasajima et al., 2010). Promoter activity of WRKY6 was stimulated around the root tip, and the transcript levels of WRKY6 were also upregulated under B deficiency. Moreover, transcriptome around the root tip of WT and *wrky6* mutant cultivated under normal or B deficient condition demonstrated that *WRKY6* mutation abolishes down- or upregulation response of part of the genes whose expression is altered upon B-deficiency in WT plants. However, the overall pattern of *WRKY6* gene expression is not widely changed in the two conditions and, even those genes whose expression is altered by the *WRKY6* mutation under B deficiency have varied functions, as to be expected for pleiotropic genes.

In addition to WRKY6, the expression of many other genes is altered upon B deprivation, which can reduce crop productivity (Camacho-Cristóbal et al., 2011; Wang et al., 2015; González-Fontes et al., 2016). Evidence has been accumulating for the role of calcium ions (Ca^{2+}) and the Ca^{2+}-related proteins as intermediates in the transduction pathway initiated by B deficiency (Quiles-Pando et al., 2013; González-Fontes et al., 2014). Remarkably, under B deficiency, the cytosolic levels of Ca^{2+} are altered concomitantly with the upregulation of several genes related to Ca^{2+} channel/transporter and Ca^{2+} sensing (Quiles-Pando et al., 2013) in Arabidopsis roots. Based on these results and in the structural roles that B and Ca^{2+} have (both participate in cross-linking pectin chains in the cell wall), a different hypothesis was raised to explain sensing and response to B deprivation (Camacho-Cristóbal et al., 2011; González-Fontes et al., 2014).

An alternative strategy that has been used to unravel B responses is the utilization of genotypes showing differential sensitivity to B deficiency. As a remarkable example, the whole genome re-sequencing of a B-deficiency-sensitive genotype, "Westar 10" ("W10"), and a B-deficiency-resistant genotype, Qingyou 10' ("QY10"), of rapeseed elucidated the responses to boron deficiency during the vegetative and reproductive development (Hua et al., 2016). B deficiency-resistant genotype "QY10" showed higher transcript levels of B transporter genes and more differentially expressed genes related to antioxidant enzymes. Given that the B-deficiency-sensitive genotype "W10" exhibited elevated ROS and malondialdehyde accumulation in the roots, these results collectively indicate that a greater ability to scavenger ROS partly explains the higher tolerance to B deficiency in rapeseed roots. Furthermore, genes related to the maintenance of cell wall and to membrane function were differentially expressed upon B deficiency in "W10" and "QY10" genotypes. In the roots of "QY10," more genes related to arabinogalactan-proteins showed higher transcript levels than that observed in "W10" roots, whereas a greater number of genes related to xyloglucan *endo*-trans-glycosylase/hydrolases and pectate lyases were higher expressed in the roots of "W10" than those of "QY10."

As well as demonstrated for B deficiency, the complete elucidation of the genetic mechanisms underlying boron toxicity responses are incomplete (Wang et al., 2020). In Arabidopsis, the rapid inhibition of root growth, one of the first and most evident symptoms exhibited by plants upon B toxicity, is likely mediated by changes in the expression of cell cycle genes (Aquea et al., 2012). In barley (*Hordeum vulgare*), a study conducted to understand the transcriptional responses of a B excess tolerant genotype grown under control (50 μM) and excessive B (1000 μM) demonstrated that some pathways associated with Ca^{2+}−calmodulin system are particularly induced in response to B toxic exposure (Tombuloglu et al., 2015). Moreover, several genes related to defense responses and cell wall modification are differentially expressed in response to excessive B (Aquea et al., 2012; Tombuloglu et al., 2015; Wang et al., 2020). A recent study proposed, among other things, to compare the transcriptional responses of Arabidopsis, a B stress-sensitive species, and its close relative *Schrenkiella parvula*, which is a B stress-tolerant extremophyte (Wang et al., 2020). Greater gene expression changes were observed in *S. parvula* in comparison to Arabidopsis. Moreover, the expression levels of some of the upregulated genes in Arabidopsis were like those exhibited by their orthologs in *S. parvula* even under control condition (Wang et al., 2020). These results suggest an altered "defense readiness" at the transcription level of *S. parvula* which is probably contributing to the differential stress response exhibited by plants that are tolerant to excess B (Wang et al., 2020). Collectively, the data described above suggest that B stress generates a series of responses at the molecular level that are only starting to be unraveled and will certainly aid in the development of crop plants that can tolerate this problem while maintaining their productive levels.

5.7 Adaptive mechanisms to boron stress in the reproductive phase

In plants, the reproductive tissues show particular sensitivity to fluctuation in the levels of B, in view, that this micronutrient plays a significant role in the germination of the pollen grain and in the growth of the pollen tube (Huang et al., 2000; Dell et al., 2002; Wang et al., 2003; Iwai et al., 2006).

Consequently, H_3BO_3 transporters and channels have evolved to rigidly regulate the uptake, distribution, redistribution and homeostasis of B for the tissues of need, and is therefore decisive in determining tolerance B deficiency and toxicity (Miwa and Fujiwara, 2010; Chatterjee et al., 2014; Durbak et al., 2014; Yoshinari and Takano, 2017; Zhang et al., 2017).

Among the membrane channels that facilitate the uptake of H_3BO_3 are the NIPs, and the family of NIPs genes appeared at the beginning of the evolution of the terrestrial plant and underwent substantial structural and functional diversification (Danielson and Johanson, 2010; Abascal et al., 2014; Roberts and Routray, 2017). NIPs are divided into three subfamilies (NIP I, NIP II, and NIP III) according to the pore structure of their channels (Roberts and Routray, 2017). The subfamily NIP II was characterized as a channel for H_3BO_3. In Arabidopsis, there are three genes (*AtNIP5;1*, *AtNIP6;1*, and *AtNIP7;1*) that encode proteins of the subfamily NIP II (Wallace and Roberts, 2004). The NIP7;1 protein presents greater evolutionary divergence in comparison to the other two proteins belonging to NIP II, being identified only in dicot species, suggesting that they have evolved more recently (Roberts and Routray, 2017).

Analyzes of the proteins NIP5;1 and NIP6;1 provide evidence for specific functions in the nutrition of B. *NIP6;1* is the most similar gene to *NIP5;1* in Arabidopsis, but NIP6;1 has different physiological roles in the transport of B. During the evolution of Arabidopsis, *NIP5;1* and *NIP6;1* apparently diversified in terms of tissue expression specificity, while maintaining their B transport activities (Takano et al., 2006). NIP5;1 is a transcript specific to the root epidermis that is induced by B limitation (Takano et al., 2006; Tanaka and Fujiwara, 2008). In contrast, *NIP6;1* is highly expressed in a specific manner in the branch, particularly within the phloem vascular tissue, in the nodal regions of the stem, developing young leaf petioles and junction areas (the base of flowers and siliquas), associated with xylem–phloem transfer (Zimmermann et al., 2004; Tanaka and Fujiwara, 2008).

Flowering is the stage of development most prone to B deficiency and subsequent production losses. NIP proteins were detected in floral tissues, contributing to the distribution of B in flowers sensitive to deficiency of this micronutrient. *NIP7;1* is predominantly expressed in the plasma membrane of tapetum cells at critical stages of development that encompass microsporogenesis and the formation of the pollen cell wall, during a narrow development window (floral stages 9–11) (Routray et al., 2018). Tapetal cells play a fundamental role in development, providing metabolites and nutrients to the developing pollen grain (Liu and Fan, 2013, Shi et al., 2015). In Arabidopsis, the development of the flower is divided into 20 stages that represent a series of remarkable events starting with the initiation of the flower and ending with the fall of the seed. In stage 10, the exine pollen cell wall generation is initiated (Liu and Fan, 2013). In stage 11, mitotic divisions of the pollen occur and the tapetum degenerates due to programmed cell death (Liu and Fan, 2013), releasing the contents of the tapetal cells, which become associated with the outer coating of mature cells from the pollen grains (Hsieh and Huang, 2007).

Given the sensitivity of microsporogenesis and the development of pollen to the limitation of B, male sterility is a common result of the limitation of this micronutrient (Huang et al., 2000; Li et al., 2016; Routray et al., 2018). NIP7;1 plays a role in B homeostasis during flower development (Li et al., 2011; Onuh and Miwa, 2021). Under the limitation of B, the pollen grains *nip7;1* have reduced viability, irregular morphology, increased incidence of agglomerated grains and deficient exine formation, with impairment of the typical cell wall (Routray et al., 2018).

The formation of the pollen cell wall is a spatially and temporally regulated process involving the synthesis, secretion, and assembly of several components, including an elaborate exine reticulated wall (Liu and Fan, 2013; Shi et al., 2015). Crosslinking of RG-II by B is also important for maintaining the mechanical strength of the cell wall and therefore the growth of the pollen tube (Iwai et al., 2006; Delmas et al., 2008). The B-mediated crosslinking of RG-II is believed to play an important role in the oscillatory growth of the pollen tube (Holdaway-Clarke et al., 2003). In addition, the following transporters, *AtNIP4;1* and *AtNIP4;2*, have been identified in Arabidopsis, which are specifically expressed in pollen and pollen tube elongation, which have been inhibited under limited B conditions (Pérez Di Giorgio et al., 2016). *AtNIP4;1* and *AtNIP4;2* possibly has a role in the efficient transport of B for the elongation of the pollen tube and subsequent fertilization (Onuh and Miwa, 2021).

In B homeostasis, the potential transporters of the BOR family can be co-express with NIP7;1 during the critical stage of pollen development by coordinating homeostasis in the anthers. Although BOR1 is predominantly expressed in roots, induction of its transcriptional activity was observed in flowers in anthesis (Pérez-Castro et al., 2012).

Monocotyledonous plants require less B for their growth compared to dicotyledonous plants. This lower B requirement correlates with lower amounts of pectic compounds in the cell wall of grassy plants. In contrast to the vegetative stage, the reproductive stage of monocots is sensitive to the low availability of B. The limited supply of B in the reproductive stage leads to the loss of yield associated with male sterility in rice (*Oryza sativa*), wheat, and barley (Wongmo et al., 2004; Yu and Bell 2002). These results suggest that B is essential for the fertility of grains in rice for supporting the germination of pollen and development of the panicle (floral organ) (Tanaka et al., 2013). The rice genome contains four genes similar to *AtBOR1*; namely *OsBOR1*, *OsBOR2*, *OsBOR3* and *OsBOR4* (Nakagawa et al., 2007). *OsBOR1*, the closest counterpart to *AtBOR1*, is the transporter of B acquisition in the roots and subsequent translocation to the shoots (Nakagawa et al., 2007). *OsBOR1* is also highly expressed in floral organs, including anthers and pollen during the maturation process (Mihara et al., 2008; Suwabe et al., 2008). *OsBOR4* is specifically expressed in pollen, the interruption of this gene causes an anomaly in pollen germination and/or pollen tube elongation (Mihara et al., 2008; Suwabe et al., 2008; Tanaka et al., 2013). Taken together, the results indicate that *OsBOR4* is a B transporter with a primary function in maintaining pollen germination and in the growth of the pollen tube. Considering the essential role of B in reproductive growth and the potential roles of *OsBORs* for B transport, members of the *OsBOR* family are involved in homeostasis B in floral organs. In addition, in barley and wheat, the contribution of BORs to high-B tolerance has been extensively investigated.

Therefore, because B is necessary for pollen germination and pollen tube growth (Iwai et al., 2006), specific BOR transporters are likely to support this process, in conjunction with NIPs, such as *AtNIP7;1* (Routray et al., 2018), *AtNIP4;1* and *AtNIP4;2* (Pérez Di Giorgio et al., 2016).

5.8 Concluding remarks

The effects of B on plant growth and development at low and high concentrations have been extensively demonstrated. Here we present further evidence that an adequate B nutrition can be efficient in the improvement of plant tolerance toward a range of abiotic and biotic stresses. This fact apart, it seems reasonable to assume that such stress should be firstly investigated separately and intensely and in the near future the combination of stress are likely able to render a much greater understanding of B aspects in plant responses. Unveiling these interactions will likely advance our understanding of B-mediated responses that affect overall plant performance under a changing environment, which is especially important as climate change expands pest ranges and places additional stresses on the plants upon which we rely for food, materials, and renewable energy. It is important to mention, however, that only a few crops have been studied thus far and as such further studies are clearly required.

As sessile organisms, plants constantly face multiple environmental stresses. While many studies have shown the plant responses to many environmental stresses, there is still a surprisingly limited understanding of the precise role played by B in modulating cell-specific plant responses. Future research work should seek to obtain more field data of B responses in crops and simultaneously explore the complex response behavior associated with other biotic and abiotic stress. A large research gap on our understanding of the physiological and molecular responses at cellular levels to different B conditions continues to exist. Due to the many links between B responses and multiple aspects of plant physiology in general, any manipulations, or changes of B fertilization may have molecular, individual, or even community-level impacts. Identifying the molecular bases of plant responses due to either high or low concentrations of B can pave the way to determine the essential steps related to the effective response of plants to these conditions.

Funding

This work was supported by funding from the Conselho Nacional de Desenvolvimento Científico e Tecnológico (CNPq-Brazil), the Fundação de Amparo Pesquisa do Estado de Minas Gerais (FAPEMIG-Brazil). Research fellowships granted by CNPq-Brazil to A.N.N. and W.L.A. and scholarship provided by CNPq to GLP and the Coordenação de Aperfeiçoamento de Pessoal de Nível Superior (CAPES-Brazil) to RPO-G are also gratefully acknowledged.

References

Abascal, F., Irisarri, I., Zardoya, R., 2014. Diversity and evolution of membrane intrinsic proteins. Biochim. Biophys. Acta. 1840, 1468–1481.

Abreu, I., Poza, L., Bonilla, I., Bolaños, L., 2014. Boron deficiency results in early repression of a cytokinin receptor gene and abnormal cell differentiation in the apical root meristem of Arabidopsis thaliana. Plant. Physiol. Biochem. 77, 117–121.

Aibara, I., Hirai, T., Kasai, K., Takano, J., Onouchi, H., Naito, S., et al., 2018. Boron-dependent translational suppression of a borate exporter BOR1 for avoidance of boron toxicity. Plant. Physiol. 177, 759–774.

Alves, M., Chicau, P., Matias, H., Passarinho, J., Pinheiro, C., Ricardo, C.P., 2011. Metabolic analysis revealed altered amino acid profiles in Lupinus albus organs as a result of boron deficiency. Physiol. Plant. 142, 224–232.

Aquea, F., Federici, F., Moscoso, C., Vega, A., Jullian, P., Haseloff, J., et al., 2012. A molecular framework for the inhibition of Arabidopsis root growth in response to boron toxicity. Plant. Cell Env. 35, 719–734.

Archana, N.P., Verma, P., 2017. Boron deficiency and toxicity and their tolerance in plants: a review. J. Global. Biosci 6, 4958–4965.

Beato, V.M., Navarro-Gochicoa, M.T., Rexach, J., Herrera-Rodríguez, M.B., Camacho-Cristóbal, J.J., Kempa, S., et al., 2011. Expression of root glutamate dehydrogenase genes in tobacco plants subjected to boron deprivation. Plant. Physiol. Biochem. 49, 1350–1354.

Bellaloui, N., Yadavc, R.C., Chern, M.S., Hu, H., Gillen, A.M., Greve, C., et al., 2003. Transgenicallyenhanced sorbitol synthesis facilitates phloem-boron mobility in rice. Physiol. Plant. 117, 79–84.

Bell, D.L., Sultan, S.E., 1999. Dynamic phenotypic plasticity for root growth in Polygonum: a comparative study. Am. J. Bot. 86, 807–819.

Ben-Gal, A., Shani, U., 2002. Yield, transpiration and growth of tomatoes under combined excess boron and salinity stress. Plant. Soil. 247, 211–221.

Blilou, I., Xu, J., Wildwater, M., Willemsen, V., Paponov, I., Friml, J., et al., 2005. The PIN auxin efflux facilitator network controls growth and patterning in Arabidopsis roots. Nature. 433, 39–44.

Bogiani, J.C., Amaro, A.C.E., Rosolem, C.A., 2013. Carbohydrate production and transport in cotton cultivars grown under boron deficiency. Sci. Agric. 7, 442–448.

Brdar-Jokanović, M., 2020. Boron toxicity and deficiency in agricultural plants. Int. J. Mol. Sci. 21, 1424.

Brown, P.H., Bellaloui, N., Hu, H., Dandekar, A., 1999. Transgenically enhanced sorbitol synthesis facilitates phloem boron transport and increases tolerance of tobacco to boron deficiency. Plant. Physiol. 119, 17–20.

Brown, P.H., Bellaloui, N., Wimmer, M.A., Bassil, E.S., Ruiz, J., Hu, H., et al., 2002. Boron in plant biology. Plant. Biol. 4, 205–223.

Brown, P.H., Shelp, B.J., 1997. Boron mobility in plants. Plant. Soil. 193, 85–101.

Brundrett, M., Ferguson, B., Gressshoff, P., Mathesius, U., Munns, Rasmussen, A., et al., 2017. 4.1-Nutrient Requir. Root Architecture .

Camacho-Cristóbal, J.J., Anzellotti, D., González-Fontes, A., 2002. Changes in phenolic metabolism of tobacco plants during short-term boron deficiency. Plant. Biol. Biochem. 40, 997–1002.

Camacho-Cristóbal, J.J., Herrera-Rodríguez, M.B., Beato, V.M., Rexach, J., Navarro-Gochicoa, M.T., Maldonado, J.M., et al., 2008. The expression of several cell wall-related genes in Arabidopsis roots is down-regulated under boron deficiency. Env. Exp. Bot. 63, 351–358.

Camacho-Cristóbal, J., Martín-Rejano, E., Rodríguez, M.B., Navarro-Gochicoa, M., Rexach, J., González-Fontes, A., 2015. Boron deficiency inhibits root cell elongation via an ethylene/auxin/ROS-dependent pathway in Arabidopsis seedlings. J. Exp. Bot. 66, 10.

Camacho-Cristóbal, J.J., Rexach, J., González-Fontes, A., 2008. Boron plant: deficiency toxicity. J. Integr. Plant. Biol. 50, 1247–1255.

Camacho-Cristóbal, J.J., Rexach, J., Herrera-Rodríguez, M.B., Navarro-Gochicoa, M.T., González-Fontes, A., 2011. Boron deficiency and transcript level changes. Plant. Sci. 181, 85–89.

Cervilla, L.M., Blasco, B., Ríos, J.J., Romero, L., Ruiz, J.M., 2007. Oxidative stress and antioxidants in tomato (*Solanum lycopersicum*) plants subjected to boron toxicity. Ann. Bot. 100, 747–756.

Cervilla, L.M., Blasco, B., Ríos, J.J., Rosales, M.A., Rubio-Wilhelmi, M.M., Sánchez-Rodríguez, E., et al., 2009. Response of nitrogen metabolism to boron toxicity in tomato plants. Plant. Biol. 11, 671–677.

Chatterjee, M., Tabi, Z., Galli, M., Malcomber, S., Buck, A., Muszynski, M., et al., 2014. The boron efflux transporter ROTTEN EAR is required for maize inflorescence development and fertility. Plant. Cell 26, 2962–2977.

Chen, L.S., Han, S., Qi, Y.P., Yang, L.T., 2012. Boron stresses and tolerance in citrus. Afr. J. Biotech. 11, 5961–5969.

Choi, E.Y., Kolesik, P., McNeill, A., Collins, H., Zhang, Q., Huynh, B.L., et al., 2007. The mechanism of boron tolerance for maintenance of root growth in barley (*Hordeum vulgare* L.). Plant. Cell Env. 30, 984–993.

Choi, E.Y., McNeill, A.M., Coventry, D., Stangoulis, J.C.R., 2006. Whole plant response of crop and weed species to high subsoil boron. Aust. J. Agr. Res. 57, 761–770.

Chormova, D., Fry, S.C., 2016. Boron bridging of rhamnogalacturonan-II is promoted in vitro by cationic chaperones, including polyhistidine and wall glycoproteins. N. Phytol. 209, 241–251.

Chormova, D., Messenger, D.J., Fry, S.C., 2014a. Boron bridging of rhamnogalacturonan II, monitored by gel electrophoresis, occurs during polysaccharide synthesis and secretion but not post-secretion. Plant. J. 77, 534–546.

Chormova, D., Messenger, D.J., Fry, S.C., 2014b. Rhamnogalacturonan-II cross-linking of plant pectins via boron bridges occurs during polysaccharide synthesis and/or secretion. Plant. Signal. Behav. 9, e28169.

Danielson, J.A., Johanson, U., 2010. Phylogeny of major intrinsic proteins. Adv. Exp. Med. Biol. 679, 19–31.

Dannel, F., Pfeffer, H., Römheld, V., 2002. Update on boron in higher plant - Uptake, primary translocation and compartmentation. Plant. Biol. 4, 193–204.

Dell, B., Huang, L., 1997. Physiological response of plants to low boron. Plant. Soil. 193, 103–120.

Dell, B., Huang, L., Bell, R.W., 2002. Boron in plant reproduction. In: Goldbach, H.E., Rerkasem, B., Wimmer, M., Brown, P.H., Thellier, M., Bell, R.W. (Eds.), Boron in Plant and Animal Nutrition. Kluwer Academic/Plenum Publishers, New York, USA, pp. 103–118.

Delmas, F., Seveno, M., Northey, J.G., Hernould, M., Lerouge, P., Mccourt, P., et al., 2008. The synthesis of the rhamnogalacturonan II component 3-deoxy-D-manno-2-octulosonic acid (Kdo) is required for pollen tube growth and elongation. J. Exp. Bot. 59, 2639–2647.

Desnos, T., 2008. Root branching responses to phosphate and nitrate. Curr. Opin. Plant. Biol. 11, 82–87.

Dong, X., Liu, G., Wu, X., Lu, X., Yan, L., Muhammad, R., et al., 2016. Different metabolite profile and metabolic pathway with leaves and roots in response to boron deficiency at the initial stage of citrus rootstock growth. Plant. Physiol. Biochem. 108, 121–131.

Dordas, C., Chrispeels, M.J., Brown, P.H., 2000. Permeability and channel-mediated transport of boric acid across membrane vesicles isolated from squash roots. Plant. Physiol. 124, 1349–1361.

Dugger, W.M., 1983. Boron in metabolism. Encyclopedia of Plant Physiol. Springer, Verlag, N.Y, p. 626.

Durbak, A.R., Phillips, K.A., Pike, S., O'Neill, M.A., Mares, J., Gallavotti, A., et al., 2014. Transport of boron by the tassel-less1 aquaporin is critical for vegetative and reproductive development in maize. Plant. Cell 26, 2978–2995.

Du, W., Pan, Z.Y., Hussain, S.B., Han, Z.X., Peng, S.A., Liu, Y.Z., 2020. Foliar supplied boron can be transported to roots as a boron-sucrose complex via phloem in citrus trees. Front. Plant. Sci. 11, 250.

Ederli, L., Reale, L., Ferranti, F., Pasqualini, S., 2004. Responses induced by high concentration of cadmium in *Phragmites australis* roots. Physiol. Plant. 121, 66–74.

Eggert, K., Wirén, N., 2017. Response of the plant hormone network to boron deficiency. N. Phytol. 216.

El-Motaium, R., Hu, H., Brown, P.H., 1994. The relative tolerance of six Prunus rootstocks to boron and salinity. J. Am. Soc. Hortic. Sci. 119, 1169–1175.

Eraslan, F., Inal, A., Gunes, A., Alpaslan, M., 2007. Impact of exogenous salicylic acid on the growth, antioxidant activity and physiology of carrot plants subjected to combined salinity and boron toxicity. Sci. Hortic. (Amst.) 113, 120–128.

Fernández, V., Sotiropoulos, T., Brown, P.H., 2013. Foliar Fertilization: Scientifc Principles and Field Practices, 1st ed. IFA, Paris.

Forde, B., Lorenzo, H., 2001. The nutritional control of root development. Plant. Soil. 232, 51–68.

Frey, A., Effroy, D., Lefebvre, V., Seo, M., Perreau, F., Berger, A., et al., 2012. Epoxycarotenoid cleavage by NCED5 fine-tunes ABA accumulation and affects seed dormancy and drought tolerance with other NCED family members. Plant. J. 70, 501–512.

Ghanati, F., Morita, A., Yokota, H., 2002. Induction of suberin and increase of lignin content by excess boron in tobacco cells. Soil. Sci. Plant. Nutr. 48, 357–364.

Ghanati, F., Morita, A., Yokota, H., 2005. Deposition of suberin in roots of soybean induced by excess boron. Plant. Sci. 168, 397–405.

Goldbach, H.E., Wimmer, M.A., 2007. Boron in plants and animals: is there a role beyond cell-wall structure? J. Plant. Nutr. Soil. Sci. 170, 39–48.

Goldbach, H.E., Yu, Q., Wingender, R., Schulz, M., Wimmer, M., Findeklee, P., et al., 2001. Rapid response reactions of roots to boron deprivation. J. Plant. Nutr. Soil. Sci. 164, 173–180.

González-Fontes, A., Herrera-Rodríguez, M.B., Martín-Rejano, E.M., Navarro-Gochicoa, M.T., Rexach, J., Camacho-Cristóbal, J.J., 2016. Root responses to boron deficiency mediated by ethylene. Front. Plant. Sci. 6, 1–6.

González-Fontes, A., Navarro-gochicoa, M.T., Camacho-Cristóbal, J.J., Quiles-pando, C., Rexach, J., Bego, M., 2014. Is Ca2 + involved in the signal transduction pathway of boron deficiency? New hypotheses for sensing boron deprivation. Plant. Sci. 218, 135–139

González-Fontes, A., Rexach, J., Quiles-Pando, C., Herrera-Rodríguez, M.B., Camacho Cristóbal, J.J., Navarro-Gochicoa, M.T., 2013. Transcription factors as potential participants in the signal transduction pathway of boron deficiency. Plant. Signal. Behav. 8, e26114.

Gruber, B.D., Giehl, R.F.H., Friedel, S., von Wirén, N., 2013. Plasticity of the Arabidopsis root system under nutrient deficiencies. Plant. Physiol. 163, 161–179.

Guo, P., Qi, Y.P., Yang, L.T., Ye, X., Jiang, H.X., Huang, J.H., et al., 2014. cDNA-AFLP analysis reveals theadaptive responses of citrus to long-term boron-toxicity. BMC Plant. Biol. 14, 284.

Hamurcu, M., Hakki, E.E., Sert, T.D., Özdemir, C., Minareci, E., Avsaroglu, Z.Z., et al., 2016. Extremely high boron tolerance in *Puccinellia distans* (Jacq.) Parl. related to root boron exclusion and a well-regulated antioxidant system. J. Biosci. 71, 273–285.

Han, S., Tang, N., Jiang, H.-X., Yang, L.-T., Li, Y., Chen, L.-S., 2009. CO2 assimilation, photosystem II photochemistry, carbohydrate metabolism and antioxidant system of citrus leaves in response to boron. Plant. Sci. 176, 43–153.

Hayes, J.E., Pallotta, M., Baumann, U., Berger, B., Langridge, P., Sutton, T., 2013. Germanium as a tool to dissect boron toxicity effects in barley and wheat. Funct. Plant. Biol. 40, 618–627.

Hayes, J.E., Reid, R.J., 2004. Boron tolerance in barley is mediated by efflux of B from the roots. Plant. Physiol. 136, 3376–3382.

Hell, R., Hillebrand, H., 2001. Plant concepts for mineral acquisition and allocation. Curr. Opin. Biotechnol. 12, 161–168.

Hokmabadi, H., Arzani, K., Gheibi, B., Hadavi, F., 2006. Growth and leaf chemical composition of three pistachio (*Pistacia vera* L.) rootstock seedlings in response to Boron excess in irrigation water. Acta Hortic. 726, 363–366.

Holdaway-Clarke, T.L., Weddle, N.M., Kim, S., Robi, A., Parris, J., Kunkel, J.G., et al., 2003. Effect of extracellular calcium, pH and borate on growth oscillations in *Lillium formosanum* pollen tubes. J. Exp. Bot. 54, 65–72.

Hsieh, K., Huang, A.H., 2007. Tapetosomes in *Brassica tapetum* accumulate endoplasmic reticulum-derived flavonoids and alkanes for delivery to the pollen surface. Plant. Cell 19, 582–596.

Huang, L., Bell, R.W., Dell, B., 2008. Evidence of phloem boron transport in response to interrupted boron supply in white lupin (*Lupinus albus* L. cv. Kiev Mutant) at the reproductive stage. J. Exp. Bot. 59, 575–583.

Huang, J.H., Lin, X.J., Zhang, L.Y., Wang, X.D., Fan, G.C., Chen, L.S., 2019. MicroRNA sequencing revealed Citrus adaptation to long-term boron toxicity through modulation of root development by miR319 and miR171. Int. J. Mol. Sci. 20, 1422.

Huang, L., Pant, J., Dell, B., Bell, R.W., 2000. Effects of boron deficiency on anther development and floret fertility in wheat (*Triticum aestivum* L. Wil-goyne). Ann. Bot. 85, 493–500.

Hua, Y., Zhou, T., Ding, G., Yang, Q., Shi, L., Xu, F., 2016. Physiological, genomic and transcriptional diversity in responses to boron deficiency in rapeseed genotypes. J. Exp. Bot. 67, 5769–5784.

Hu, H., Brown, P.H., 1997. Absorption of boron by plant roots. Plant. Soil. 193, 49–58.

Hu, H., Penn, S.G., Lebrilla, C.B., Brown, P.H., 1997. Isolation and characterization of soluble boron complexes in higher plants. Plant. Physiol. 113, 649–655.

Ishibashi, K., Kondo, S., Hara, S., Morishita, Y., 2011. The evolutionary aspects of aquaporin family. Am. J. Integr. Comp. Physiol. 300, R566–R576.

Iuchi, S., Kobayashi, M., Taji, T., Naramoto, M., Seki, M., Kato, T., et al., 2001. Regulation of drought tolerance by gene manipulation of 9-cisepoxycarotenoid dioxygenase, a key enzyme in abscisic acid biosynthesis in Arabidopsis. Plant. J. 27, 325–333.

Iwai, H., Hokura, A., Oishi, M., Chiba, H., Ishii, T., Sakai, S., et al., 2006. The gene responsible for borate cross-linking of pectin rhamnogalactronan-II is required for plant reproduction tissue development and fertilization. Proc. Natl. Acad. Sci. U.S.A. 103, 16592–16597.

Jefferies, S., Pallotta, M., Paull, J.G., Karakousis, A., Kretschmer, J., Manning, S., et al., 2000. Mapping and validation of chromosome regions conferring boron toxicity tolerance in wheat (*Triticum aestivum*). Theor. Appl. Genet. 101, 767–777.

Kajikawa, M., Fujibe, T., Uraguchi, S., Miwa, K., Fujiwara, T., 2011. Expression of the the Arabidopsis borate efflux transporter gene, AtBOR4, in rice affects the xylem loading of boron and tolerance to excess boron. Biosci. Biotechnol. Biochem. 75, 2421–2423.

Kasai, K., Takano, J., Miwa, K., Toyoda, A., Fujiwara, T., 2011. High boron-induced ubiquitination regulates vacuolar sorting of the BOR1 borate transporter in Arabidopsis thaliana. J. Biol. Chem. 286, 6175–6183.

Kasajima, I., Ide, Y., Yokota, M., Fujiwara, T., 2010. WRKY6 is involved in the response to boron deficiency in *Arabidopsis thaliana*. Physiol. Plant. 139, 80–92.
Kaur, S., Nicolas, M.E., Ford, R., Norton, R.M., Taylor, P.W.J., 2006. Selection of *Brassica rapa* genotypes for tolerance to boron toxicity. Plant. Soil. 285, 115–123.
Kayıhan, D.S., Kayıhan, C., Çiftçi, Y.Ö., 2016. Excess boron responsive regulations of antioxidative mechanism at physio-biochemical and molecular levels in *Arabidopsis thaliana*. Plant. Physiol. Biochem. 109, 337–345.
Kayıhan, C., Öz, M.T., Eyidoğan, F., Yucel, M., Öktem, H.A., 2017. Physiological, biochemical, and transcriptomic responses to boron toxicity in leaf and root tissues of contrasting wheat cultivars. Plant. Mol. Biol. Rep. 35, 97–109.
Kobayashi, M., Miyamoto, M., Matoh, T., Kitajima, S., Hanano, S., Sumerta, I.N., et al., 2017. Mechanism underlying rapid responses to boron deprivation in Arabidopsis roots. Soil. Sci. Plant. Nutr. 64, 1–10.
Kobayashi, M., Mutoh, T., Matoh, T., 2004. Boron nutrition of cultured tobacco BY-2 cells. IV. Genes induced under low boron supply. J. Exp. Bot. 55, 1441–1443.
Kocabek, T., Svoboda, Z., Al-Zwi, A., Rolfe, S., Fellner, M., 2009. Boron-regulated hypocotyl elongation is affected in Arabidopsis mutants with defects in light signalling pathways. Environ. Exp. Bot. 67, 101–111.
Konuk, M., Liman, R., Ciğerci, İ., 2007. Determination of genotoxic effect of boron on *Allium cepa* root meristematic cells. Pak. J. Bot. 39.
Kováčik, J., Klejdus, B., 2008. Dynamics of phenolic acids and lignin accumulation in metal-treated *Matricaria chamomilla* roots. Plant. Cell Rep. 27, 605–615.
Kutschera, U., Niklas, K.J., 2017. Boron and the evolutionary development of roots. Plant. Signal. Behav. 12, 1–4.
Lambers, H., Shane, M.W., Cramer, M.D., Pearse, S.J., Veneklaas, E.J., 2006. Root structure and functioning for efficient acquisition of phosphorus: matching morphological and physiological traits. Ann. Bot. 98, 693–713.
Landi, M., Deg'Innocenti, E., Pardossi, A., Guidi, L., 2012. Antioxidant and photosynthetic responses in plants under boron toxicity: a review. Am. J. Agric. Biol. Sci. 7, 255–270.
Landi, M., Margaritopoulou, T., Papadakis, I.E., Araniti, F., 2019. Boron toxicity in higher plants: an update. Planta. 250, 1–22.
Lehto, T., Raisanen, M., Lavola, A., Julkunen-Tiitto, R., Aphalo, P.J., 2004. Boron mobility in deciduous forest trees in relation to their polyols. N. Phytol. 163, 333–339.
Lewis, D.H., 1980. Boron, lignification and the origin of vascular plants—a unified hypothesis. N. Phytol. 84, 209–229.
Le, J., Vandenbussche, F., Van Der Straeten, D., Verbelen, J.P., 2001. In the early response of Arabidopsis roots to ethylene, cell elongation is up and down regulated and uncoupled from differentiation. Plant. Physiol. 125, 519–522.
Liakopoulos, G., Stavrianakou, S., Nikolopoulos, D., Karvonis, E., Vekkos, K.A., Psaroudi, V., et al., 2009. Quantitative relationships between boron and mannitol concentrations in phloem exudates of *Olea europaea* leaves under contrasting boron supply conditions. Plant. Soil. 323, 177–186.
Liu, G., Dong, X., Liu, L., Wu, L., Jiang, C., 2014. Boron deficiency is correlated with changes in cell wall structure that lead to growth defects in the leaves of navel orange plants. Sci. Hortic. 176, 54–62.
Liu, L., Fan, X.D., 2013. Tapetum: regulation and role in sporopollenin biosynthesis in Arabidopsis. Plant. Mol. Biol. 83, 165–175.
Liu, D., Jiang, W., Zhang, L., Li, L., 2000. Effects of boron ions on root growth and cell division of broad bean (*Vicia faba* L.). Isr. J. Plant. Sci. 48, 47–51.

Liu, G.D., Wang, R.D., Wu, L.S., Peng, S.A., Wang, Y.H., Jiang, C., 2012. Boron distribution and mobility in navel orange grafted on citrange and trifoliate orange. Plant. Soil. 360, 123–133.

Li, T., Choi, W.G., Wallace, I.S., Baudry, J., Roberts, D.M., 2011. Arabidopsis thaliana NIP7;1: an anther-specific boric acid transporter of the aquaporin superfamily regulated by an unusual tyrosine in helix 2 of the transport pore. Biochemistry. 50, 6633–6641.

Li, K., Kamiya, T., Fujiwara, T., 2015. Differential roles of PIN1 and PIN2 in root meristem maintenance under low-B conditions in *Arabidopsis thaliana*. Plant. Cell Physiol. 56, 1205–1214.

Li, Q., Liu, Y., Pan, Z., Xie, S., Peng, S., 2016. Boron deficiency alters root growth and development and interacts with auxin metabolism by influencing the expression of auxin synthesis and transport genes. Biotechnol. Biotechnol. Equip. 30, 661–668.

Li, Z., Zhang, P., Lv, J., Cheng, Y., Cui, J., Zhao, H., et al., 2016. Global dynamic transcriptome programming of rapeseed (*Brassica napus* L.) anther at different development stages. PLoS ONE 11, e0154039.

Lv, Q., Wang, L., Wang, J.Z., Li, P., Chen, Y., Du, J., et al., 2017. SHB1/HY1 alleviates excess boron stress by increasing BOR4 expression level and maintaining boron homeostasis in Arabidopsis roots. Front. Plant. Sci. 8, 790.

López-Búcio, J., Cruz-Ramírez, A., Herrera-Estrella, L., 2003. The role of nutrient availability in regulating root architecture. Curr. Opin. Plant. Biol. 6, 280–287.

Macho-Rivero, M.A., Camacho-Cristóbal, J.J., Herrera-Rodríguez, M.B., Müller, M., Munné-Bosch, S., González-Fontes, A., 2017. Abscisic acid and transpiration rate are involved in the response to boron toxicity in Arabidopsis plants. Physiol. Plant. 160, 21–32.

Macho-Rivero, M.A., Herrera-Rodríguez, M.B., Brejcha, R., Schäffner, A.R., Tanaka, N., Fujiwara, T., et al., 2018. Boron toxicity reduces water transport from root to shoot in Arabidopsis plants. Evidence for a reduced transpiration rate and expression of major PIP aquaporin genes. Plant. Cell Physiol. 59, 841–849.

Marschner, P., 2012. Marchner's Mineral Nutrition of Higher Plants, Ed 3 Academic Press, San Diego.

Martinez-Cuenca, M.R., Martinez-Alcantara, B., Quiñones, A., Ruiz, M., Iglesias, D.J., Primo-Millo, E., et al., 2015. Physiological and molecular responses to excess boron in *Citrus macrophylla* W. PLoS One 10 (7), e0134372.

Martín-Rejano, E., Camacho-Cristóbal, J., Rodríguez, M.B., Rexach, J., Navarro-Gochicoa, M., González-Fontes, A., 2011. Auxin and ethylene are involved in the responses of root system architecture to low boron supply in Arabidopsis. Physiol. Plant. 142, 170–178.

Matthes, M.S., Robil, J.M., Tran, T., Kimble, A., Mcsteen, P., 2018. Increased transpiration is correlated with reduced boron deficiency symptoms in the maize tassel-less1 mutant. Physiol. Plant. 163, 344–355.

Matthes, M., Torres-Ruiz, R.A., 2017. Boronic acids as tools to study (plant) developmental processes? Plant. Signal. Behav. 12, e1321190.

Maurel, C., Boursiac, Y., Luu, D.T., Santoni, V., Shahzad, Z., Verdoucq, L., 2015. Aquaporins in plants. Physiol. Rev. 95, 1321–1358.

Mei, L., Sheng, O., Peng, S., Zhou, G., Wei, Q., Li, Q., 2011. Growth, root morphology and boron uptake by citrus rootstock seedlings differing in boron- deficiency responses. Sci. Hortic. 129, 426–432.

Meriño-Gergichevich, C., Reyes-Díaz, M., Guerrero, J., Ondrasek, G., 2017. Physiological and nutritional responses in two highbush blueberry cultivars exposed to deficiency and excess of boron. J. Soil. Sci. Plant. Nutr. 17, 307–318.

Mesquita, G.L., Zambrosi, F.C.B., Tanaka, F.A.O., Boaretto, R.M., Quaggio, J.A., Ribeiro, R.V., et al., 2016. Anatomical and physiological responses of Citrus trees to varying boron availability are dependent on rootstock. Front. Plant. Sci. 7, 224.
Metwally, A., El-Shazoly, R., Hamada, A.M., 2017. Physiological responses to excess boron in wheat cultivars. Eur. J. Biol. Res. 7, 1−8.
Mihara, M., Itoh, T., Izawa, T., 2008. In silico identification of short nucleotide sequences associated with gene expression of pollen development in rice. Plant. Cell Physiol. 49, 1451−1464.
Miwa, K., Aibara, I., Fujiwara, T., 2014. Arabidopsis thaliana BOR4 is upregulated under high boron conditions and confers tolerance to high boron. Soil. Sci. Plant. Nutr. 60, 349−355.
Miwa, K., Fujiwara, T., 2010. Boron transport in plants: co-ordinated regulation of transporters. Ann. Bot. 105, 1103−1108.
Miwa, K., Fujiwara, T., 2011. Role of overexpressed BOR4, a boron exporter, in tolerance to high level of boron in shoots. Soil. Sci. Plant. Nutr. 57, 558−565.
Miwa, K., Takano, J., Omori, H., Seki, M., Shinozaki, K., Fujiwara, T., 2007. Plants tolerant of high boron levels. Science. 318, 1417.
Miwa, K., Wakuta, S., Takada, S., Ide, K., Takano, J., Naito, S., et al., 2013. Roles of BOR2, a boron exporter, in cross linking of rhamnogalacturonan II and root elongation under boron limitation in Arabidopsis. Plant. Physiol. 163, 1699−1709.
Moghadam, N.K., Lajayer, B.A., Ghorbanpour, M., 2020. Boron tolerance in plants. Metalloids Plants. 301−314.
Mosa, K., Kumar, K., Chhikara, S., Musante, C., White, J., Dhankher, O.P., 2016. Enhanced boron tolerance in plants mediated by bidirectional transport through plasma membrane intrinsic proteins. Sci. Rep. 6.
Nakagawa, Y., Hanaoka, H., Kobayashi, M., Miyoshi, K., Miwa, K., Fujiwara, T., 2007. Cell-type specificity of the expression of OsBOR1, a rice efflux boron transporter gene, is regulated in response to boron availability for efficient boron uptake and xylem loading. Plant. Cell 19, 2624−2635.
Nielsen, F.H., Eckhert, C.D., 2020. Boron. Adv. Nutr. 11, 461−462.
O'Neill, M.A., Ishii, T., Albersheim, P., Darvill, A.G., 2004. Rhamnogalacturonan II: structure and function of a borate cross-linked cell wall pectic polysaccharide. Annu. Rev. Plant. Biol. 55, 109−139.
Ochiai, K., Uemura, S., Shimizu, A., Okumoto, Y., Matoh, T., 2008. Boron toxicity in rice (Oryza sativa L.). I. Quantitative trait locus (QTL) analysis of tolerance to boron toxicity. TAG. Theoretical and applied genetics. Theoretische und Angew. Genetik. 117, 125−133.
Oertli, J.J., 1993. The mobility of boron in plants. Plant. Soil. 155, 301−304.
Oikonomou, A., Ladikou, E.-V., Chatziperou, G., Margaritopoulou, T., Landi, M., Sotiropoulos, T., et al., 2019. Boron excess imbalances root/shoot allometry, photosynthetic and chlorophyll fluorescence parameters and sugar metabolism in apple plants. Agronomy. 9, 731.
Oiwa, Y., Kitayama, K., Kobayashi, M., Matoh, T., 2013. Boron deprivation immediately causes cell death in growing roots of *Arabidopsis thaliana* (L.) Heynh. Soil. Sci. Plant. Nutr. 59, 621−627.
Onuh, A., Miwa, K., 2021. Regulation, diversity, and evolution of boron transporters in plants. Plant. Cell Physiol.
Ou, Y., Wu, X., Gao, Y., Wu, Y., Yao, Y., 2019. Analysis of physiological responses and expression profiling of boron transporter-like genes in response to excess boron in *Populus russkii*. Chemosphere. 224, 369−378.
Ozhuner, E., Eldem, V., Ipek, A., Okay, S., Sakcali, S., Zhang, B., et al., 2013. Boron stress responsive microRNAs and their targets in barley. PLoS ONE 8, e59543.

Ozturk, M., Sakcali, S., Gucel, S., Rombuloglu, H., 2010. Boron and plants. In: Ashrat, M., Ozturk, M., Ahmad, M.S.A. (Eds.), Plant Adaptation and Phytoremediation. Springer, Berlin, pp. 275–311.

Ozyigit, I.I., Filiz, E., Saracoglu, I.A., Karadeniz, S., 2020. Exploration of two major boron transport genes BOR1 and NIP5;1 in the genomes of different plants. Biotechnol. Biotechnol. Equip. 34, 455–468.

O'Neill, M.A., Eberhard, S., Albersheim, P., Darvill, A.G., 2001. Requirement of borate cross-linking of cell wall rhamnogalacturonan II for Arabidopsis growth. Science. 294, 846–849.

Pallotta, M., Schnurbusch, T., Hayes, J., Hay, A., Baumann, U., Paul, J., et al., 2014. Molecularbasis of adaptation to high soil boron in wheat landraces and elite cultivars. Nature.

Pandey, N., Archana, 2013. Antioxidant responses and water status in Brassica seedlings subjected to boron stress. Acta Physiol. Plant. 35, 697–706.

Pang, J., Ryan, M.H., Tibbett, M., Cawthray, G.R., Siddique, K.H., Bolland, M.D., et al., 2010. Variation in morphological and physiological parameters in herbaceous perennial legumes in response to phosphorus supply. Plant. Soil. 331, 241–255.

Papadakis, I.E., Dimassi, K.N., Therios, I.N., 2003. Response of two citrus genotypes to six boron concentrations: concentration and distribution of nutrients, total absorption, and nutrient use efficiency. Aust. J. Agric. Res. 54, 571–580.

Papadakis, I.E., Tsiantas, P.I., Tsaniklidis, G., Landi, M., Psychoyou, M., Fasseas, C., 2018. Changes in sugar metabolism associated to stem bark thickening partially assist young tissues of *Eriobotrya japonica* seedlings under boron stress. J. Plant. Physiol. 231, 337–345.

Pommerrening, B., Diehn, T.A., Bienert, G.P., 2015. Metalloido-porins: Essentiality of nodulin 26-like intrinsic proteins in metalloid transport. Plant. Sci. 238, 212–227.

Poss, J.A., Grattan, S.R., Suarez, D.L., Grieve, C.M., 2000. Stable carbon isotope discrimination: an indicator of cumulative salinity and boron stress in *Eucalyptus camaldulensis*. Tree Physiol. 20, 1121–1127.

Potocka, I., Szymanowska-Pułka, J., 2018. Morphological responses of plant roots to mechanical stress. Ann. Bot. 122, 711–723.

Princi, M.P., Lupini, A., Araniti, F., Sunseri, F., Abenavoli, M.R., 2013. Short-term effects of boron excess on root morphological and functional traits in tomato. XVII International Plant Nutrition Colloquium- Boron Satellite Meeting–Proceedings Book. 2013. Istanbul, Turkey, pp. 1150–1151.

Pereira, G.L., Siqueira, J.A., Batista-Silva, W, Cardoso, F.B., Nunes-Nesi, A, Araújo, W. L., 2021. Boron: more than an essential element for land plants? Front. Plant Sci. 11, 610307. Available from: https://doi.org/10.3389/fpls.2020.610307.

Pérez Di Giorgio, J., Bienert, G.P., Ayub, N., Yaneff, A., Barberini, M.L., Mecchia, M.A., et al., 2016. Pollen-specific aquaporins NIP4;1 and NIP4;2 are required for pollen development and pollination in *Arabidopsis thaliana*. Plant. Cell 28, 1053–1077.

Pérez-Castro, R., Kasai, K., Gainza-Cortés, F., Ruiz-Lara, S., Casaretto, J.A., Peña-Cortés, H., et al., 2012. VvBOR1, the grapevine ortholog of AtBOR1, encodes an efflux boron transporter that is differentially expressed throughout reproductive development of *Vitis vinifera* L. Plant. Cell Physiol. 53, 485–494.

Quiles-Pando, C., Rexach, J., Navarro-Gochicoa, M.T., Camacho-Cristóbal, J.J., Herrera-Rodríguez, M.B., González-Fontes, A., 2013. Boron deficiency increases the levels of cytosolic Ca2 + and expression of Ca2 + -related genes in Arabidopsis thaliana roots. Plant. Physiol. Biochem. 65, 55–60.

Rao, I.M., Miles, J.W., Beebe, S.E., Horst, W.J., 2016. Root adaptations to soils with low fertility and aluminium toxicity. Ann. Bot. 118, 593–605.

Reid, R., Fitzpatrick, K., 2009. Influence of leaf tolerance mechanisms and rain on boron toxicity in barley and wheat. Plant. Physiol. 151, 413–420.

Reid, R.J., Hayes, J.E., Post, A., Stangoulis, J.C.R., Graham, R.D., 2004. A critical analysis of the causes of boron toxicity in plants. Plant. Cell Env. 27, 1405–1414.

Reid, R., 2010. Can we really increase yields by making crop plants tolerant to boron toxicity? Plant. Sci. 178, 9–11.

Reid, R., 2014. Understanding the boron transport network in plants. Plant. Soil. 385, 1–13.

Rerkasem, B., Jamjod, S., 2004. Boron deficiency in wheat: a review. Field Crop. Res. 89, 173–186.

Riaz, M., Yan, L., Wu, X.W., Hussain, S., Aziz, O., Imran, M., et al., 2018. Boron reduces aluminum-induced growth inhibition, oxidative damage and alterations in the cell wall components in the roots of trifoliate orange. Ecotoxicol. Environ. Saf. 153, 107–115.

Roberts, D.M., Routray, P., 2017. The nodulin 26 intrinsic protein subfamily. In: Chaumont, F., Tyerman, S.D. (Eds.), Plant Aquaporins: From Transport to Signaling. Springer International Publishing AG, Cham, Switzerland, pp. 267–296.

Roessner, U., Patterson, J.H., Forbes, M.G., Fincher, G.B., Langridge, P., Bacic, A., 2006. An investigation of boron toxicity in barley using metabolomics. Plant. Physiol. 142, 1087–1101.

Rosolem, C.A., Bogiani, J.C., 2011. Physiology of boron stress in cotton. In: Oosterhuis, D. M. (Ed.), Stress physiology in cotton. The Cotton Foundation, Cordova. pp. 113–124.

Routray, P., Li, T., Yamasaki, A., Yoshinari, A., Takano, J., Choi, W.G., et al., 2018. Nodulin Intrinsic Protein 7;1 is a tapetal boric acid channel involved in pollen cell wall formation. Plant. Physiol. 178, 1269–1283.

Ryden, P., Sugimoto-Shirasu, K., Smith, A.C., Findlay, K., Reiter, W.D., McCann, M.C., 2003. Tensile properties of Arabidopsis cell walls depend on both xyloglucan cross-linked microfibrillar network and rhamnogalacturonan II-borate complexes. Plant. Physiol. 132, 1033–1040.

Sakamoto, T., Tsujimoto-Inui, Y., Sotta, N., Hirakawa, T., Matsunaga, T., Fukao, Y., et al., 2018. Proteasomal degradation of BRAHMA promotes boron tolerance in Arabidopsis. Nat. Commun. 9.

Sakamoto, T., Tsujimoto-Inui, Y., Uraguchi, S., Yoshizumi, T., Matsunaga, S., Mastui, M., et al., 2011. Condensin II alleviates DNA damage and is essential fortolerance of boron overload stress in Arabidopsis. Plant. Cell 23, 3533–3546.

Sang, W., Huang, Z.R., Qi, Y.P., Yang, L.T., Guo, P., Chen, L.S., 2015. An investigation of boron-toxicity in leaves of two citrus species differing in boron-tolerance using comparative proteomics. J. Proteomics. 123, 128–146.

Sang, W., Huang, Z.-R., Yang, L.-T., Guo, P., Ye, X., Chen, L.-S., 2017. Effects of high toxic boron concentration on protein profiles in roots of two citrus species differing in boron-tolerance. Front. Plant. Sci. 8, 180.

Scheres, B., Benfey, P., Dolan, L., 2002. Root development. The Arabidopsis Book. American Society of Plant Biologists.

Schopfer, P., 2001. Hydroxyl radical-induced cell-wall loosening in vitro and in vivo: implications for the control of elongation growth. Plant. J. 28, 679–688.

Scorei, R., 2012. Is boron a prebiotic element? A mini-review of the essentiality of boron for the appearance of life on earth. Orig. Life Evol. Biosph. 42, 3–17.

Shelp, B.J., Marentes, E., Kitheka, A.M., Vivekanandan, P., 1995. Boron mobility in plants. Physiol. Plant. 94, 356–361.

Shimotohno, A., Sotta, N., Sato, T., De Ruvo, M., Marée, A.F., Grieneisen, V.A., et al., 2015. Mathematical modeling and experimental validation of the spatial distribution of boron in the root of *Arabidopsis thaliana* identify high boron accumulation in the tip and predict a distinct root tip uptake function. Plant. Cell Physiol. 56, 620–630.

Shireen, F., Nawaz, M.A., Chen, C., Zhang, Q., Zheng, Z., Sohail, H., et al., 2018. Boron: functions and approaches to enhance its availability in plants for sustainable agriculture. Int. J. Mol. Sci. 19, 1856.

Shi, J., Cui, M., Yang, L., Kim, Y.J., Zhang, D., 2015. Genetic and biochemical mechanisms of pollen wall development. Trends Plant. Sci. 20, 741–753.

Shorrocks, V.M., 1997. The occurrence and correction of boron deficiency. Boron Agric. 17, 3–5.

Smith, T.E., Grattan, S.R., Grieve, C.M., Poss, J.A., Läuchli, A.E., Suarez, D.L., 2013. pH dependent salinity-boron interactions impact yield, biomass, evapotranspiration and boron uptake in broccoli (*Brassica oleracea* L.). Plant. Soil. 370, 541–554.

Sorgonà, A., Abenavoli, M.R., Cacco, G., 2005. A comparative study between two citrus rootstocks: effect of nitrate on the root morpho-topology and net nitrate uptake. Plant. Soil. 270, 257–260.

Sotta, N., Duncan, S., Tanaka, M., Sato, T., Marée, A.F., Fujiwara, T., et al., 2017. Rapid transporter regulation prevents substrate flow traffic jams in boron transport. eLife 6, e27038.

Stangoulis, J.C.R., Brown, P.H., Bellaloui, N., Reid, R.J., Graham, R.D., 2001. The efficiency of boron utilisation in canola. Aust. J. Plant. Physiol. 28, 1109–1114.

Stangoulis, J., Tate, M., Graham, R., Bucknall, M., Palmer, L., Boughton, B., et al., 2010. The mechanism of boron mobility in wheat and canola phloem. Plant. Physiol. 153, 876–881.

Steingrobe, B., Schmid, H., Claassen, N., 2001. Root production and root mortality of winter barley and its implication with regard to phosphate acquisition. Plant. Soil. 237, 239–248.

Stepanova, A.N., Yun, J., Likhacheva, A.V., Alonso, J.M., 2007. Multilevel interactions between ethylene and auxin in Arabidopsis roots. Plant. Cell 19, 2169–2185.

Sutton, T., Baumann, U., Hayes, J., Collins, N., Shi, B.-J., Schnurbusch, T., et al., 2007. Boron-toxicity tolerance in barley arising from efflux transporter amplification. Science 318, 1446–1449.

Suwabe, K., Suzuki, G., Takahashi, H., Shiono, K., Endo, M., Yano, K., et al., 2008. Separated transcriptomes of male gametophyte and tapetum in rice: validity of a laser microdissection (LM) microarray. Plant. Cell Physiol. 49, 1407–1416.

Swarup, R., Perry, P., Hagenbeek, D., Van Der Straeten, D., Beemster, G.T., Sandberg, G., et al., 2007. Ethylene upregulates auxin biosynthesis in Arabidopsis seedlings to enhance inhibition of root cell elongation. Plant. Cell 19, 2186–2196.

Takano, J., Miwa, K., Fujiwara, T., 2008. Boron transport mechanisms: collaboration of channels and transporters. Trends Plant. Sci. 13, 451–457.

Takano, J., Miwa, K., Yuan, L., von Wirén, N., Fujiwara, T., 2005. Endocytosis and degradation of BOR1, a boron transporter of *Arabidopsis thaliana*, regulated by boron availability. Proc. Natl. Acad. Sci. USA. 102, 12276–12281.

Takano, J., Noguchi, K., Yasumori, M., Kobayashi, M., Gajdos, Z., Miwa, K., et al., 2002. Arabidopsis boron transporter for xylem loading. Nature 420, 337–340.

Takano, J., Tanaka, M., Toyoda, A., Miwa, K., Kasai, K., Fuji, K., et al., 2010. Polar localization and degradation of Arabidopsis boron transporters through distinct trafficking pathways. Proc. Natl. Acad. Sci. USA 107, 5220–5225.

Takano, J., Wada, M., Ludewig, U., Schaaf, G., von Wirén, N., Fujiwara, T., 2006. The Arabidopsis major intrinsic protein NIP5;1 is essential for efficient boron uptake and plant development under boron limitation. Plant. Cell 18, 1498–1509.

Tanaka, H., Dhonukshe, P., Brewer, P.B., Friml, J., 2006. Spatiotemporal asymmetric auxin distribution: a means to coordinate plant development. Cell Mol. Life Sci. 63, 2738–2754.

Tanaka, M., Fujiwara, T., 2008. Physiological roles and transport mechanisms of boron: perspectives from plants. Eur. J. Physiol. 456, 671–677.
Tanaka, M., Sotta, N., Yamazumi, Y., Yamashita, Y., Miwa, K., Murota, K., et al., 2016. The minimum open reading frame, AUG-stop, induces boron-dependent ribosome stalling and mRNA degradation. Plant. Cell 28, 2830–2849.
Tanaka, M., Takano, J., Chiba, Y., Lombardo, F., Ogasawara, Y., Onouchi, H., et al., 2011. Boron-dependent degradation of NIP5;1 mRNA for acclimation to excess boron condition in Arabidopsis. Plant. Cell 23, 3547–3559.
Tanaka, N., Uraguchi, S., Saito, A., Kajikawa, M., Kasai, K., Sato, Y., et al., 2013. Roles of pollen-specific boron efflux transporter, OsBOR4, in the rice fertilization process. Plant. Cell Physiol. 54, 2011–2019.
Tewari, R.K., Kumar, P., Sharma, P.N., 2010. Morphology and oxidative physiology of boron-deficient mulberry plants. Tree Physiol. 30, 68–77.
Tombuloglu, G., Tombuloglu, H., Sakcali, M.S., Unver, T., 2015. High-throughput transcriptome analysis of barley (*Hordeum vulgare*) exposed to excessive boron. Gene. 557, 71–81.
Torun, A.A., Yazici, A., Erdem, H., 2006. Genotypic variation in tolerance to boron toxicity in 70 durum wheat genotypes. Turk. J. Agric. For. 30.
Tsiantas, P.I., Papadakis, I.E., Tsaniklidis, G., Landi, M., Psychoyou, M., 2019. Allocation pattern, nutrient partitioning, sugar metabolism, and pigment composition in hydroponically grown loquat seedlings subjected to increasing boron concentrations. J. Soil. Sci. Plant. Nut. 19, 556–564.
Voxeur, A., Gilbert, L., Rihouey, C., Driouich, A., Rothan, C., Baldet, P., et al., 2011. Silencing of the GDP-D-mannose 3,5-epimerase affects the structure and cross-linking of the pectic polysaccharide rhamnogalacturonan II and plant growth in tomato. J. Biol. Chem. 286, 8014–8020.
Wallace, I.S., Roberts, D.M., 2004. Homology modeling of representative subfamilies of Arabidopsis major intrinsic proteins: classification based on the aromatic/arginine selectivity filter. Plant. Physiol. 135, 1059–1068.
Wang, G., DiTusa, S.F., Oh, D.H., Herrmann, A.D., Mendoza-Cozatl, D.G., O'Neill, M.A., et al., 2020. Cross species multi-omics reveals cell wall sequestration and elevated global transcription as mechanisms of boron tolerance in plants. bioRxiv .
Wang, Q., Lu, L., Wu, X., Li, Y., Lin, J., 2003. Boron influences pollen germination and pollen tube growth in *Picea meyeri*. Tree physiol. 23, 345–351.
Wang, N., Yang, C., Pan, Z., Liu, Y., Peng, S., 2015. Boron deficiency in woody plants: Various responses and tolerance mechanisms. Front. Plant. Sci. 6, 1–14.
Wang, N.N., Yan, T.S., Fu, L.N., Zhou, G.F., Liu, Y.Z., Peng, S.A., 2014. Differences in boron distribution and forms in four citrus scion-rootstock combinations with contrasting boron efficiency under boron-deficient conditions. Trees 28, 1589–1598.
Warington, K., 1923. The effect of boric acid and borax on the broad bean and certain other plants. Ann. Bot. 37, 629–672.
Will, S., Eichert, T., Fernández, V., Möhring, J., Müller, T., Römheld, V., 2011. Absorption and mobility of foliar-applied boron in soybean as affected by plant boron status and application as a polyol complex. Plant. Soil. 344, 283–293.
Wimmer, M.A., Eichert, T., 2013. Mechanisms for boron deficiency mediated changes in plant water relations. Plant. Sci. 203, 25–32.
Wongmo, J., Jamjod, S., Rerkasem, B., 2004. Contrasting responses to boron deficiency in barley and wheat. Plant. Soil. 259, 103–110.
Wu, X.-Li, Ou, Y.-B., Yuan, G.-H., Chen, Y.-F., Wang, Y., Yao, Y.-A., 2015. Physiological responses of two poplar species to high boron stress[J]. Chin. J. Plan. Ecolo 39, 407–415.

Wu, X., Riaz, M., Yan, L., Du, C., Liu, Y., Jiang, C., 2017. Boron deficiency in trifoliate orange induces changes in pectin composition and architecture of components in root cell walls. Front. Plant. Sci. 8, 1882.

Yan, X., Wu, P., Ling, H., Xu, G., Xu, F., Zhang, Q., 2006. Plant nutriomics in China: an overview. Ann. Bot. 98, 473–482.

Yoshinari, A., Hosokawa, T., Amano, T., Beier, M.P., Kunieda, T., Shimada, T., et al., 2019. Polar localization of the borate exporter BOR1 requires AP2-dependent endocytosis. Plant. Physiol. 179, 1569–1580.

Yoshinari, A., Takano, J., 2017. Insights into the mechanisms underlying boron homeostasis in plants. Front. Plant. Sci. 8, 195.

Yuai, Y., Jianming, X., Zhengqiang, Y., Ke, W., 1993. Responses of rape genotypes to boron application. Plant. Soil. 155, 321–324.

Yu, X., Bell, P.F., 2002. Boron and lime effects on yield and deficiency symptoms of rice grown in greenhouse on acid typic glossaqualf. J. Plant. Nutri. 25, 2591–2602.

Zhang, Q., Chen, H., He, M., Zhao, Z., Cai, H., Ding, G., et al., 2017. The boron transporter BnaC4.BOR1;1c is critical for inflorescence development and fertility under boron limitation in *Brassica napus*. Plant. Cell. Environ. 40, 1819–1833.

Zhao, D., Oosterhuis, D., 2003. Cotton growth and physiological responses to boron deficiency. J. Plant. Nutr. 26, 855–867.

Zhou, G.F., Liu, Y.Z., Sheng, O., Wei, Q.J., Yang, C.Q., Peng, S.A., 2015. Transcription profiles of boron-deficiency-responsive genes in citrus rootstock root by suppression subtractive hybridization and cDNA microarray. Front. Plant. Sci. 5, 795.

Zimmermann, P., Hirsch-Hoffmann, M., Hennig, L., Gruissem, W., 2004. GENEVESTIGATOR. Arabidopsis microarray database and Analysis Toolbox. Plant. Physiol. 136, 2621–2632.

CHAPTER 6

Physiological and biochemical mechanisms and adaptation strategies of plants under boron deficiency conditions

Milan Kumar Lal[1,2], Rahul Kumar Tiwari[1,2], Arvind Jaiswal[3], Biswaranjan Behera[4], Aalok Shiv[5], Awadhesh Kumar[6], Sudhir Kumar[2], Vijay Paul[2], Madan Pal Singh[2], Brajesh Singh[1] and Prakash Kumar Jha[7]
[1]ICAR-Central Potato Research Institute, Shimla, India
[2]ICAR-Indian Agricultural Research Institute, New Delhi, India
[3]ICAR-Central Potato Research Station, Jalandhar, India
[4]ICAR-Indian Institute of Water Management, Bhubaneswar, India
[5]ICAR-Indian Institute of Sugarcane Research, Lucknow, India
[6]ICAR-National Rice Research Institute, Cuttack, India
[7]Feed the Future Sustainable Intensification Innovation Lab, Department of Agronomy, Kansas State University, Manhattan, KS, United States

6.1 Introduction

Boron (B) is a unique micronutrient that belongs to group 13 (IIIa) of the periodic table and possesses both metallic and nonmetallic properties. It is an essential element required for the growth and development of the plants (Brdar-Jokanović, 2020; Dehors et al., 2019; Matthes, Robil, and McSteen, 2020; Persson et al., 2007; Ramos et al., 2021; Rerkasem and Jamjod, 2004; Schnurbusch et al., 2010; Stavrianakou et al., 2006; Tanaka and Fujiwara, 2008; Vera et al., 2021; Wang et al., 2003). It is involved in plant physiological, biochemical, and metabolic processes (Mousavi and Motesharezadeh, 2019; Seth and Aery, 2017; Stavrianakou et al., 2006; Tavallali, 2017). The major functions in the plant where boron is essentially involved are respiration, sugar transport, lignification, structural stability of cell wall, and metabolism of carbohydrates, phenolics, auxin, and RNA (Ahmed et al., 2011). The early report on the importance of B suggests that it is essentially involved in plant growth and development (Landi et al., 2012; Mukhopadhyay et al., 2013; Nooli et al., 2020; Padbhushan and Kumar, 2015; Stewart et al., 2020, 2021;

Boron in Plants and Agriculture
DOI: https://doi.org/10.1016/B978-0-323-90857-3.00005-9

Wang et al., 2003). Brandenburg (1931) reported heart rot in sugar beet and dry rot in mangolds due to B deficiency. The application of boron inhibits scab disease in potatoes due to improved skin resistance and fruit dropping in grape and frost resistance in fruit trees due to better carbohydrate and protein metabolism (Mousavi and Motesharezadeh, 2019; Tariq and Mott, 2007). Boron has a high potency, it can cause damages like inhibition of germination, root growth and chlorosis and necrosis of shoots (Chatterjee et al., 2005; Emamverdian et al., 2015). Therefore, it is also regarded as a hazardous element when present in higher concentrations.

In soil, boron concentration usually ranges between 20 and 200 ppm and the availability of boron in the soil varies from different kinds of soil (Brdar-Jokanović, 2020). The soil factors that affect boron uptake are pH, temperature, moisture, texture, soil calcium carbonate content and soil organic matter (Goldberg, 1997). With an increase in soil pH, B gets adsorbed on soil particles and becomes less mobile (Lehto, 1995). Although B exists in several forms in the soil solution and the dominant form in the pH range most suitable for plant growth (5.5−7.5) is undissociated boric acid (H_3BO_3) (Yan, Riaz, Wu, Wang, et al., 2018b). The uncharged boric acid is prone to leaching in light-textured soil under high rainfall regions leading to boron deficiency. In heavy textured soil under less rainfall, boron may accumulate in the soil, leading to B toxicity (Brdar-Jokanović, 2020; Camacho-cristóbal, Rexach, and González-Fontes, 2008). During the dry summer season, boron deficiency is severe under high temperature and low moisture conditions. In soil, calcium carbonate directly acts as a sink for boron by adsorbing it on the surface and indirectly by increasing soil pH, thus restricting boron mobility in soil (Welch, 1995). Soil organic matter also adsorbs boron on it and affects its availability in soil (Yermiyahu et al., 2001).

Boron is the only mineral plant nutrient that is taken up by the plant from the soil in a nonionic undissociated form, that is, boric acid (Marschner, 1995). The movement of boron from soil to plant system occurs through the mass flow process, and it is distributed in the plant system through the transpiration stream (Landi et al., 2012; Matthes et al., 2018; Mukhopadhyay et al., 2013). The undissociated boric acid has a strong ability to form complexes with cis-diols inside the plant system (Cakmak and Römheld, 1997a; Camacho-Cristóbal et al., 2011). Boron is immobile in the plant system as it has been found that a part of a leaf exhibits B deficiency even if the other part of the same leaf contains

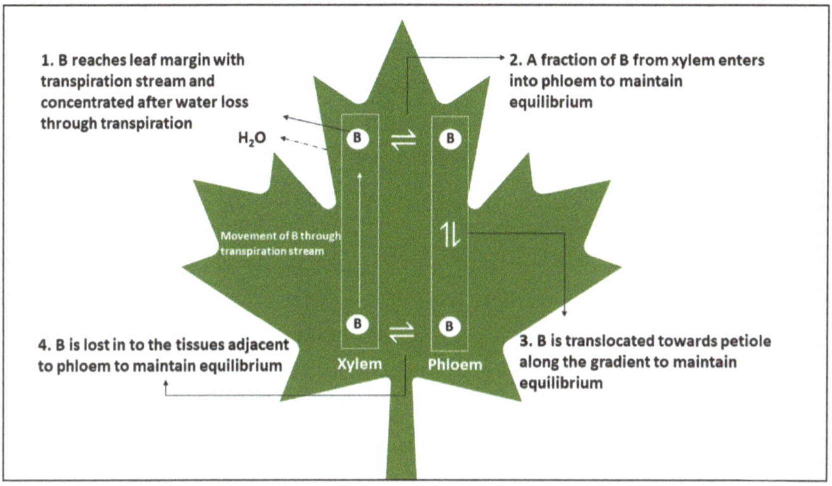

Figure 6.1 Model for boron immobility in the plant. *Idea adapted from Oertli and Richardson (1970).*

excess boron (Chatzissavvidis and Antonopoulou, 2019). Oertli and Richardson (1970) suggested the possible reason for the immobility of boron in a plant is high cyclic local mobility (Fig. 6.1). The boron tends to maintain equilibrium in concentration in all tissues over a small region. Such tendency leads to a highly localized cyclic movement of boron in the plant (Alloway, 2013). Therefore, it is the extremely high local mobility of boron that makes it immobile in the plant system.

6.2 Impact of boron stress on crop productivity

Since the first report of B deficiency in faba bean (*Vicia faba*) by Warington (1923), at least 132 crops in over 80 countries have shown positive responses to B applications (Ravikumar et al., 2021). In general, crops of the Gramineae family require less B due to the lower quantity of B-complexing pectin in the primary cell wall. However, other monocots like onion, asparagus, coconut, and oil palm are pectin rich and have higher B requirements (Corrales et al., 2008). It is established that the boron requirement of dicot species is 4–7 times higher than that of monocot species (Marschner, 1995). Among dicot families, Cruciferae and Chenopodiaceae are most sensitive to B deficiency. Shorrocks (1997) hypothesized that the evolution of a family-like Chenopodiaceae from a B-rich environment resulted in high B requirements of the cultivated crops of the family. Although cereals are often considered low B requiring crops, many researchers have reported the response of cereal crops to boron application (Göçmen Taşkın et al., 2021; Rerkasem and Jamjod, 2004; Schnurbusch et al., 2010). The deficiency of B results in failure in the grain set and reduced productivity in cereals like wheat (Nadeem et al., 2021) and barley (Wongmo et al., 2004) due to depressed male fertility. However, the extent of such effects is dependent on the genetic makeup of cultivars, and the B-efficient genotypes grow and yield better than other genotypes (Rerkasem and Jamjod, 1997). Besides, crop growth phases also decide the variation in B deficiency response. For example, Rerkasem and Jamjod (1997) reported the genotypic variation to B deficiency in wheat during the reproductive phase, whereas, in barley, it was reported in both vegetative and reproductive phases. The B deficiency in barley has been reported to either suppress or enhance crop growth during the vegetative phase (Mousavi and Motesharezadeh, 2019; Shireen et al., 2018). In a sand culture without added boron, Wongmo et al. (2004) observed 0%–90% and 0%–67% grain set index (percentage grain set, in the first two florets of 10 central spikelets) in wheat and barley, respectively. Under B deficiency in barley, they noticed a complex phenotypic response like increased tillering and dry weight straw, rat tail symptom on terminal spikelet, reduced no. of grains/spike, and spikes/plant. Under boron deficiency, the productivity of crop like alfalfa is severely affected due to nutritional disorders in which flowers drop before setting any seeds (Chen et al., 2020). The response of different crops to boron application in terms of crop productivity has been listed in Table 6.1. The positive response of crops to B application is broadly due to increased male fertility, better pollination, and higher grain/seed setting (Ahmed et al., 2011).

6.3 Physiological response of crops under boron stress

Boron deficiency in the plant has numerous anatomical, physiological and biochemical changes which are difficult to identify with primary and secondary effects and deprivation (Mukhopadhyay et al., 2013). The physiological changes that altered under B deficiency condition include reduction in photosynthesis, unbalance in source and sink relationship, disproportionate partitioning of carbon, premature senescence, reduction in biomass accumulation, reduction in male fertility rate, nutritional imbalance, degradation of embryogenesis and ultimately lowering of fruit

Table 6.1 Response of various crops to external boron application.

Crop	Boron level	Productivity	Reference
Maize	No boron (B_0) and Boron 20 μM (B_{20}) in sand culture	B_0 plants produced 0.4 grains/ear, whereas B_{20} plants produced 410 grains/ear	Lordkaew et al. (2011)
Rice	No boron and B at 1 kg/ha in field experiment	14%–30% yield increase depending on cultivars	Rashid et al. (2009)
Wheat	No boron and B at 1 kg/ha	19.9% increase in grain yield under 1 kg/ha B application over control	Khan et al. (2006)
Rice	No boron (B_0) and Boron 10 μM (B_{10}) in pot culture	Yield decrease ranged from 28% (in var SPR1) to 79% (in var CNT1) in B_0 as compared to B_{10}	Lordkaew et al. (2013)
Pigeon pea	Soil application of boron @2.5 kg/ha + foliar application @0.1% at flowering and pod filling stage	32% higher seed yield as compared to that of no B application	Nooli et al. (2020)
Green gram	0.5 mg/kg of soil in pot culture	Mean seed weight per pot increased by 33.5% over control	Padbhushan and Kumar (2015)
Mustard	Soil application of B at 1 kg/ha	8.3% increase in grain yield over no boron application	Hussain et al. (2008)
Soyabean	B at 1 kg/ha	76% increase in seed dry matter; 80% less hollow heart as compared to control	Rerkasem et al. (1993)
Pea nut		61%; 80% less hollow heart as compared to control	
Black gram		5 times more seed dry weight; no hollow heart symptoms	
Cotton	Soil application of B at 1.1 kg/ha	Increase cotton seed yield by more than 560 kg/ha	Borax (2002)

set and seed formation (Brdar-Jokanović, 2020; Cheng and Rerkasem, 1993; Lordkaew et al., 2013). Moreover, the metabolic activities in the plant are regulated by B, and its deficiency can lead to disruption of the membrane system of chloroplast, mitochondria and plasma membrane (PM) (Brdar-Jokanović, 2020; Camacho-cristóbal et al., 2008). It also plays an important role in the transport of sugar, synthesis of lignin and flavonoid, metabolism of nitrogenous and phenolic compounds (Seth and Aery, 2017). The detail about the physiological response of B deficiency in the plant is subsequently discussed.

6.3.1 Photosynthesis

Photosynthesis is one of the major metabolic events of the plant which is severely affected either by B stress or toxicity (Landi et al., 2012; Mukhopadhyay et al., 2013). Boron deficiency in *Citrus sinensis* reported a reduction in stomatal conductance and carbon dioxide assimilation in leaves with an increased level of intercellular carbon dioxide concentration (Han et al., 2008). Under B deficiency condition there was lowering of stomatal conductance along with CO_2 assimilation with an increased level of CO_2 concentration in *Citrus sinensis* leaves. The increased starch accumulation and reduction in nonreducing sugar such as sucrose were also reported which indicates that there might be feedback regulation of CO_2 assimilation which ultimately increased hexose sugar accumulation (Han et al., 2009). The carbon balance under B deficiency is disturbed which is further maintained by the respiration, organic-acid metabolism, amino-acid biosynthesis and anapleurotic pathway which is upregulated in leaves and downregulated in roots (Lu et al., 2014). Similarly, previous studies have suggested that heavy metal decreases the photosynthetic capacity of the plant (Tavallali, 2017). There was a reduction in the Hill reaction activity and low intercellular CO_2 concentration in mustard (*Brassica campestris*) which is the result of lowered CO_2 assimilation rate (Wimmer and Eichert, 2013). The reduction in photosynthetic rate could be correlated with the reduction in chlorophyll content, electron transport rate in thylakoid and phosphorylation of various rate-limiting enzymes involved in photosynthesis (Mukhopadhyay et al., 2013).

Boron deficiency has a detrimental effect on lowering photosynthesis and related parameters. Li et al. (2017) reported a significant lowering of photosynthetic rate by 25.1% in cotton leaves which resulted in the reduction of leaf area and stomatal conductance ultimately leading to growth inhibition. Moreover, B deficiency in the plant may affect the phloem sieve element and thereby hindering the transport of photoassimilate from source to sink. As discussed earlier in this section, starch accumulation was suggested to be increased under B stress conditions which might further lead to the reduction of synthesis of photoassimilate in leaves and its transport to sink tissues (Li et al., 2017). However, the exact mechanism of the detrimental effect of B deficiency on the photosynthetic process is still lacking which needs to be elucidated. However, the possible role of B stress affecting photosynthesis might be due to structural damage of thylakoid membrane leading to altered electron transport rate and accumulation of carbohydrate in leaf, which ultimately leads to lowering of photosynthetic rate (Landi et al., 2012).

6.3.2 Carbon partitioning and source-sink relationship

The relationship between source and sink has a key role in the carbohydrate assimilation and distribution of photoassimilate during the growth and development of plants (Hastilestari et al., 2018; Lemoine et al., 2013; Shi et al., 2013). The production and productivity of the plant are dependent on source activity (photosynthesis and nutrient remobilization rate) and sink strength (conversion of simple sugar to complex sugar and protein formation along with storage in sink tissue) (Albacete et al., 2014; Hastilestari et al., 2018). Under B stress conditions, carbon partitioning of photoassimilates was reported to be significantly affected in many plant species (Brdar-Jokanović, 2020; García-Sánchez et al., 2020; Li et al., 2017; Mousavi and Motesharezadeh, 2019; Rerkasem and Jamjod, 2004; Wongmo et al., 2004). Moreover, the lowering of photoassimilate synthesis in the source (leaf) and its transportation in the sink tissue (reproductive, root and other developing parts) affects its relationship under B deficiency condition (Cakmak and Römheld, 1997b; Han et al., 2008; Mousavi and Motesharezadeh, 2019). Chatterjee et al. (2005) reported under B stress conditions the sink activity was suggested to be lowered which results in decreased seed weight of gram (*Cicer arietinum* L.). The loss in the seed weight and size might be due to the formation of a lesser number of endosperm cells per seed and decreased capacity of stored carbohydrates in grams. The lower photosynthetic rate in leaves and followed by a reduction of accumulation of carbohydrate and protein resulted in lower storage and subsequently affected the seed size under B deficiency (Camacho-Cristóbal and González-Fontes, 1999). Boron stress can lead to lowering of sink activity which results in premature flower and fruit drops which is correlated with the requirement of B for development of reproductive organ formation and development (Hegazi et al., 2018).

6.3.3 Nitrogen metabolism

Boron is the important micronutrient that is involved in nitrogen (N) metabolism (Gülüt, 2021). Under boron stress conditions, the regulation of N metabolism in the plant occurs by enhancement of the level of nitrate and reduction in the activity of nitrate reductase (Bhardwaj et al., 2021). The bioavailability and absorption of plant nutrients are affected by B availability. Ahmed et al. (2011) reported that the application of B in cotton plants showed increased absorption and translocation of nitrogen, phosphorus, potassium, zinc, iron and copper in the leaves, bud and seeds. Asparagine synthetase (AS) transcript-level was positively regulated in both leaves and roots by carbon and nitrogen content. However, under B deficiency the expression of AS was higher in roots compared to the control *Nicotiana tabacum* plant (Beato et al., 2010). Moreover, studies suggest that an enhanced or restricted level of B supply reduces the nitrate level by modulating the expression of a nitrate transporter and preventing PMA2 expression in the root. Thus resulting in the decreased activity of H + ATPase of the PM (Cervilla et al., 2020). Nitrogenase activity is regulated by B status in the plant and its deficiency can lead to lowering of nitrogen assimilation in the plant (Bellaloui et al., 2010; Cervilla et al., 2009). Boron also regulates the activity of glutathione reductase which is an important enzyme that he alleviated under oxidative stress conditions (Aydin et al., 2019). In legume crops such as pea (*Pisum sativum*) and soybean (*Glycine max*) under B stress condition, the weight of nodule and nitrogen-fixing capacity was reported to be decreased significantly (Bolanos et al., 1994; Yamagishi and Yamamoto, 1994). The inhibition of nitrogenase activity and deterioration in nodule development might be due to degradation of the PM and cell wall, including the peribacteriod membrane which is the major symptom of B deficiency in the plant (Matamoros and Becana, 2020).

6.3.4 Interaction other nutrients

Boron uptake in the plant might be affected by other minerals available in the soil (Dinh et al., 2021). The deficiency and toxicity of B may affect the plant metabolic processes and it acts as a regulator and inhibitor of accumulation and utilization of nutrients essential for plants (Eraslan et al., 2007; Lordkaew et al., 2013; Mousavi and Motesharezadeh, 2019). Studies suggest that B deficiency may reduce the level of plant nutrients directly involved in the metabolic regulation of plants. Moreover, the acidification of soil or nutrient solution can reduce the accumulation and concentration of B, N, K, Mg, and Fe in the shoot of *Brassica napus* (Dinh et al., 2021). Boron is involved in the maintenance of permeability of plasmalemma of the root surface which leads to the increase in the P uptake. The uptake of Fe and Cu in the plant was reported to be positively correlated with the B content in the plant. The uptake of K in the plant is enhanced due to the synergistic effect of B, whereas Ca shows an antagonistic effect (Cui et al., 2021). Similarly, the uptake of Mn and Zn was reported to be negatively correlated with the application of B in the soil. A study on pomegranate suggests that the boron deficiency can adversely affect the absorption of macronutrients (Ca) and micronutrients (Cu Fe, Mn, and Zn) in the plant which revealed that B is essential for homeostasis of nutrients and their interaction (Marathe et al., 2016). It can be concluded that B is involved in different physiological and biochemical processes in the plant which upon changing (deficiency or toxicity) can lead to change in concentration and translocation of other nutrients.

6.3.5 Transpiration

Boron uptake in the plant is generally considered a passive process that is absorbed in the plant from the soil in the form of not dissociated boric acid and further, it is transported to the leaves via xylem (Hua et al., 2020). The long-distance transport of nutrients in the plant occurs in the vascular system which consists of the xylem and phloem. Most importantly the upward movement of nutrients in the plant occurs via the xylem which is mainly driven through water potential gradient resulting from surface transpiration during daytime (Dahal et al., 2019; Li et al., 2017; Tsukaguchi et al., 2003). The movement of B is generally restricted in the phloem tissues, therefore higher accumulation of B occurs in the highly-transpiring matured leaves (Montanaro et al., 2014). The accumulation of B from roots to leaves occurs in the terminal part of the leaf veins and thus under B deficiency or toxicity these tissue show early symptoms such as chlorosis and necrosis. A higher amount of B was reported in the leaf tissues as compared to the phloem sap as B is not efficiently remobilized in the plant (Landi et al., 2012).

6.3.6 Pollen tube formation

Boron deficiency is detrimental for pollen and anther development which leads to male sterility and ultimately failure in grain set (Cheng and Rerkasem, 1993; Fang et al., 2016; Lordkaew et al., 2013; Rerkasem and Jamjod, 2004; Shorrocks, 1997). Cellulose is the major component of the pollen tube wall under and callose can be synthesized in the normal pollen tuber walls (Persson et al., 2007; Polko and Kieber, 2019). Under B deficiency the formation of pollen tuber was reported to be degraded due to inhibition of cellulose deposition in pollen tuber. Boron deficiency can alter the formation of cellulose and callose in the tip of pollen tubes which ultimately leads to an abnormal condition of sterility condition (Dehors et al., 2019; Yan, Riaz, Wu, Du, et al., 2018a). Pectin is another important constituent of the pollen tube wall which is composed of methoxy-ester and later de-esterified by the enzyme pectin methylesterase (PME). Boron and calcium are essential minerals that are required for the formation of a pectic polysaccharide such as pectin in the cell wall. In *Picea meyeri* plant, the germination of pollen tube under boron deficient condition ranged from 18% to 24% as compared to control which was reported to be 61% (Wang et al., 2003). The aforementioned studies suggest that B is crucial for fruit set and which ultimately result in the yield of food crops.

6.4 Biochemical response of crops under boron stress

Boron deficiency has been shown to affect the different developmental and biochemical processes in the plant which includes reduction of cell development and growth, shoot cell expansion, production in the reactive oxygen species (ROS), lowering of oxidant scavenging capacity and ultimately reduction in the grain and fruit size (Cervilla et al., 2020; Fang et al., 2016; Li et al., 2017; Lu et al., 2014; Nooli et al., 2020; Rerkasem and Jamjod, 2004; Tavallali, 2017). The detailed biochemical responses of boron stress conditions in the plant are discussed in the subsequent section.

6.4.1 Cell wall and membrane permeability

Boron is an essential micronutrient that is required for the normal growth and developmental process of higher plants (Matthes et al., 2020). Boron deficiency could inhibit the of the plant due to reduction in cell elongation and cell division in both roots and shoot that shown to have an imbalance of shoot to root biomass (Hermans et al., 2006). The major role of B in the synthesis of the cell wall is linking borate-diol ester bonds to rhamnogalacturonan-II (RG-II) pectic subunits via their apiosyl residues (Li et al., 2017; Mukhopadhyay et al., 2013). Moreover, the recent studies suggest that B is also involved in the cross-linking of glycosylinositol phosphorylceramides of the cell membrane with arabinogalactan proteins (AGPs) of the cell wall (Fang et al., 2019). Boron is also involved in the function and localization of PM protein, transport process across the membrane and maintenance of the integrity of PM (Matthes and Torres-Ruiz, 2016). Therefore, B is well known for the mechanical properties of the cell wall.

6.4.2 Boron transporters

Boron uptake in plants involves both passive diffusion and active process in the root. Plant root uptake B in the form of boric acid which is an uncharged molecule and relatively permeable across PMs (Matthes et al., 2020). Under B sufficient condition in soil, the passive process of B uptake is prevalent which satisfy the plant demand. However, when the B is limited in soil, the boric acid channels activates that is a major intrinsic protein (MIP) and BOR family (borate exporters for transport of B) (Yoshinari and Takano, 2017). Boric acid transporter channel belongs to the MIP family which are the homologs of aquaporin transporter (Li et al., 2017). It was well reported that members of this family transport water and small uncharged molecules in plants (Javot and Maurel, 2002; Kapilan et al., 2018; Tanaka and Fujiwara, 2008). The MIP family of transporters are divided into five plasma membrane intrinsic protein (PIP), tonoplast intrinsic proteins, small basic intrinsic proteins, nodulin 26-like intrinsic proteins (NIPs), and X intrinsic proteins (XIPs) (Kapilan et al., 2018; Matthes et al., 2020; Onuh and Miwa, 2021). Out of these transporters, NIPS are further divided into subclasses I, II, and III depending upon the channel pore structure. Yoshinari and Takano (2017) reported that NIP IIs are physiological significant for B transport in the plant. Moreover, under excess B condition, BOR transporter activated and exports B from tissues to the soil which has a similar structure to anion transporter (Schnurbusch et al., 2010). Under luxury consumption of B, BOR1 was reported to be regulated at the posttranscriptional level and degradation of protein takes place for homeostasis of B in variable growth conditions (Matthes et al., 2020; Yoshinari and Takano, 2017). Controlling the expression of the aforementioned transporter proteins in the plant can help to regulate B in the plant.

6.4.3 Reactive oxygen species and antioxidant system

ROS are generated in the plant which is the output of cellular metabolism. However, under biotic and abiotic stress conditions the level of ROS in the plant is increased many folds which leads to degradation of lipids, protein and nucleic acid and ultimately leading to senescence and death of the plant (Arnao and Hernández-Ruiz, 2019; García-Sánchez et al., 2020; Landi et al., 2012; Lu et al., 2014; Mousavi and Motesharezadeh, 2019; Tavallali, 2017; Tiwari et al., 2020a,b). Studies suggest that B stress inhibit the root cell elongation in *Arabidopsis* seedling via ROS dependent pathway (Camacho-Cristóbal et al., 2015). The stress condition might have triggered the formation of ROS in the root cell which impairs the cell wall integrity and ultimately leads to the reduction in the root elongation under the B deficiency condition (Tsang et al., 2011). The imbalance of photosynthesis and carbon assimilation in the plant under biotic and abiotic stress condition can lead to accelerated production of ROS viz., oxygen ($^1O_2$), superoxide anion (O_2^-), hydrogen peroxide (H_2O_2) and hydroxyl radicals ($^{\bullet}OH$) which damage the membrane system, nucleic acid, lipid and protein components of the cell (Lal et al., 2021; Olson, 2020; Sharif et al., 2018; Sharma et al., 2012; Suzuki and Mittler, 2006). However, to minimize the cellular damage caused by ROS in the plant, there is a ROS scavenging mechanism that is composed of an enzymatic and nonenzymatic pathway (Tiwari et al., 2020a, b). Boron deficiency in the plant can activate the ROS scavenging mechanism and ameliorate the negative impact of these molecules in the plant system (Emamverdian et al., 2015).

To scavenge cellular damage antioxidant molecules involved is mainly ascorbate (AsA) and reduced glutathione. Whereas, the antioxidant enzyme such as superoxide dismutase (SOD), ascorbate peroxidase (APX), glutathione reductase (GR), monodehydroascorbate reductase (MDAR) and dehydroascorbate reductase (DHAR) and catalase (CAT) were involved in the scavenging mechanism (Asthir, 2015; Han et al., 2008; Kumar et al., 2021; Neelambari et al., 2019; Sadak et al., 2020; Sun et al., 2019). Han et al. (2008) reported that under B stress conditions, there was higher activity of SOD, APX, MDAR and GR, whereas the lower activity of DHAR and CAT was observed in leaves of *Citrus* seedlings. The young tea plantlet when subjected to boron stress condition increased the level of $^1O_2$, O_2^-, H_2O_2 and $^{\bullet}OH$ which leads to electrolyte leakage. Moreover, the activities of APX, CAT, POD and SOD was reported to be increased under B deprived tea plants (Mukhopadhyay et al., 2013). In soybean (*Glycine max*) leaves, low B resulted in decreased activities of SOD, APX and CAT (Bellaloui et al., 2010). Antioxidant system activated under B stress/ deficiency condition to maintain the metabolic processes in the plant.

6.4.4 Carbohydrate metabolism

Biochemical responses such as carbohydrate metabolism are altered under B deficient conditions. Previous reports suggest that there was an increased soluble sugar content in the leaf of citrus under B stress (Han et al., 2008). Most importantly, glucose and fructose content was higher in B deficient leaves as compared to control. Starch content in leaves was also reported to be increased significantly under boron stress in plants which leads to disruption of chloroplast structure and lowering of CO_2 assimilation rate and chlorophyll content (Han et al., 2009). Higher starch accumulation and downregulation of photosynthesis together limits the sink capacity and unbalance the source-sink relationship under B stress (Yu et al., 2015).

6.5 Conclusion

Boron is an essential element that is required by the plant as a nutrient for growth and development which ultimately affect production and productivity. It is a micronutrient which that is required in small quantities and any imbalance in the plant might affect its vegetative and reproductive growth stages. The uptake of B takes place through passive (along with water) and active transport (involves various transporter). Boron is transported in the plant through transpiration stress and the homeostasis maintained by various transporter exporting B out from the plant system. There are many transporters such as BOR, PIPs, NIPs, and XIPs which are involved in maintaining B homeostasis. The physiological and biochemical response in the plant occurs due to B deficiency which can be observed in the morphological and cellular levels. The significant process has been made worldwide to understand the B homeostasis and understand its mechanism in the plant that severely affect the yield and quality of grain, fruits and vegetables.

References

Ahmed, N., Abid, M., Ahmad, F., Ullah, M.A., Javaid, Q., Ali, M.A., 2011. Impact of boron fertilization on dry matter production and mineral constitution of irrigated cotton. Pak. J. Bot. 43 (6), 2903–2910.

Albacete, A.A., Martínez-Andújar, C., Pérez-Alfocea, F., 2014. Hormonal and metabolic regulation of source-sink relations under salinity and drought: from plant survival to crop yield stability. Biotechnol. Adv. 32 (1), 12–30. Available from: https://doi.org/10.1016/j.biotechadv.2013.10.005.

Alloway, B.J., 2013. Bioavailability of elements in soil. Essentials of Medical Geology: Revised Edition. Springer, Netherlands, pp. 351–373.

Arnao, M.B., Hernández-Ruiz, J., 2019. Melatonin: a new plant hormone and/or a plant master regulator? Trends Plant. Sci. 24 (1), 38–48. Available from: https://doi.org/10.1016/j.tplants.2018.10.010.

Asthir, B., 2015. Protective mechanisms of heat tolerance in crop plants. J. Plant. Interact. 10 (1), 202–210.

Aydin, M., Tombuloglu, G., Sakcali, M.S., Hakeem, K.R., Tombuloglu, H., 2019. Boron alleviates drought stress by enhancing gene expression and antioxidant enzyme activity. J. Soil. Sci. Plant. Nutr. 19 (3), 545–555. Available from: https://doi.org/10.1007/s42729-019-00053-8.

Beato, V.M., Rexach, J., Navarro-Gochicoa, M.T., Camacho-Cristóbal, J.J., Herrera-Rodríguez, M.B., Maldonado, J.M., et al., 2010. A tobacco asparagine synthetase gene responds to carbon and nitrogen status and its root expression is affected under boron stress. Plant. Sci. 178 (3), 289–298. Available from: https://doi.org/10.1016/j.plantsci.2009.12.008.

Bellaloui, N., Reddy, K.N., Gillen, A.M., Abel, C.A., 2010. Nitrogen metabolism and seed composition as influenced by foliar boron application in soybean. Plant. Soil. 336 (1), 143–155. Available from: https://doi.org/10.1007/s11104-010-0455-6.

Bhardwaj, S., Kapoor, D., Singh, S., Gautam, V., Dhanjal, D.S., Jan, S., et al., 2021. Nitric oxide: a ubiquitous signal molecule for enhancing plant tolerance to salinity stress and their molecular mechanisms. J. Plant. Growth Regul. 1–13. Available from: https://doi.org/10.1007/s00344-021-10394-3.

Bolanos, L., Esteban, E., de Lorenzo, C., Fernandez-Pascual, M., de Felipe, M.R., Garate, A., et al., 1994. Essentiality of boron for symbiotic dinitrogen fixation in pea (*Pisum sativum*) rhizobium nodules. Plant. Physiol. 104 (1), 85–90. Available from: https://doi.org/10.1104/pp.104.1.85.

Borax. 2002. Boron application for improved cotton yield. LT-AC10-AN104. Available at: borax.coaleurcitron.ath.cx/www.borax.com fr/agriculture/files/an104.pdf (accessed 31 March 2022).

Brandenburg, E., 1931. The heart and dry rot of beets as a symptom of boron deficiency. Phytopathol. Z. 3, 499–517.

Brdar-Jokanović, M., 2020. Boron toxicity and deficiency in agricultural plants. Int. J. Mol. Sci. 21 (4), 1424.

Cakmak, I., Römheld, V., 1997a. Boron deficiency-induced impairments of cellular functions in plants. Plant. Soil. 193 (2), 71–83. Available from: https://doi.org/10.1023/A:1004259808322.

Cakmak, I., Römheld, V., 1997b. Boron deficiency-induced impairments of cellular functions in plants, Plant and Soil, Vol. 193. Kluwer Academic Publishers, pp. 71–83.

Camacho-Cristóbal, J.J., González-Fontes, A., 1999. Boron deficiency causes a drastic decrease in nitrate content and nitrate reductase activity, and increases the content of carbohydrates in leaves from tobacco plants. Planta 209 (4), 528–536. Available from: https://doi.org/10.1007/s004250050757.

Camacho-Cristóbal, J.J., Rexach, J., González-Fontes, A., 2008. Boron in plants: deficiency and toxicity. J. Integr. Plant. Biol. 50 (10), 1247–1255.

Camacho-Cristóbal, J.J., Rexach, J., Herrera-Rodríguez, M.B., Navarro-Gochicoa, M.T., González-Fontes, A., 2011. Boron deficiency and transcript level changes. Plant. Sci. 181 (2), 85–89.

Camacho-Cristóbal, J.J., Martín-Rejano, E.M., Herrera-Rodríguez, M.B., Navarro-Gochicoa, M.T., Rexach, J., González-Fontes, A., 2015. Boron deficiency inhibits root cell elongation via an ethylene/auxin/ROS-dependent pathway in arabidopsis seedlings. J. Exp. Botany 66 (13), 3831–3840. Available from: https://doi.org/10.1093/jxb/erv186.

Cervilla, L.M., Blasco, B., Ríos, J.J., Rosales, M.A., Rubio-Wilhelmi, M.M., Sánchez-Rodríguez, E., et al., 2009. Response of nitrogen metabolism to boron toxicity in tomato plants. Plant. Biol. (Stuttgart, Ger.) 11 (5), 671–677. Available from: https://doi.org/10.1111/j.1438-8677.2008.00167.x.

Cervilla, L.M., Blasco, B., Ríos, J.J., Rosales, M.A., Rubio-Wilhelmi, M.M., Sá Nchez-Rodríguez, E., et al., 2020. Response of nitrogen metabolism to boron toxicity in tomato plants. Plants 9 (3), 322. Available from: https://doi.org/10.1111/j.1438-8677.2008.00167.x. 2020, Vol. 9, Page 322.

Chatterjee, C., Sinha, P., Dube, B.K., 2005. Biochemical changes, yield, and quality of gram under boron stress. Commun. Soil. Sci. Plant. Anal. 36 (13–14), 1763–1771. Available from: https://doi.org/10.1081/CSS-200062440.

Chatzissavvidis, C., Antonopoulou, C., 2019. Boron toxicity in fruit crops: agronomic and physiological implications. Fruit Crops: Diagnosis and Management of Nutrient Constraints. Elsevier, pp. 211–221.

Chen, L., Xia, F., Wang, M., Wang, W., Mao, P., 2020. Metabolomic analyses of alfalfa (*Medicago sativa* L. Cv. 'Aohan') reproductive organs under boron deficiency and surplus conditions. Ecotoxicol. Environ. Saf. 202, 111011. Available from: https://doi.org/10.1016/j.ecoenv.2020.111011.

Cheng, C., Rerkasem, B., 1993. Effects of boron on pollen viability in wheat. Plant. Soil. 155–156 (1), 313–315. Available from: https://doi.org/10.1007/BF00025045.

Corrales, I., Poschenrieder, C., Barceló, J., 2008. Boron-induced amelioration of aluminium toxicity in a monocot and a dicot species. J. Plant. Physiol. 165 (5), 504–513. Available from: https://doi.org/10.1016/j.jplph.2007.03.014.

Cui, J., Davanture, M., Lamade, E., Zivy, M., Tcherkez, G., 2021. Plant low-K responses are partly due to Ca prevalence and the low-K biomarker putrescine does not protect from Ca side effects but acts as a metabolic regulator. Plant. Cell Environ. 44 (5), 1565–1579. Available from: https://doi.org/10.1111/pce.14017.

Dahal, K., Li, X.-Q., Tai, H., Creelman, A., Bizimungu, B., 2019. Improving potato stress tolerance and tuber yield under a climate change scenario – a current overview. Front. Plant. Sci. 10 (May), 563. Available from: https://doi.org/10.3389/fpls.2019.00563.

Dehors, J., Mareck, A., Kiefer-Meyer, M.C., Menu-Bouaouiche, L., Lehner, A., Mollet, J.C., 2019. Evolution of cell wall polymers in tip-growing land plant gametophytes: composition, distribution, functional aspects and their remodeling. Front. Plant. Sci. 10, 441.

Dinh, A.Q., Naeem, A., Sagervanshi, A., Wimmer, M.A., Mühling, K.H., 2021. boron uptake and distribution by oilseed rape (*Brassica napus* L.) as affected by different nitrogen forms under low and high boron supply. Plant. Physiol. Biochem. 161, 156–165. Available from: https://doi.org/10.1016/j.plaphy.2021.02.009.

Emamverdian, A., Ding, Y., Mokhberdoran, F., Xie, Y., 2015. Heavy metal stress and some mechanisms of plant defense response. Sci. World J. 2015.

Eraslan, F., Inal, A., Gunes, A., Alpaslan, M., 2007. Boron toxicity alters nitrate reductase activity, proline accumulation, membrane permeability, and mineral constituents of tomato and pepper plants. J. Plant. Nutr. 30 (6), 981–994. Available from: https://doi.org/10.1080/15226510701373221.

Fang, K., Zhang, W., Xing, Y., Zhang, Q., Yang, L., Cao, Q., et al., 2016. Boron toxicity causes multiple effects on malus domestica pollen tube growth. Front. Plant. Sci. 7 (FEB2016), 208. Available from: https://doi.org/10.3389/fpls.2016.00208.

Fang, K.F., Du, B.S., Zhang, Q., Xing, Y., Cao, Q.Q., Qin, L., 2019. Boron deficiency alters cytosolic Ca 2 + concentration and affects the cell wall components of pollen tubes in malus domestica. Plant. Biol. 21 (2), 343–351. Available from: https://doi.org/10.1111/plb.12941.

García-Sánchez, F., Simón-Grao, S., Martínez-Nicolás, J.J., Alfosea-Simón, M., Liu, C., Chatzissavvidis, C., et al., 2020. Multiple stresses occurring with boron toxicity and deficiency in plants. J. Hazard. Mater. 397, 122713. Available from: https://doi.org/10.1016/j.jhazmat.2020.122713.

Göçmen Taşkın, B., Özbek, Ö., Şan, S.K., Nachit, M.M., Kaya, Z., 2021. The assessment of boron toxicity tolerance in F6 RILs of durum wheat [*Triticum turgidum* (L.) Tell. Convar. Durum (Desf.) Mackey]. Cereal Res. Commun. 1–9. Available from: https://doi.org/10.1007/s42976-021-00174-w.

Goldberg, S., 1997. Reactions of boron with soils. Plant. Soil. 193 (2), 35–48. Available from: https://doi.org/10.1023/A:1004203723343.

Gülüt, K.Y., 2021. Nitrogen and boron nutrition in grafted watermelon i: impact on pomological attributes, yield and fruit qualityedited by S. FarooqPLOS ONE 16 (5), e0252396. Available from: https://doi.org/10.1371/journal.pone.0252396.

Han, S., Chen, L.S., Jiang, H.X., Smith, B.R., Yang, L.T., Xie, C.Y., 2008. Boron deficiency decreases growth and photosynthesis, and increases starch and hexoses in leaves of citrus seedlings. J. Plant. Physiol. 165 (13), 1331–1341. Available from: https://doi.org/10.1016/j.jplph.2007.11.002.

Han, S., Tang, N., Jiang, H.X., Yang, L.T., Li, Y., Chen, L.S., 2009. CO_2 assimilation, photosystem II photochemistry, carbohydrate metabolism and antioxidant system of citrus leaves in response to boron stress. Plant. Sci. 176 (1), 143–153. Available from: https://doi.org/10.1016/j.plantsci.2008.10.004.

Hastilestari, B.R., Lorenz, J., Reid, S., Hofmann, J., Pscheidt, D., Sonnewald, U., et al., 2018. Deciphering source and sink responses of potato plants (*Solanum tuberosum* L.) to elevated temperatures. Plant. Cell Environ. 41 (11), 2600–2616. Available from: https://doi.org/10.1111/pce.13366.

Hegazi, E.S., El-Motaium, R.A., Yehia, T.A., Hashim, M.E., 2018. Effect of foliar boron application on boron, chlorophyll, phenol, sugars and hormones concentration of olive (*Olea europaea* L.) buds, leaves, and fruits. J. Plant. Nutr. 41 (6), 749–765. Available from: https://doi.org/10.1080/01904167.2018.1425438.

Hermans, C., Hammond, J.P., White, P.J., Verbruggen, N., 2006. How do plants respond to nutrient shortage by biomass allocation? Trends Plant. Sci. 11 (12), 610–617.

Hua, T., Zhang, R., Sun, H., Liu, C., 2020. Alleviation of boron toxicity in plants: mechanisms and approaches. Crit. Rev. Environ. Sci. Technol. Available from: https://doi.org/10.1080/10643389.2020.1807451.

Hussain, M.J., Sarker, M.M.R., Sarker, M.H., Ali, M., Salim, M.M.R., 2008. Effect of different levels of boron on the yield and yield attributes of mustard in Surma-Kushiara flood plain soil (AEZ 20). J. Soil. Nature 2 (3), 6–9.

Javot, H., Maurel, C., 2002. The role of aquaporins in root water uptake. Ann. Botany 90 (3), 301–313.

Kapilan, R., Vaziri, M., Zwiazek, J.J., 2018. Regulation of aquaporins in plants under stress. Biol. Res. 51 (1), 4.

Khan, R.A.H.M.A.T.U.L.L.A.H., Gurmani, A.H., Gurmani, A.R., Zia, M.S., 2006. Effect of boron application on rice yield under wheat rice system. International Journal of Agriculture and Biology 8 (6), 805–808.

Kumar, A., Singh, B., Raigond, P., Sahu, C., Mishra, U.N., Sharma, S., et al., 2021. Phytic acid: blessing in disguise, a prime compound required for both plant and human nutrition. Food Res. Int. 142 (June 2020), 110193. Available from: https://doi.org/10.1016/j.foodres.2021.110193.

Lal, M.K., Tiwari, R.K., Kumar, R., Naga, K.C., Kumar, A., Singh, B., et al., 2021. Effect of potato apical leaf curl disease on glycemic index and resistant starch of potato (*Solanum tuberosum* L.) tubers. Food Chem. 359, 129939. Available from: https://doi.org/10.1016/j.foodchem.2021.129939.

Landi, M., Degl'Innocenti, E., Pardossi, A., Guidi, L., 2012. Antioxidant and photosynthetic responses in plants under boron toxicity: a review. Am. J. Agric. Biol. Sci. 7 (3), 255–270. Available from: https://doi.org/10.3844/ajabssp.2012.255.270.

Lehto, T., 1995. Boron retention in limed forest mor. For. Ecol. Manag. 78 (1–3), 11–20. Available from: https://doi.org/10.1016/0378-1127(95)03599-7.

Lemoine, R., Camera, S.L., Atanassova, R., Dédaldéchamp, F., Allario, T., Pourtau, N., et al., 2013. Source-to-sink transport of sugar and regulation by environmental factors. Front. Plant. Sci. 4 (JUL), 272.

Li, M., Zhao, Z., Zhang, Z., Zhang, W., Zhou, J., Xu, F., et al., 2017. Effect of boron deficiency on anatomical structure and chemical composition of petioles and photosynthesis of leaves in cotton (*Gossypium hirsutum* L.). Sci. Rep. 7 (1), 1–9. Available from: https://doi.org/10.1038/s41598-017-04655-z.

Lordkaew, S., Dell, B., Jamjod, S., Rerkasem, B., 2011. Boron deficiency in maize. Plant and Soil 342 (1), 207–220.

Lordkaew, S., Konsaeng, S., Jongjaidee, J., Dell, B., Rerkasem, B., Jamjod, S., 2013. Variation in responses to boron in rice. Plant. Soil. 363 (1–2), 287–295. Available from: https://doi.org/10.1007/s11104-012-1323-3.

Lu, Y.B., Yang, L.T., Li, Y., Xu, J., Liao, T.T., Chen, Y.B., et al., 2014. Effects of boron deficiency on major metabolites, key enzymes and gas exchange in leaves and roots of citrus sinensis seedlings. Tree Physiol. 34 (6), 608–618. Available from: https://doi.org/10.1093/treephys/tpu047.

Marathe, R.A., Murkute, A.A., Dhinesh Babu, K., 2016. Mineral nutrient deficiencies and nutrient interactions in pomegranate. Natl Acad. Sci. Lett. 39 (6), 407–410. Available from: https://doi.org/10.1007/s40009-016-0487-4.

Marschner, H., 1995. Mineral Nutrition of Higher Plants, 2nd Edition Academic, Great Britain.

Matamoros, M.A., Becana, M., 2020. Redox control of the legume-rhizobium symbiosis. Adv. Botanical Res. 94, 67–96. Available from: https://doi.org/10.1016/bs.abr.2019.09.011.

Matthes, M., Torres-Ruiz, R.A., 2016. Boronic acid treatment phenocopies monopteros by affecting PIN1 membrane stability and polar auxin transport in *Arabidopsis thaliana* embryos. Dev. (Camb.) 143 (21), 4053–4062. Available from: https://doi.org/10.1242/dev.131375.

Matthes, M.S., Robil, J.M., Tran, T., Kimble, A., McSteen, P., 2018. Increased transpiration is correlated with reduced boron deficiency symptoms in the maize tassel-less1 mutant. Physiol Plant. 163 (3), 344–355. Available from: https://doi.org/10.1111/ppl.12717.

Matthes, M.S., Robil, J.M., McSteen, P., 2020. From element to development: the power of the essential micronutrient boron to shape morphological processes in plants. J. Exp. Botany 71 (5), 1681–1693.

Montanaro, G., Dichio, B., Lang, A., Mininni, A.N., Nuzzo, V., Clearwater, M.J., et al., 2014. Internal vs external control of calcium nutrition in kiwifruit. J. Plant. Nutr. Soil. Sci. 177 (6), 819–830.

Mousavi, S.M., Motesharezadeh, B., 2019. Boron deficiency in fruit crops. Fruit Crops: Diagnosis and Management of Nutrient Constraints. Elsevier, pp. 191–209.

Mukhopadhyay, M., Ghosh, P.D., Mondal, T.K., 2013. Effect of boron deficiency on photosynthesis and antioxidant responses of young tea plantlets. Russian J. Plant. Physiol. 60 (5), 633–639. Available from: https://doi.org/10.1134/S1021443713030096.

Nadeem, F., Farooq, M., Mustafa, B., Nawaz, A., 2021. Influence of soil residual boron on rice performance and soil properties under conventional and conservation rice–wheat cropping systems. Crop. Pasture Sci. 72 (5), 335. Available from: https://doi.org/10.1071/CP20339.

Neelambari, R.K., Mandavia, C., Mandavia, M., Jha, P.K., 2019. Curative effect of ascorbic acid and gibberellic acid on physiological and biochemical mechanism of wheat (*Triticum aestivum* L.) under saline condition. J. Biotechnol. Crop. Sci. 8 (13), 78–84.

Nooli, S.S., Wali, S.Y., Patil, M.B., Bevinakatti, H.P., 2020. Response of pigeonpea to boron nutrition. J. Pharmacognosy Phytochemistry 9 (2), 372–374.

Oertli, J.J., Richardson, W.F., 1970. The mechanism of boron immobility in plants. Physiologia Plant. 23 (1), 108–116. Available from: https://doi.org/10.1111/j.1399-3054.1970.tb06397.x.

Olson, K.R., 2020. Reactive oxygen species or reactive sulfur species: why we should consider the latter. J. Exp. Biol. 223 (4).

Onuh, A.F., Miwa, K., 2021. Regulation, diversity and evolution of boron transporters in plants. Plant. Cell Physiol. Available from: https://doi.org/10.1093/pcp/pcab025.

Padbhushan, R., Kumar, D., 2015. Soil boron fractions and response of green gram in calcareous soils. J. Plant. Nutr. 38 (8), 1143–1157. Available from: https://doi.org/10.1080/01904167.2015.1009106.

Persson, S., Paredez, A., Carroll, A., Palsdottir, H., Doblin, M., Poindexter, P., et al., 2007. Genetic evidence for three unique components in primary cell-wall cellulose synthase complexes in arabidopsis. Proc. Natl Acad. Sci. U S Am. 104 (39), 15566–15571. Available from: https://doi.org/10.1073/pnas.0706592104.

Polko, J.K., Kieber, J.J., 2019. The regulation of cellulose biosynthesis in plants. Plant. Cell 31 (2), 282–296.

Ramos, Q., Armienta, M.A., Aguayo, A., Cruz, O., 2021. Evaluation of the interactions of arsenic (As), boron (B), and lead (Pb) from geothermal production wells with agricultural soils. Ecotoxicol. Environ. Saf. 211, 111843. Available from: https://doi.org/10.1016/j.ecoenv.2020.111843.

Rashid, A., Yasin, M., Ali, M.A., Ahmad, Z., Ullah, R., 2009. Boron deficiency in rice in Pakistan: A serious constraint to productivity and grain quality. In: Salinity and water stress, Springer, Dordrecht, pp. 213–219.

Ravikumar, C., Karthikeyan, A., Senthilvalavan, P., Manivannan, R., 2021. Effect of sulphur, zinc and boron on the growth and yield enhancement of sunflower (*Helianthus annuus* L.). J. Appl. Nat. Sci. 13 (1), 295–300. Available from: https://doi.org/10.31018/jans.v13i1.2569.

Rerkasem, B., Bell, R.W., Lodkaew, S., Loneragan, J.F., 1993. Boron deficiency in soybean [Glycine max (L.) Merr.] peanut (Arachis hypogaea L.) and black gram [Vigna mungo (L.) Hepper]: Symptoms in seeds and differences among soybean cultivars in susceptibility to boron deficiency. Plant and Soil 150 (2), 289–294.

Rerkasem, B., Jamjod, S., 1997. Genotypic variation in plant response to low boron and implications for plant breeding. Plant. Soil. 193 (2), 169–180. Available from: https://doi.org/10.1023/A:1004220226977.

Rerkasem, B., Jamjod, S., 2004. Boron deficiency in wheat: a review. Field Crop. Res. 89 (2–3), 173–186. Available from: https://doi.org/10.1016/j.fcr.2004.01.022.

Sadak, M.S., Abdalla, A.M., Abd Elhamid, E.M., Ezzo, M.I., Abdalla-Bulletin, A.M., Abdalla, A.M., et al., 2020. Role of melatonin in improving growth, yield quantity and quality of *Moringa oleifera* L. plant under drought stress. Bull. Natl Res. Cent. 44 (1), 18. Available from: https://doi.org/10.1186/s42269-020-0275-7.

Schnurbusch, T., Hayes, J., Hrmova, M., Baumann, U., Ramesh, S.A., Tyerman, S.D., et al., 2010. Boron toxicity tolerance in barley through reduced expression of the multifunctional aquaporin HvNIP2;1. Plant. Physiol. 153 (4), 1706–1715. Available from: https://doi.org/10.1104/pp.110.158832.

Seth, K., Aery, N.C., 2017. Boron induced changes in biochemical constituents, enzymatic activities, and growth performance of wheat. Acta Physiologiae Plant. 39 (11), 1–10. Available from: https://doi.org/10.1007/s11738-017-2541-3.

Sharif, R., Xie Id, C., Zhang, H., Arnao Muhammad Ali Id, M.B., Arnao Muhammad Ali Id, M.B., Ali Id, Q., et al., 2018. Melatonin and its effects on plant systems. Molecules 23 (9), 1–20. Available from: https://doi.org/10.3390/molecules23092352.

Sharma, P., Jha, A.B., Dubey, R.S., Pessarakli, M., 2012. Reactive oxygen species, oxidative damage, and antioxidative defense mechanism in plants under stressful conditions. J. Botany 2012, 1–26. Available from: https://doi.org/10.1155/2012/217037.

Shi, W., Muthurajan, R., Rahman, H., Selvam, J., Peng, S., Zou, Y., et al., 2013. Source-sink dynamics and proteomic reprogramming under elevated night temperature and their impact on rice yield and grain quality. N. Phytologist 197 (3), 825–837. Available from: https://doi.org/10.1111/nph.12088.

Shireen, F., Nawaz, M.A., Chen, C., Zhang, Q., Zheng, Z., Sohail, H., et al., 2018. Molecular sciences boron: functions and approaches to enhance its availability in plants for sustainable agriculture. Int. J. Mol. Sci. 19 (7), 1856. Available from: https://doi.org/10.3390/ijms19071856. 2018, Vol. 19, Page 1856.

Shorrocks, V.M., 1997. The occurrence and correction of boron deficiency, Plant and Soil, Vol. 193. Kluwer Academic Publishers, pp. 121–148.

Stavrianakou, S., Liakopoulos, G., Karabourniotis, G., 2006. Boron deficiency effects on growth, photosynthesis and relative concentrations of phenolics of *Dittrichia viscosa* (Asteraceae). Environ. Exp. Botany 56 (3), 293–300. Available from: https://doi.org/10.1016/j.envexpbot.2005.03.007.

Stewart, Z.P., Paparozzi, E.T., Wortmann, C.S., Jha, P.K., Shapiro, C.A., 2020. Foliar micronutrient application for high-yield maize. Agronomy 10 (12), 1946. Available from: https://doi.org/10.3390/agronomy10121946.

Stewart, Z.P., Paparozzi, E.T., Wortmann, C.S., Jha, P.K., Shapiro, C.A., 2021. Effect of foliar micronutrients (B, Mn, Fe, Zn) on maize grain yield, micronutrient recovery, uptake, and partitioning. Plants 10 (3), 528. Available from: https://doi.org/10.3390/plants10030528.

Sun, Y., Liu, Z., Lan, G., Jiao, C., Sun, Y., 2019. Effect ofexogenous melatonin on resistance of cucumber to downy mildew. Sci. Horticulturae 255, 231–241. Available from: https://doi.org/10.1016/j.scienta.2019.04.057.

Suzuki, N., Mittler, R., 2006. Reactive oxygen species and temperature stresses: a delicate balance between signaling and destruction. Physiologia Plant. 126 (1), 45–51.

Tanaka, M., Fujiwara, T., 2008. Physiological roles and transport mechanisms of boron: perspectives from plants. Pflug. Arch. Eur. J. Physiol. 456 (4), 671–677.

Tariq, M., Mott, C.J.B., 2007. The significance of boron in plant nutrition and environment - a review. J. Agron. 6 (1), 1–10. Available from: https://doi.org/10.3923/ja.2007.1.10.

Tavallali, V., 2017. Interactive effects of zinc and boron on growth, photosynthesis, and water relations in pistachio. J. Plant. Nutr. 40 (11), 1588–1603. Available from: https://doi.org/10.1080/01904167.2016.1270308.

Tiwari, R.K., Lal, M.K., Kumar, R., Chourasia, K.N., Naga, K.C., Kumar, D., et al., 2020a. Mechanistic insights on melatonin-mediated drought stress mitigation in plants. Physiologia Plant. 13307. Available from: https://doi.org/10.1111/ppl.13307. ppl.
Tiwari, R.K., Lal, M.K., Naga, K.C., Kumar, R., Chourasia, K.N., Subhash, S., et al., 2020b. Emerging roles of melatonin in mitigating abiotic and biotic stresses of horticultural crops. Sci. Horticulturae 272 (July), 109592. Available from: https://doi.org/10.1016/j.scienta.2020.109592.
Tsang, D.L., Edmond, C., Harrington, J.L., Nühse, T.S., 2011. Cell wall integrity controls root elongation via a general 1-aminocyclopropane-1-carboxylic acid-dependent, ethylene-independent pathway. Plant. Physiol. 156 (2), 596–604. Available from: https://doi.org/10.1104/pp.111.175372.
Tsukaguchi, T., Kawamitsu, Y., Takeda, H., Suzuki, K., Egawa, Y., 2003. Water status of flower buds and leaves as affected by high temperature in heat-tolerant and heat-sensitive cultivars of snap bean (*Phaseolus vulgaris* L.). Plant. Prod. Sci. 6 (1), 24–27. Available from: https://doi.org/10.1626/pps.6.24.
Vera, A., Moreno, J.L., Siles, J.A., López-Mondejar, R., Zhou, Y., Li, Y., et al., 2021. Interactive impacts of boron and organic amendments in plant-soil microbial relationships. J. Hazard. Mater. 408, 124939. Available from: https://doi.org/10.1016/j.jhazmat.2020.124939.
Wang, Q., Lu, L., Wu, X., Li, Y., Lin, J., 2003. Boron influences pollen germination and pollen tube growth in *Picea meyeri*. Tree Physiol. 23 (5), 345–351. Available from: https://doi.org/10.1093/treephys/23.5.345.
Warington, K., 1923. The effect of boric acid and borax on the broad bean and certain other plants. Ann. Botany 37 (148), 629–672.
Welch, R.M., 1995. Micronutrient nutrition of plants. Crit. Rev. Plant. Sci. 14 (1), 49–82. Available from: https://doi.org/10.1080/07352689509701922.
Wimmer, M.A., Eichert, T., 2013. Review: mechanisms for boron deficiency-mediated changes in plant water relations. Plant. Sci. 203–204, 25–32.
Wongmo, J., Jamjod, S., Rerkasem, B., 2004. Contrasting responses to boron deficiency in barley and wheat. Plant. Soil. 259 (1–2), 103–110. Available from: https://doi.org/10.1023/B:PLSO.0000020946.76017.20.
Yamagishi, M., Yamamoto, Y., 1994. Effects of boron on nodule development and symbiotic nitrogen fixation in soybean plants. Soil. Sci. Plant. Nutr. 40 (2), 265–274. Available from: https://doi.org/10.1080/00380768.1994.10413300.
Yan, L., Riaz, M., Wu, X., Du, C., Liu, Y., Jiang, C., 2018a. Ameliorative effects of boron on aluminum induced variations of cell wall cellulose and pectin components in trifoliate orange (*Poncirus trifoliate* (L.) Raf.) rootstock. Environ. Pollut. 240, 764–774. Available from: https://doi.org/10.1016/j.envpol.2018.05.022.
Yan, L., Riaz, M., Wu, X., Wang, Y., Du, C., Jiang, C., 2018b. Interaction of boron and aluminum on the physiological characteristics of rape (*Brassica napus* L.) seedlings. Acta Physiologiae Plant. 40 (2), 33. Available from: https://doi.org/10.1007/s11738-018-2614-y.
Yermiyahu, U., Keren, R., Chen, Y., 2001. Effect of composted organic matter on boron uptake by plants. Soil. Sci. Soc. Am. J. 65 (5), 1436–1441. Available from: https://doi.org/10.2136/sssaj2001.6551436x.
Yoshinari, A., Takano, J., 2017. Insights into the mechanisms underlying boron homeostasis in plants. Front. Plant. Sci. 8, 1951.
Yu, S.M., Lo, S.F., David Ho, T.H., 2015. Source-sink communication: regulated by hormone, nutrient, and stress cross-signaling. Trends Plant. Sci. 20 (12), 844–857.

CHAPTER 7

Role of physical and chemical agents in plants for tolerance to boron nutrition

Shiva Aliyari Rad[1], Zahra Dehghanian[1], Behnam Asgari Lajayer[2], Muazzez Gürgan[3] and Tess Astatkie[4]
[1]Department of Biotechnology, Faculty of Agriculture, Azarbaijan Shahid Madani University, Tabriz, Iran
[2]Department of Soil Science, Faculty of Agriculture, University of Tabriz, Tabriz, Iran
[3]Department of Biology, Tekirdağ Namık Kemal University, Tekirdağ, Turkey
[4]Faculty of Agriculture, Dalhousie University, Truro, NS, Canada

7.1 Introduction

Plants need necessary macronutrients and micronutrients to expand and mature normally (Asgari Lajayer et al., 2019). Inadequate nutrient provision results in nutrient scarcity or toxicity, which has an impact on plant development resulting in a decline in the function and quality of plants. Boron (B) is one of the required micronutrients for suitable plant development, but it restricts plant development when in deficit or at high levels (Davies et al., 2011; Khadem Moghadam et al., 2020). Boric acid is water-soluble and B is mostly attracted by the soil solution via the root (Hu and Brown, 1997). An appropriate amount of boron in a soil solution is crucial to plant accessibility to facilitate boron absorption to plants (Majidi et al., 2010). Since the difference between the insufficient and toxic levels of boron for a plant is so slight, each shift in B concentration dramatically affects plant development, and this must be considered when adjusting B nutrition for plants via fecundation (Matula, 2009). B availability to plants is influenced by soil solution amount, pH, soil type, moisture levels, temperature, oxide, carbonate, aluminum, iron hydroxides, and clay minerals. Eggert and von Wirén (2016) reported that B supply boosts nutrient, water absorption, and biomass creation. Plant dry matter output is influenced by the amount of available P and B (Davies et al., 2011). It has been discovered that a higher amount of B reduces the amount of P in plant leaves (Kaya et al., 2009). It was demonstrated that as P rate increases, the amount of B in the plant decreases (Yamanouchi,

Boron in Plants and Agriculture
DOI: https://doi.org/10.1016/B978-0-323-90857-3.00013-8

1980). These findings were also confirmed by a study that reported supplementing with P might minimize the harmful effects of high B levels on tomato plant fruit function and development (Kaya et al., 2009). The study's goal was to evaluate the effects of increasing P values on (1) B value and absorption via barley, as well as the relationships between these two traits; (2) variation in B value assessed via 2 soil experiments (Mehlich 3 and NH4-acetate); and (3) potential communication among P and B values in soil.

7.2 Public attributes and chemistry of boron

Boron is a metalloid that has an atomic number and atomic mass of 5 and 10.811 g/mol, respectively; and is in group 13 of the periodic table. Boron was discovered as an element in 1808 by Sir Davy, Gay-Lussac, and Thenard (Bolaños et al., 2004). Boron has several stable isotopes, of which the most stables are 10B and 11B. It does not exist in elemental form in nature, but it does form organoboron structures with Na and O_2, which are boron's physiological forms in organisms (Hunt, 2003). Despite its low concentration in comparison to H, C, N, or O_2, boron is abundant in the environment and can be discovered naturally in minerals, soil, and water (Kot, 2009). Organoboron complexes, such as orthoborates, have B—O or B—N bonds and are essential in a biological system. Plants, animals, and human tissues all produce organoboron structures. Approximately 96% of the boron in an organism is in the form of inert boric acid $B(OH)_3$ and a slight volume of borate-anion B $(OH)^{4-}$ (Bolaños et al., 2004; Hunt, 2003; Devirian and Volpe, 2003). Boric acid is a feeble Lewis acid by a pKa of 9.25. As boron constitutes covalent bonds with its ligands, its pKa value drops, and it transforms into the negative bars foursquare borate-anion $B(OH)^{4-}$ (Hunt, 2003). Both borate and boric acid interact reversibly with biomolecules such as riboflavin, adenosine monophosphate, pyridoxine, ascorbic acid, pyrimidine nucleotides, and five-carbon sugar compounds such as apiose, ribose, and polysaccharides (Bolaños et al., 2004). H_3BO_3 and its Na salts (borax, di-sodium tetra-borate) are often employed as antiseptics, bactericides, washing agents like detergents, preservatives, fertilizers, insecticides, and herbicides.

7.3 Higher plants require boron as a micronutrient

B has metallic and nonmetallic properties and can produce acids and bases. It is primarily absorbed via the roots of the plants as tiny inert boric acid molecules, which quickly permeate through the phospholipid bilayer (Bolaños et al., 2004). Boron is a micronutrient for plants, which means it is required in concentrations as low as 1 mg/g for optimum development and output. Plant creatures have been recognized to include boron since the mid-nineteenth century (Kot, 2009). Its presence in plants species, including in wheat, has been verified by scientific studies, but no claims have been made that the element is vital (Devirian and Volpe, 2003). Mazéin investigated boron as a micronutrient in 1915, but the author's experimental methods were later criticized. As a result, Warington is credited with being the first to demonstrate the element's importance in plant metabolism. Boron is an important micronutrient for multiple monocotyledonous and dicotyledonous species, as well as for conifers, ferns, and phytoplankton species (Kim et al., 2003). Boron is a necessary or at minimum, highly helpful micronutrient for animals, too. It affects the metabolism of P, Ca, and Mg minerals, triglycerides, sugars, proteins, amino acids, ROS, and steroid hormones. It is suggested for humans to consume 1–13 mg/day, with the primary sources being juices, vegetables, nuts, and legumes (Chen et al., 2002).

7.4 Toxicity of boron in plants and boron nutrition

B is an essential nutritional element for plant development, and boric acid is the most readily available source of B in the soil. It is a weak Lewis-acid that produces the anion-borate as in the following reaction:

$$B(OH)_3 + H_2O \rightarrow \leftarrow B(OH)^{4-} + H^+ (pKa = 9.24).$$

Boric acid is a comparatively soluble compound in the soil that is easily washed away by rain. Thus, B shortage is common in areas with a lot of rain, such as in Southeast Asia and Southeast China (Shorrocks, 1997). Borate creates a network in the cell wall of plants by covalently crosslinking two chains of pectin at rhamnogalacturonan II (RG-II) sites (Funakawa and Miwa, 2015). Pectin is a polysaccharide found in abundance in the cell wall of higher plants being vital in defining cell size and form. The requirements of plant species for B are closely related to pectin material and B have also been proposed to be essential for cytoskeleton and membrane functions (Voxeur and Fry, 2014; Bassil et al., 2004). B toxicity impacts certain facets of cellular metabolism, damages DNA, and often induces tissue damage (Sakamoto et al., 2011; Reid et al., 2004).

Plants need B transport systems that are regulated by B situations to prevent B deficiency and toxicity.

7.5 Boron pptake and transport mechanisms

Boric acid is the available form in the soil and is absorbed via plant roots as it is a tiny, inert molecule that may pass through cell envelope (Stangoulis et al., 2001; Dordas and Brown, 2000). Plants control the boron transfer owing to the major-intrinsic-protein (MIP) family channels and borate exporters (BORs), the latter is used in the case of boron overabundance to exclude B in tissues (Takano et al., 2008; Miwa et al., 2007; Schnurbusch et al., 2010; Sutton et al., 2007). The structure of the BOR family is analogous to that of anion transporters (Coudray et al., 2017; Thurtle-Schmidt and Stroud, 2016). As shown in Fig. 7.1, Boron plays key roles in essential processes of higher plants including sugar transport and carbohydrate metabolism; photosynthesis; lignification and cell wall synthesis; keeping the cell membrane intact; nucleic acids metabolism; indole acetic acid metabolism; the ascorbate/glutathione cycle; metabolism of phenolic compounds; nitrogen metabolism; pollen tube formation; and several enzymatic activities (Rehman et al., 2006).

The amount of boron transferred within or outside of the cell must be controlled via a cellular mechanism to sustain boron homeostasis. Boron's effects could be seen first in plants because they are in contact with boron directly in the soil. So, multiple plant genes linked to boron transportation and resistance pathways have been detected. Boron is absorbed by the roots and transported to shoots through the xylem. It pursues transpiration flow and gathers at the edges of adult leafage during this long-distance transfer (Brown and Shelp, 1997). Passive and active transportation in the absorption of boron from the soil has been documented. Boric acid is an uncharged, tiny molecule with 71.5 a volume, almost identical to urea (75.3) (Kot, 2009). The boric acid penetrance coefficient for lipid bilayer is 8×10^{-6} cm/s, according to studies performed with the root of squash (*Cucurbita pepo*) and charophyte alga (*Chara corallina*) (Stangoulis et al., 2001) cells (Dordas and Brown, 2000). So, under great boron reserve, inactive diffusion demonstrates an important path for boron to pass the epidermis of the root. When boron access in the soil is limited, membrane proteins are required to absorb boron, indicating the existence of active transfer mechanisms.

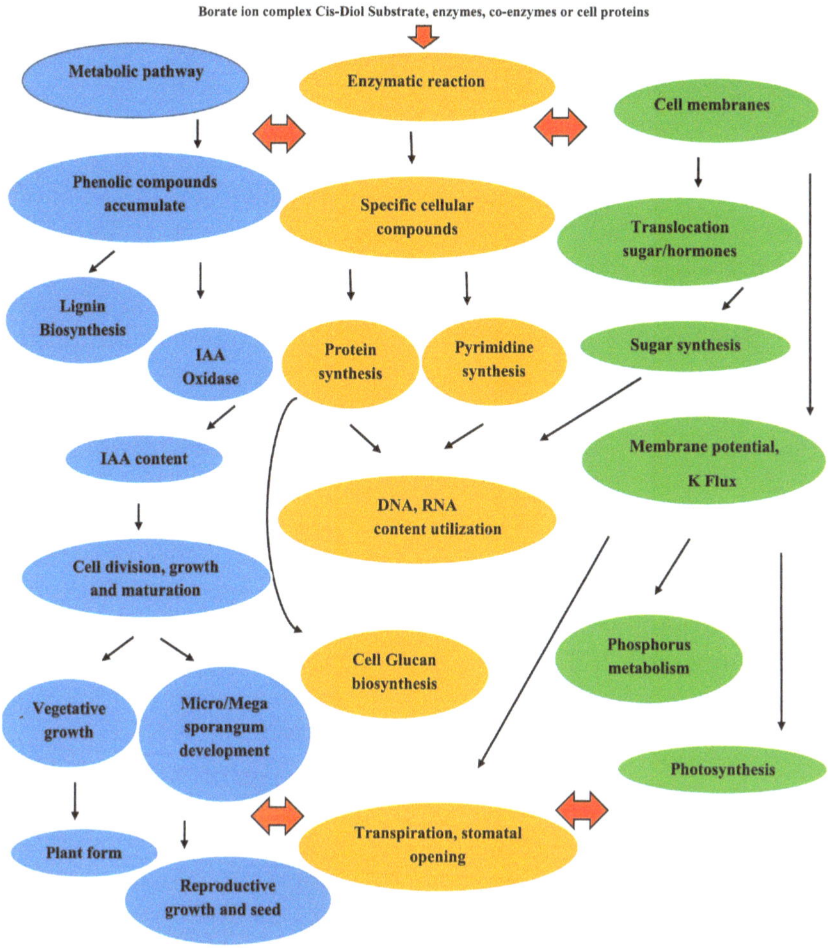

Figure 7.1 Physiological and molecular processes where boron is involved.

7.6 Boric acid channels, functions and regulation

Boric acid channels belong to the aquaporin homologs MIP family. Water/tiny uncharged molecules are transported by members of this family. Different MIPs carry boric acid and other tiny, uncharged molecules in plants, with their different substrate specifications (Reid, 2014). Determined by the structure of channel pore, NIPs are more apportioned to subclasses I, II, and III (Roberts and Routray, 2017). NIP I proteins have a physiologic function that is unknown. As silicic acid channels, NIP III proteins have been specified (Ma and Yamaji, 2006).

NIP5;1 in Arabidopsis was the primary boric acid channel recognized (Takano et al., 2006). Under B-limiting situation, NIP5;1 function-lost mutant displays intense developmental decrease handled by little B absorption to roots. When B is limiting, NIP5;1 is expressed in the root tip rhizodermis and the mature root endodermis (Takano et al., 2010; Wang et al., 2017). NIP5;1 indicates polar restriction into the soil in the PM of these cell kinds. The NIP5;1 function in B absorption is necessary when B is scarce, and the level of NIP5;1 mRNA is controlled by post-transcriptional regulation in response to cellular B concentrations (Takano et al., 2006; Tanaka et al., 2011). The minimum upstream open reading frame in the mRNA 5′-untranslated region is responsible for this regulation (Tanaka et al., 2016). Elevated B situations cause delaying ribosome at the AUG-stop, which inhibits mRNA destruction and its translation. This downadjustment is essential for plant acclimatization to achieve the B situation (Tanaka et al., 2011).

The nearest similarity to AtNIP5;1 is AtNIP6;1. Under a B-limiting situation, young leaf expansion and apical dominance are disrupted in NIP6;1 function-lost mutant (Tanaka et al., 2008). NIP6;1 has a role in B transport from xylem to phloem and is expressed in the cells of phloem companion, sieve elements in nodes and parenchyma cells (Tanaka et al., 2008). Initially, AtNIP7;1 was identified as an arsenite transporter but later it was specified as another boric acid channel (Isayenkov and Maathuis, 2008; Li et al., 2011). In Xenopus oocytes, the activity of NIP7;1 B transportation is smaller than that of NIP5;1 and NIP6;1. Nevertheless, the replacement of Tyr-81 in the transportation pore by Cys that occurs on NIP7;1 boosts B transportation activity than that of NIP5;1 or NIP6;1. Li et al. (2011) suggested that the NIP7;1 pore size may be handled by protein—protein interactions, pH, and phosphorylation. The NIP7;1 physiological yield in pollen still needs to be studied.

The NIP II proteins ZmTLS1 and OsNIP3;1 act like the channels of boric acid in rice and maize, and are needed in B transportation during vegetative and reproductive growth under B-limiting situations (Leonard et al., 2014; Hanaoka et al., 2014).

7.7 The borate exporters: physiological functions

The genome of *Arabidopsis thaliana* houses seven genes that encode seven different BOR-type borate transporters (Takano et al., 2002). The BORs could be split into clade I and clade II evolutionary clades in the angiosperms (Wakuta et al., 2015). BOR genes are expressed in different organs in *Arabidopsis* at all growth periods, according to microarray analysis (Winter et al., 2007; Schmid et al., 2005).

A mutant of bor1−1 was used to identify AtBOR1, the clade I prototype (Takano et al., 2002). The bor1−1 mutant exhibited a strong decrease in shoot development under boron limiting situations in the normal-B situations (Takano et al., 2001). These flaws were linked to decreased B aggregation in the shoot tissues; however, they could be corrected by increasing B supply. BOR1 is expressed in the roots, cotyledons, and hypocotyls of *Arabidopsis thaliana* seedlings (Yoshinari et al., 2016; Takano et al., 2010). It is expressed in the root tips and displays rather an intense expression in provascular, endodermal, and epidermal cells. It is also expressed in the endodermal cells in hypocotyls and mature roots, and in epidermal cells in cotyledons. In cooperation with NIP5;1, the BOR1 polar localization is possibly significant for the B effective transportation from the root to the xylem. Though localization is unknown in the shoot, AtBOR1 has evidently a role in preferential B transition in the nodes and basal leafages in cooperation with NIP6;1 (Tanaka et al., 2008; Takano et al., 2001). AtBOR1 may have a role in B absorption at the cotyledon's surface. BOR1;1c, ZmRTE, and OsBOR1 have been shaped as B transporter, and needed for B transportation in B-restricted situations in plants and, thus regarded to be AtBOR1 orthologs in rapeseed, maize, and rice, respectively (Nakagawa et al., 2007; Zhang et al., 2017; Chatterjee et al., 2014). BOR1 homologs have also been found to have expression (Pérez-Castro et al., 2012; Leaungthitikanchana et al., 2013; Cañon et al., 2013).

AtBOR2, with 90% of identity to AtBOR1, is dispensed in the plasma membrane (PM) with a stele-sided polarity like AtBOR1; however, unlike AtBOR1, AtBOR2 is evidently produced in the cap of the root and the epidermal cells more than in the root tip's internal cell layers (Miwa et al., 2013). In the roots of mutants (called bor2−1 and bor2−2) in which AtBOR2 was deleted, the B−RG-II crosslinking ratios and cell elongation were considerably reduced compared with the bor1−3 mutants that lack AtBOR1 under B-scarce situations (Miwa et al., 2013). To support root development, BOR2 enhanced B−RG-II cross-link in the cells of the roots. The second addition of boric acid to Rosa cells cultivated in B-free medium resulted in RG-II dimer, which is a progressive formation without observable lack of available monomers in the wall of a cell (Chormova et al., 2014). However, the RG-II monomer in the wall of the cell was not cross-linked permanently when polysaccharides de novo biosynthesis was inhibited. These kinds of results suggest that only newly synthesized RG-II are cross-linked by borate during secretion from the Golgi apparatus to the wall of the cell. BOR2–GFP was not constantly localized in the PM but displayed cycling among the endomembrane sections and the PM (Miwa et al., 2013). BOR2 also transports B to secretory vesicles to upgrade RG-II crosslinking. Excess B resistance is involved by AtBOR4 which belongs to clade II. BOR4 is produced in columella and epidermal cells in the root tip, as well as in endodermal root cells (Miwa et al., 2007, 2014). When B is in excess, BOR4 mRNA starts to aggregate (Miwa et al., 2014). Unlike BOR1/2, BOR4 displays poor localization (polar) unto the soil sides in roots' epidermal cells (Miwa et al., 2007). Lv et al. (2017) and Miwa et al. (2014) reported that the incorporation of BOR4 T-DNA caused the accumulation of higher B in the shoot and in the root, and it displayed a larger decrease in development than wild-type plants under excess B situations. Therefore, BOR4 is mainly involved in the elimination of B from tissues.

In wheat and barley, HvBot1 as well as its homologs are regarded as orthologs of AtBOR4 and have been determined as essential agents for tolerating excess B (Sutton et al., 2007; Reid, 2014; Pallotta et al., 2014). In rice, preanthesis anthers and mature pollens encode OsBOR4, a clade II gene (Tanaka et al., 2013). Pollen germinations and elongations were disrupted when OsBOR4 was disrupted, showing that OsBOR4 retained intracellular B levels in pollen. According to microarray results, the stamen has the highest degree of AtBOR4 mRNA aggregation (Schmid et al., 2005). Besides the deletion of B for excess B tolerance, AtBOR4 may play a role in B homeostasis for fertilization. Bryophytes and lycophytes have been shown to carry BOR genes (Wakuta et al., 2015). BORs in the bryophyte *Physcomitrella patens* were classified into clades I and II, whereas they were separated into clades I and II in *Selaginella moellendorffii*. In the *Physcomitrella patens*, the B transportation function of BORs is unknown. When expressed in yeast cells, SmBOR1 (clade I), and SmBOR3 and 4 (clade II) act as B exporters (Wakuta et al., 2015). Though the substantial importance of them is yet to be determined, it is possible that the joint ancestors of vascular plants also had two forms of BOR for reduced and excessive B toleration.

7.8 Boron roles in plants

Warington (1923) reported that the growth of *Vicia faba* (field bean) was slower in a medium without boron, but growth was recovered once boron was added to the medium. Boron has been embraced as an essential and helpful microelement for a variety of organisms since that time. In plants, boron has a role in the metabolism of nucleic acid, carbohydrate, protein, and phenol; cell wall synthesis and construction, and membrane integrity and activities (Goldbach and Wimmer, 2007; Miwa et al., 2007). However, the molecular mechanisms underlying these metabolisms are mostly unclear. Boron's main molecular role is to cross-link pectins in the plant cell wall. Pectin is a complex polysaccharide that contributes to the structure of the plant cell wall. Borate forms a cross-link with pectin RG-II apiose residues that are needed for cell wall stabilization and natural leave expansion (O'Neill et al., 2004; Kobayashi et al., 2004).

RG-II is a pectic polysaccharide with cis−diol bounds that covalently binds to borate esters (York et al., 1985). Borate crosslinking in these pectic systems physically reinforces the cell walls and has an important role in cell adhesion (Ishii and Matsunaga, 2003; Scorei, 2012). Boron plays an important function in extending organs in plants, so boron shortage could be mostly seen in expanding organs. Boron deficiency results in aberrant cell wall formation, membrane permeability and cytoskeletal polymerization alterations. Furthermore, leaf expansions, root elongation, flowers and fruit expansions were hindered (Kot, 2009; Dell and Huang, 1997). Boron shortage interferes with the metabolism of sugar, starch, phenol, and RNA, and the composition of the cell wall, among other physiological activities (Kot, 2009). In 2004, under boron restriction, the expression of glucosyltransferase and glutathione S-transferase in tobacco (*Nicotiana tabacum*) BY-2 cells were induced (Kobayashi et al., 2004). Another research using the same cells in 2009 found that oxidative harm specifically resulted in cell death in the presence of boron deficiency and that it was unlikely to be a normal PCD (programmed cell death) (Koshiba et al., 2009).

7.9 Roles of boron in plant metabolism

Despite the fact that boron is accepted as an essential element for almost a century, its role in plant metabolism is still debatable (Amin et al., 2007; Wolkenstein et al., 2010). The element's primary role is to aid in the synthesis of cell walls as well as to safeguard the structure and integrity of these structures. The value of boron in the wall pertains to the species, organs, and boron accumulation of the plant (Anderson and Jordan, 1961). Plants absorb boron in a form of the molecules of small uncharged boric acid (around 96%), and to a lesser extent, borate anions (Ahmed et al., 2007). Borate helps the cell wall's construction and function by forming ester linkages among the apiose remainders of RGII monomers. Dicotyledonous plants have more pectin in their cell walls than monocotyledonous plants, so their boron needs are less and their tolerance to excess boron is higher (Scorei, 2012).

Boron's main role in the metabolism of plants, according to one theory (Ahmed et al., 2007), might be the fixation of molecules with cis–diol groups, regardless of their purpose. Boron maintains the stability of the plasma membrane by forming a cis–diol complex in combination with glycoproteins and glycolipids. It is engaged in enzyme reactions, ion, metabolites, and hormones' transfer, and hence engages in membrane metabolism and activities (Hunt, 2003; Kobayashi et al., 2004). It takes place in sugar transport across the membrane, lignin and flavonoid production, auxin, nitrogen combinations, and phenol metabolism, and it agitates, inhibits, or stabilizes enzymes (Koshiba et al., 2009; Pahl et al., 2001). Hampering of root development was detected in situations of inadequate boron amount due to the hindrance of DNA synthesis resulting in the shortage of pyrimidine base.

In legumes and mycorrhiza, the element is engaged in mitosis (WHO, 1996; Clarke et al., 1987), tissue vegetative development and differentiation (Eckhert, 1998), translation and transcription (Fort et al., 1998), and nitrogen fixation (Ahmed et al., 2007). Boron scarcity affects photosynthetic valence and photosynthetic product transport (Kobayashi et al., 2004). There was a reduction in stomatal conductance and CO_2 assimilation, as well as a rise in intercellular CO_2 concentration, in boron-deficient orange leaves. The aggregation of hexose and starch was enhanced, but sucrose aggregation was reduced, indicating feedback regulation in CO_2 absorption.

Antioxidant enzymes persisted to be active, but the system altogether was not effective enough to protect against oxidative harm (Spielberg et al., 1979). Expression of genes of the respiration, aminoacid biosynthesis, organicacid metabolism, an anaplerotic pathway increased in leaf cells but decreased in root cells to maintain net carbon balance during a carbon deficit. Since insufficient boron supply induces various metabolic, physiological, and anatomic modifications, distinguishing the primary and secondary efficacy of the deficiency is exceedingly difficult. The most noticeable effect was the disruption of cell division in the apical meristem, which leads to a decline or termination of root development. Boron shortage reduces male reproduction by declining microsporogenesis, germination, and pollen tube elongation. The nourishment derangement disturbs embryogenesis after fertilization, which causes seed decay or the development of imperfect or harmed embryos (Goldbach and Wimmer, 2007; Miwa et al., 2007; Benderdour et al., 2000).

7.10 Plant tolerance to boron

It shows that soil, not the crust of the earth, seas, oceans (Wolska and Bryjak, 2013), or surface streams (Türker et al., 2014), has the greatest average concentration of B. As a result, toxicity concerns are prevalent, particularly in dry and semiarid areas. In some regions, B agglomeration in the soil as a result of water evaporating from a shallow water table or from the soil surface results in B poisoning symptoms. In most plants, the symptoms of B poisoning are identical. They start with marginal and tip chlorosis and quickly progress to necrosis (Shorrocks, 1982). The venation of the leaf determines the pattern of chlorosis and necrosis. Monocotyledons have tip necrosis rather than marginal necrosis. Many plant species can tolerate B concentrations of 0.5−2 ppm in the soil solution, but when they approach 20 ppm in plant tissues, indications of B toxicity (Marschner, 2011) including chlorosis commencing at the leaf tips or edges of developed leaves, as well as a decrease in development (Reid et al., 2004; Yıldırım, 2017). Boron extraction using hot water is the most precise technique for determining soil B access. When the concentration is greater than 5 ppm, the soil might not be appropriate for the growth of plenty of plants and might induce visual B poisoning symptoms (Nable et al., 1997). Observable B poisoning indications are mentioned by Gupta (1980) for a variety of crops including forages, vegetables, and cereals (Table 7.1). Plants have systems in place to combat the toxicity of B in the soil, one of which is extruding B from plant cells.

According to Hayes and Reid (2004), *Hordeum vulgare* L., Sahara cultivar, which is B-tolerant, uses an extraction system to retain 50% lower B content in the root than a B-sensitive cultivar. B extracted from the root of Sahara is thought to be the reason for the B tolerance in Sahara, which indicates that the borate anion efflux transporter has a role.

B is recognized to bind a variety of biological molecules, especially those with two hydroxyl groups. Ribose, particularly the *cis*-hydroxyl on the ribose side of electron carriers; for example, ATP, NADH, and NADPH is suggested to be the location of B binding.

Table 7.1 The observable symptoms of boron toxicity in some crops (Gupta, 1980).

No.	Plant species		Toxicity symptoms of B
1	Vegetable crops	Rutabaga	Yellow leaf margins and tends to curl and wrinkle
		Turnip	Marginal whitening of seedlings of cotyledons and initial leaves
		Peas	The margins of old leaves burn, and the extent of burning depends on the severity of B toxicity
		Onions	Burning leaves' tips, progressing to the root, and no bulb growth
		Radishes	Burning of lower leaves
		Tomatoes	Drying out of older leaves
2	Forage crops	Alfalfa	The oldest leaves have burnt edges
		Red Clover	Burning and yellowing of older leaf margins
		Bird's-foot trefoil	Tiny and dark leaves, with weak stalks, and the growth is weak
3	Cereal crops	Wheat	Older leaf tips become light brown
		Barley	Older leaves have elongated dark-brown spots at the tips
		Oats	Light-yellow bleached leaf tips
		Corn	Marginal burning of tips and yellowing between the veins
4	Other crops	Tobacco	Stunted growth and brown circular patches at the edge of the leaves
		Beans	Mottled and necrotic patches on the leaves, mainly at the leaf edges
		Black-eyed bean	Yellow or dead areas occur between veins without any pattern
		Indian bean	Slight chlorosis follows dark-brown spots on the leaves; some burning on the border of the leaves, and development is inhibited

Therefore, the intensity of photosynthesis and respiration as a cause of B toxicity is disproved (Princi et al., 2016; Reid et al., 2004). Several studies have found the significant roles of B efflux transporters like BOR1 in resistance to B toxicity (Nagarajan et al., 2016).

7.11 Primary considerations

For a long time, soil amendments like washing away B with water and adding organic compounds that can immobilize or inactivate B in soil were thought to be the best strategy to deal with B toxicity. Because of economic and logistical considerations, it does not appear to be practicable on a broad range in B-toxic locations at the moment. In contrast, selecting B excess tolerant genotypes may be the most feasible and practical approach to increasing outputs in B-rich soils (Nable et al., 1997). Since the 1980s, mutations for tolerance to high B in a variety of agricultural plants has been studied (Bogacki et al., 2013; Schnurbusch et al., 2010). As a result, more resistant cultivars may be well-produced, providing a very promising method to avoiding agricultural output declines in places with high soil B content.

7.12 Revisiting tolerant mechanisms

Physiological processes relevant to B excess tolerance are still unknown. B absorption from soil and its movement through the plant, a buildup of B at the end of the transpiration stream, B content in tissues, and a concentration gradient among a leaf are all examples of tolerance mechanisms in vascular plants (Reid et al., 2004). B tolerant models imply the presence of binding B combinations when intercellular hazardous amounts are reached; B compartmentalization; and active B outflow via transporters (Hayes and Reid, 2004). Furthermore, decreased B agglomeration in cultivars that are B tolerant compared to sensitive cultivars highlighted the importance of efflux-type borate transporters in roots over internal tolerance mechanisms such as B binding complexes or B compartmentalization in vacuoles. Therefore, the foundation of resistance to B excesses posits a more restricted tissue B concentration including both B absorption decrease and active efflux of the micronutrients from the roots, at least partially (Reid, 2014).

As stated earlier, when their access to the soil is restricted, AtBOR1 and AtNIP5;1 are essential for effective B absorption (Takano et al., 2002, 2006). Therefore, when B absorption reaches an excessive level, it is mostly controlled by the transcriptional control of AtNIP5;1 or by the endocytosis and destruction of AtBOR1 (Takano et al., 2005, 2006). Furthermore, overexpression of AtBOR1 did not improve plant development in the presence of hazardous B amounts (Miwa et al., 2007).

Furthermore, the destruction of AtNIP5; 1 mRNA in the presence of B excess is regulated via the 50 untranslated regions of AtNIP5; 1, implying that both genes do not seem to be implicated in B tolerant (Tanaka et al., 2011). In contrast, BOR1 and NIP5; 1 activities were hypothesized to be employed as markers for plant genotypes with greater resistance to B tension in barley (Mishra et al., 2015). Various more B transporters, also aquaporins were lately found in a variety of plants, with some of them implicated in B tolerant processes (Miwa and Fujiwara, 2010).

Using a green fluorescent protein construct, BOR4 protein was discovered on the outside root cell membranes of *Arabidopsis*. This localization is critical for B directed exports to the soils, since it prevents excessive B concentrations in developing cells and xylem. Overexpression of AtBOR4 increased plant development greatly in the presence of excess B, suggesting that it prevents the posttranslational degradation process (reported for AtBOR1) and is, in contrast, highly inductive (Miwa et al., 2007, 2014). Furthermore, rice plants with genetic modifications to express AtBOR4 demonstrated a great resistance to harmful B levels (Kajikawa et al., 2011). The increased development was linked to the efficient B exports of the root, which occurred by reason of an ideal B content managed inside the plant. Therefore, the discrepancy in BOR1 and BOR4 gene control indicates the attendance of complicated processes for sensing and regulating the B state in the cell. Aquaporin isoforms have recently been found to promote tolerance to a variety of abiotic conditions (Pang et al., 2010). Furthermore, high expression of AtTIP5; 1, which is a tonoplast aquaporin, leads to greater tolerance to a moderately high B level in the growth mediums, indicating that it is involved in vacuolar borate compart-mentation (Pang et al., 2010). Furthermore, OsPIP2; 4 and OsPIP2; 7 (two aquaporin rice genes) were found to be linked to B transmittance and resistance (Kumar et al., 2014).

Both genes, which are important for transporting B from roots in the presence of an excess of B, were expressed lower in shoots but highly overexpressed in roots, where their increased production prevented the damage of B. Moreover, a root discharge B test revealed that 10B was eliminated from roots after 1 h of exposure in *Arabidopsis* genetic variants to overexpress OsPIP2; 4orOsPIP2; 7 (Kumar et al., 2014). Using recombinant inbred lines, researchers have found a gene that codes for a NAC-like TF with a single nucleotide variation among susceptible and resistant rice (Ochiai et al., 2011). By disrupting the gene known as BET1, a one-nucleotide mutation turned out to produce resistance to B damage in rice (boron excess tolerant 1). Because there were no variations in the root or shoot B contents among susceptible and tolerant rice, this process might be independent of B efflux (Ochiai et al., 2011). Two *Arabidopsis* mutants with defective genes associated with B uptake were also examined to find novel mechanisms involved in B resistance (Sakamoto et al., 2011). Therefore, heb1−1 and heb2−1, which are mutants to be hypersensitive to excess B that showed development abnormalities in the presence of excess B, was deficient in the capacity to express two components of the chromosomal protein complex recognized as condensin II. Despite having lower B than the original plant, both heb mutants were substantially more sensitive to B excess. These data supported the presence of resistance mechanisms other than B efflux. Condensin II appeared to improve dsDNA breaks and sustain the replication process, both of which are thought to be necessary for B resistance (Sakamoto et al., 2011). Differential transcripts in reaction to the high amount of B resistant barley using RNA-Seq were analyzed and profiled by Tombuloglu et al. (2015). The researchers discovered that genes involved in cell envelope formation, the Ca2/calmodulin pathway, phospholipase activities, and signal transduction all play an important role in the case of high B amount.

7.13 Plant genetic modifications for boron susceptibility and resilience

Tissue types and developmental periods determine the tolerable B concentration ranges, and the limits between B shortage and the harmful amount is frequently rather small. As a result, within each plant species, genotype assessment for acute B content is vital for crop yields and resistant B choice. Tolerance has been classified into three broad categories: sensitive, semitolerant, and resistant (Ayvaz, 2002).

Boron sensitive plants can withstand up to 0.5 mg/L, but the resistant species may withstand up to 4 mg/L (Batar et al., 2009). Fruits such as apples, beans, figs, grapes, and peaches are the most susceptible to excess B; while plants like barley, peas, maize, potato, tomato, and tobacco are semiresistant; and those like alfalfa, cotton, carrot, turnip, and sugar beet turn out to be the most resistant (Mengel et al., 2001). Furthermore, plants that are sensitive to B show a considerable response to varying amounts of the microelement, whilst tolerant plants demonstrate flexibility to a larger range of B contents with no discernible growth declines (Ozturk et al., 2010). To increase a plant's B surplus tolerance, newer genotypes capable of adapting to places with high B amounts must be selected within the genetic variation.

According to the physiological basis of B tolerance, tolerant breeds have lower B levels in leaves than sensitive breeds, which is most likely due to lower B absorption in both roots and shoots (Nable et al., 1990). Boron excess tolerance is frequently associated with minimal or nonexistent B toxicity evidence, low B in tissues, and good development or output in the presence of soil B excess. As a result, effective phenotypic tests to scan B tolerance contain measures of leaf symptom expression, shoot dry weight, relative root length, and B content in these tissues (Schnurbusch et al., 2008; Jefferies et al., 2000). The measures of B amount and content vary with genotypes for B excess tolerance. Although tolerant barley genotypes with lower tissue B were reported in B surplus (Rehman et al., 2006), high B in tissues of tolerant wheat were also reported (Torun et al., 2006). IAC287 demonstrated B excess tolerance in their trials, as a usual demonstration of toxicity was not detected up to 32.4 mM B in the growth medium. Genetic variation in reaction to B excess has lately been exploited for QTL studies in crops including wheat and barley (Hayes et al., 2013; Torun et al., 2006; Pallotta et al., 2014), lentil (Yau and Erskine, 2000), rice (Ochiai et al., 2008), citrus (Guo et al., 2014), and alfalfa (Bogacki et al., 2013).

7.14 Conclusions

B is a micro plant nutrient that is necessary for plant development. Excess and shortage of B, on the other hand, can have negative impacts on plants and plant productivity. In arid/semiarid and humid environments, B toxicity and scarcity were reported. Boron is involved in protein synthesis, sugar transfer, cell division, carbohydrate metabolism, seed development, and the metabolism of plant hormones similar to IAA. Cell envelope construction and lignification are also well-known roles of B. The pectic polysaccharide RGII is a main structural element in cell walls that preserves the cells, and it has been shown that RGII with crosslinking by the creation of boratediol eater bonds is required for plant development. Dimerization of RG-II reduces when B shortage develops, and the structure of cell walls changes as a result. B channels like NIP5; 1 and BOR1 that are required for efficient B transport across the cell membrane in B shortage, can help in B uptake when there is B deficiency. The NIP5; 1 is a water channel and a boric acid, while BOR1 is a B transporter involved in the xylem loading mechanism, as previously noted. The BOR1 protein could transfer B outside of cells in pericycle cells and load xylem in B toxicity situations to the soil solution. An anion efflux transporter's presence on the cell membrane could play a role in B toxicity tolerance.

References

Ahmed, I., Yokota, A., Fujiwara, T., 2007. A novel highly boron tolerant bacterium, Bacillus boroniphilus sp. nov., isolated from soil, that requires boron for its growth. Extremophiles 11 (2), 217–224.

Amin, S.A., Küpper, F.C., Green, D.H., Harris, W.R., Carrano, C.J., 2007. Boron binding by a siderophore isolated from marine bacteria associated with the toxic dinoflagellate Gymnodinium catenatum. J. Am. Chem. Soc. 129 (3), 478–479.

Anderson, G.R., Jordan, J.V., 1961. Boron: a non-essential growth factor for *Azotobacter chroococcum*. Soil. Sci. 92 (2), 113–116.

Asgari Lajayer, B., Najafi, N., Moghiseh, E., Mosaferi, M., Hadian, J., 2019. Effects of gamma irradiated and non-irradiated sewage sludge on growth characteristics, leaf chlorophyll index, and macronutrients concentrations in basil. J. Soil. Sci. Plant. Nutr. 19, 580–591.

Ayvaz, M., 2002. Bazı arpa çeşitlerinde borun büyüme ve gelişme üzerine etkileri, Yüksek Lisans Tezi, Biyoloji Anabilim Dalı.

Bassil, E., Hu, H., Brown, P.H., 2004. Use of phenylboronic acids to investigate boron function in plants. Possible role of boron in transvacuolar cytoplasmic strands and cell-to-wall adhesion. Plant. Physiol. 136 (2), 3383–3395.

Batar, T., Köksal, N.S., Yersel, Ş.E., 2009. Atık bor, atık kâğıt ve perlit katkılı sıva malzemesinin üretimi ve karakterizasyonu. Ekoloji 18 (72), 45–53.

Benderdour, M., Bui, Van, Hess, T., Dicko, K., Belleville, A., Dousset, B., F., 2000. Effects of boron derivatives on extracellular matrix formation. J. Trace Elem. Med. Biol. 14 (3), 168–173.

Bogacki, P., Peck, D.M., Nair, R.M., Howie, J., Oldach, K.H., 2013. Genetic analysis of tolerance to boron toxicity in the legume *Medicago truncatula*. BMC plant. Biol. 13 (1), 1–11.

Bolaños, I., Lukaszewski, K., Bonilla, I., Blevins, D., 2004. Why boron? Plant. Physiol. Biochem. 42 (11), 907–912.

Brown, P.H., Shelp, B.J., 1997. Boron mobility in plants. Plant. Soil. 193 (1), 85–101.

Cañon, P., Aquea, F., Rodríguez-Hoces de la Guardia, A., Arce-Johnson, P., 2013. Functional characterization of *Citrus macrophylla* BOR1 as a boron transporter. Physiologia Plant. 149 (3), 329–339.

Chatterjee, M., Tabi, Z., Galli, M., Malcomber, S., Buck, A., Muszynski, M., et al., 2014. The boron efflux transporter ROTTEN EAR is required for maize inflorescence development and fertility. Plant. Cell 26 (7), 2962–2977.

Chen, X., Schauder, S., Potier, N., Van Dorsselaer, A., Pelczer, I., Bassler, B.L., et al., 2002. Structural identification of a bacterial quorum-sensing signal containing boron. Nature 415 (6871), 545–549.

Chormova, D., Messenger, D.J., Fry, S.C., 2014. Boron bridging of rhamnogalacturonan-II, monitored by gel electrophoresis, occurs during polysaccharide synthesis and secretion but not post-secretion. Plant. J. 77 (4), 534–546.

Clarke, W.B., Webber, C.E., Koekebakker, M., Barr, R.D., 1987. Lithium and boron in human blood. J. Lab Clin. Med. 109 (2), 155–158.

Coudray, N., Seyler, S.L., Lasala, R., Zhang, Z., Clark, K.M., Dumont, M.E., et al., 2017. Structure of the SLC4 transporter Bor1p in an inward-facing conformation. Protein Sci. 26 (1), 130–145.

Davies, M., Atkinson, C., Burns, C., Arroo, R., Woolley, J., 2011. Increases in leaf artemisinin concentration in *Artemisia annua* in response to the application of phosphorus and boron. Ind. Crop. Products 34 (3), 1465–1473.

Dell, B., Huang, L.B., 1997. Physiological response of plants to low boron. Plant. Soil. 193 (1–2), 103–120.

Devirian, T.A., Volpe, S.L., 2003. The physiological effects of dietary boron. Crit. Rev. Food Sci. Nutr. 43, 219–231.

Dordas, C., Brown, P., 2000. Permeability of boric acid across lipid bilayers and factors affecting it. J. Membr. Biol. 175 (2), 95–105.

Eckhert, C.D., 1998. Boron stimulates embryonic trout growth. J. Nutr. 128 (12), 2488–2493.

Eggert, K., von Wirén, N., 2016. The role of boron nutrition in seed vigour of oilseed rape (Brassica napus L.). Plant. Soil. 402 (1–2), 63–76.

Fort, D.J., Propst, T.L., Stover, E.L., Strong, P.L., Murray, F.J., 1998. Adverse reproductive and developmental effects in Xenopus from insufficient boron. Biol. trace Elem. Res. 66 (1), 237–259.

Funakawa, H., Miwa, K., 2015. Synthesis of borate cross-linked rhamnogalacturonan II. Front. Plant. Sci. 6, 223.

Goldbach, H.E., Wimmer, M.A., 2007. Boron in plants and animals: is there a role beyond cell-wall structure? J. Plant. Nutr. Soil. Sci. 170 (1), 39–48.

Guo, P., Qi, Y.-P., Yang, L.-T., Ye, X., Jiang, H.-X., Huang, J.-H., et al., 2014. cDNA-AFLP analysis reveals the adaptive responses of citrus to long-term boron-toxicity. BMC Plant. Biol. 14 (1), 1–22.

Gupta, U.C., 1980. Boron nutrition of crops. Adv. Agron. 31, 273–307.

Hanaoka, H., Uraguchi, S., Takano, J., Tanaka, M., Fujiwara, T., 2014. O s NIP 3; 1, a rice boric acid channel, regulates boron distribution and is essential for growth under boron-deficient conditions. Plant. J. 78 (5), 890–902.

Hayes, J.E., Reid, R.J., 2004. Boron tolerance in barley is mediated by efflux of boron from the roots. Plant. Physiol. 136 (2), 3376–3382.

Hayes, J.E., Pallotta, M., Baumann, U., Berger, B., Langridge, P., Sutton, T., 2013. Germanium as a tool to dissect boron toxicity effects in barley and wheat. Funct. Plant. Biol. 40 (6), 618–627.

Hu, H., Brown, P.H., 1997. Absorption of boron by plant roots. Plant. Soil. 193 (1), 49–58.

Hunt, C.D., 2003. Dietary boron: an overview of the evidence for its role in immune function. J. Trace Elem. Exp. Medicine: Off. Publ. Int. Soc. Trace Elem. Res. Hum. 16 (4), 291–306.

Isayenkov, S.V., Maathuis, F.J., 2008. The *Arabidopsis thaliana* aquaglyceroporin AtNIP7; 1 is a pathway for arsenite uptake. Febs Lett. 582 (1), 1625–1628.

Ishii, T., Matsunaga, T., 2003. Chemical structure and function of the plant cell wall polysaccharide rhamnogalacturonan II. J. Jpn. Wood Res. Soc. (Jpn.) 49 (3), 153–160.

Jefferies, S., Pallotta, M., Paull, J., Karakousis, A., Kretschmer, J., Manning, S., et al., 2000. Mapping and validation of chromosome regions conferring boron toxicity tolerance in wheat (*Triticum aestivum*). Theor. Appl. Genet. 101 (5–6), 767–777.

Kajikawa, M., Fujibe, T., Uraguchi, S., Miwa, K., Fujiwara, T., 2011. Expression of the Arabidopsis borate efflux transporter gene, AtBOR4, in rice affects the xylem loading of boron and tolerance to excess boron. Biosci, Biotechnol, Biochem. 75 (12), 2421–2423.

Kaya, C., Tuna, A.L., Dikilitas, M., Ashraf, M., Koskeroglu, S., Guneri, M., 2009. Supplementary phosphorus can alleviate boron toxicity in tomato. Sci. Horticulturae 121 (3), 284–288.

Khadem Moghadam, K., Asgari Lajayer, B., Ghorbanpour, M., 2020. Boron tolerance in plants: physiological roles and transport mechanisms. In: Deshmukh, R., Tripathi, D. K., Guerriero, G. (Eds.), Metalloids in Plants: Advances and Future Prospects. John Wiley & Sons Ltd, pp. 301–314.

Kim, D.H., Marbois, B.N., Faull, K.F., Eckhert, C.D., 2003. Esterification of borate with NAD + and NADH as studied by electrospray ionization mass spectrometry and 11B NMR spectroscopy. J. Mass. Spectrometry 38 (6), 632–640.

Kobayashi, M., Mutoh, T., Matoh, T., 2004. Boron nutrition of cultured tobacco BY-2 cells. IV. Genes induced under low boron supply. J. Exp. Botany 55 (4), 1441–1443.

Koshiba, T., Kobayashi, M., Matoh, T., 2009. Boron nutrition of tobacco BY-2 cells. V. Oxidative damage is the major cause of cell death induced by boron deprivation. Plant. Cell Physiol. 50 (1), 26–36.

Kot, F.S., 2009. Boron sources. speciation and its potential impact on health. Rev. Environ. Sci. Bio/Technol 8 (1), 3–28.

Kumar, K., Mosa, K.A., Chhikara, S., Musante, C., White, J.C., Dhankher, O.P., 2014. Two rice plasma membrane intrinsic proteins, OsPIP2; 4 and OsPIP2; 7, are involved in transport and providing tolerance to boron toxicity. Planta 239 (1), 187–198.

Leaungthitikanchana, S., Fujibe, T., Tanaka, M., Wang, S., Sotta, N., Takano, J., et al., 2013. Differential expression of three BOR1 genes corresponding to different genomes in response to boron conditions in hexaploid wheat (*Triticum aestivum* L.). Plant. Cell Physiol. 54 (7), 1056–1063.

Leonard, A., Holloway, B., Guo, M., Rupe, M., Yu, G., Beatty, M., et al., 2014. tassel-less1 encodes a boron channel protein required for inflorescence development in maize. Plant. Cell Physiol. 55 (6), 1044–1054.

Li, T., Choi, W.G., Wallace, I.S., Baudry, J., Roberts, D.M., 2011. *Arabidopsis thaliana* NIP7; 1: an anther-specific boric acid transporter of the aquaporin superfamily regulated by an unusual tyrosine in helix 2 of the transport pore. Biochemistry 50 (31), 6633–6641.

Lv, Q., Wang, L., Wang, J.Z., Li, P., Chen, Y.L., Du, J., et al., 2017. SHB1/HY1 alleviates excess boron stress by increasing BOR4 expression level and maintaining boron homeostasis in Arabidopsis roots. Front. Plant. Sci. 8, 790.

Ma, J.F., Yamaji, N., 2006. Silicon uptake and accumulation in higher plants. Trends Plant. Sci. 11 (8), 392–397.

Majidi, A., Rahnemaie, R., Hassani, A., Malakouti, M.J., 2010. Adsorption and desorption processes of boron in calcareous soils. Chemosphere 80 (7), 733–739.

Marschner, H., 2011. Marschner's Mineral Nutrition of Higher Plants. Academic Press, London, UK.

Matula, J., 2009. Boron sorption in soils and its extractability by soil tests (Mehlich 3, ammonium acetate and water extraction). Plant, Soil. Environ. 55 (1), 42–49.

Mengel, K., Kirkby, E.A., Kosegarten, H., Appel, T., 2001. Zinc, Principles of Plant Nutrition. Springer, pp. 585–597.

Mishra, S., Heckathorn, S., Krause, C., 2015. The levels of boron-uptake proteins in roots are correlated with tolerance to boron stress in barley. Crop. Sci. 55 (4), 1741–1748.

Miwa, K., Fujiwara, T., 2010. Boron transport in plants: co-ordinated regulation of transporters. Ann. Botany 105 (7), 1103–1108.

Miwa, K., Takano, J., Omori, H., Seki, M., Shinozaki, K., Fujiwara, T., 2007. Plants tolerant of high boron levels. Science 318 (5855), 1417-1417.

Miwa, K., Wakuta, S., Takada, S., Ide, K., Takano, J., Naito, S., et al., 2013. Roles of BOR2, a boron exporter, in cross linking of rhamnogalacturonan II and root elongation under boron limitation in Arabidopsis. Plant. Physiol. 163 (4), 1699–1709.

Miwa, K., Aibara, I., Fujiwara, T., 2014. *Arabidopsis thaliana* BOR4 is upregulated under high boron conditions and confers tolerance to high boron. Soil. Sci. Plant. Nutr. 60 (3), 349–355.
Nable, R.O., Cartwright, B., Lance, R.C., 1990. Genotypic differences in boron accumulation in barley: relative susceptibilities to boron deficiency and toxicity. Genetic aspects of plant mineral nutrition. Springer, pp. 243–251.
Nable, R.O., Bañuelos, G.S., Paull, J.G., 1997. Boron toxicity. Plant. Soil. 193 (1), 181–198.
Nagarajan, Y., Rongala, J., Luang, S., Singh, A., Shadiac, N., Hayes, J., et al., 2016. A barley efflux transporter operates in a Na + -dependent manner, as revealed by a multidisciplinary platform. Plant. Cell 28 (1), 202–218.
Nakagawa, Y., Hanaoka, H., Kobayashi, M., Miyoshi, K., Miwa, K., Fujiwara, T., 2007. Cell-type specificity of the expression of Os BOR1, a rice efflux boron transporter gene, is regulated in response to boron availability for efficient boron uptake and xylem loading. Plant. Cell 19 (8), 2624–2635.
Ochiai, K., Uemura, S., Shimizu, A., Okumoto, Y., Matoh, T., 2008. Boron toxicity in rice (*Oryza sativa* L.). I. Quantitative trait locus (QTL) analysis of tolerance to boron toxicity. Theor. Appl. Genet. 117 (1), 125–133.
Ochiai, K., Shimizu, A., Okumoto, Y., Fujiwara, T., Matoh, T., 2011. Suppression of a NAC-like transcription factor gene improves boron-toxicity tolerance in rice. Plant. Physiol. 156 (3), 1457–1463.
O'Neill, M.A., Ishii, T., Albersheim, P., Darvill, A.G., 2004. Rhamnogalacturonan II: structure and function of a borate cross-linked cell wall pectic polysaccharide. The. Annu. Rev. Plant. Biol. 55, 109–139.
Ozturk, M., Sakcali, S., Gucel, S., Tombuloglu, H., 2010. Boron and plants. In: Ashraf, M., Ozturk, M., Ahmad, M.S.A. (Eds.), Plant Adaptation and Phytoremediation. Springer, The Netherlands, pp. 275–312.
Pahl, M.V., Culver, B.D., Strong, P.L., Murray, F.J., Vaziri, N.D., 2001. The effect of pregnancy on renal clearance of boron in humans: a study based on normal dietary intake of boron. Toxicol Sci. 60 (2), 252–256.
Pallotta, M., Schnurbusch, T., Hayes, J., Hay, A., Baumann, U., Paull, J., et al., 2014. Molecular basis of adaptation to high soil boron in wheat landraces and elite cultivars. Nature 514 (7520), 88–91.
Pang, Y., Li, L., Ren, F., Lu, P., Wei, P., Cai, J., et al., 2010. Overexpression of the tonoplast aquaporin AtTIP5; 1 conferred tolerance to boron toxicity in Arabidopsis. J. Genet. Genomics 37 (6), 389–397.
Pérez-Castro, R., Kasai, K., Gainza-Cortés, F., Ruiz-Lara, S., Casaretto, J.A., Peña-Cortés, H., et al., 2012. VvBOR1, the grapevine ortholog of AtBOR1, encodes an efflux boron transporter that is differentially expressed throughout reproductive development of Vitis vinifera L. Plant. Cell Physiol. 53 (2), 485–494.
Princi, M.P., Lupini, A., Araniti, F., Longo, C., Mauceri, A., Sunseri, F., et al., 2016. Boron toxicity and tolerance in plants: recent advances and future perspectives. Plant Metal Interaction. Elsevier, Amsterdam, pp. 115–147.
Rehman, S., Park, T.I., Kim, Y.J., Seo, Y.W., Yun, S.J., 2006. Inverse relationship between boron toxicity tolerance and boron contents of barley seed and root. J. Plant. Nutr. 29 (10), 1779–1789.
Reid, R., 2014. Understanding the boron transport network in plants. Plant. Soil. 385 (1), 1–13.
Reid, R.J., Hayes, J.E., Post, A., Stangoulis, J.C.R., Graham, R.D., 2004. A critical analysis of the causes of boron toxicity in plants. Plant, Cell Environ. 27 (11), 1405–1414.
Roberts, D.M., Routray, P., 2017. The nodulin 26 intrinsic protein subfamily. Plant Aquaporins. Springer, pp. 267–296.

Sakamoto, T., Inui, Y.T., Uraguchi, S., Yoshizumi, T., Matsunaga, S., Mastui, M., et al., 2011. Condensin II alleviates DNA damage and is essential for tolerance of boron overload stress in Arabidopsis. Plant. Cell 23 (9), 3533–3546.

Schmid, M., Davison, T., Henz, S., Pape, U., Demar, M., Vingron, M., et al., 2005. A gene expression map of *Arabidopsis thaliana* development. Nat. Genet. 37, 501–506.

Schnurbusch, T., Langridge, P., Sutton, T., 2008. The Bo1-specific PCR marker AWW5L7 is predictive of boron tolerance status in a range of exotic durum and bread wheats. Genome 51 (12), 963–971.

Schnurbusch, T., Hayes, J., Hrmova, M., Baumann, U., Ramesh, S.A., Tyerman, S.D., et al., 2010. Boron toxicity tolerance in barley through reduced expression of the multifunctional aquaporin HvNIP2; 1. Plant. Physiol. 153 (4), 1706–1715.

Scorei, R., 2012. Is boron a prebiotic element? A mini-review of the essentiality of boron for the appearance of life on earth. Orig. Life Evolution Biospheres 42 (1), 3–17.

Shorrocks, V.M., 1982. Boron Deficiency Its Prevention and Cure. Borax Holdings.

Shorrocks, V.M., 1997. The occurrence and correction of boron deficiency. Plant. Soil. 193 (1), 121–148.

Spielberg, S.P., Butler, J.D., MacDermot, K., Schulman, J.D., 1979. Treatment of glutathione synthetase deficient fibroblasts by inhibiting γ-glutamyl transpeptidase activity with serine and borate. Biochem. Biophys. Res. Commun. 89 (2), 504–511.

Stangoulis, J.C., Reid, R.J., Brown, P.H., Graham, R.D., 2001. Kinetic analysis of boron transport in Chara. Planta 213 (1), 142–146.

Sutton, T., Baumann, U., Hayes, J., Collins, N.C., Shi, B.J., Schnurbusch, T., et al., 2007. Boron-toxicity tolerance in barley arising from efflux transporter amplification. Science 318 (5855), 1446–1449.

Takano, J., Yamagami, M., Noguchi, K., Hayashi, H., Fujiwara, T., 2001. Preferential translocation of boron to young leaves in *Arabidopsis thaliana* regulated by the BOR1 gene. Soil. Sci. Plant. Nutr. 47 (2), 345–357.

Takano, J., Noguchi, K., Yasumori, M., Kobayashi, M., Gajdos, Z., Miwa, K., et al., 2002. Arabidopsis boron transporter for xylem loading. Nature 420 (6913), 337–340.

Takano, J., Miwa, K., Yuan, L., von Wirén, N., Fujiwara, T., 2005. Endocytosis and degradation of BOR1, a boron transporter of *Arabidopsis thaliana*, regulated by boron availability. Proc. Natl Acad. Sci. 102 (34), 12276–12281.

Takano, J., Wada, M., Ludewig, U., Schaaf, G., Von Wirén, N., Fujiwara, T., 2006. The Arabidopsis major intrinsic protein NIP5; 1 is essential for efficient boron uptake and plant development under boron limitation. Plant. Cell 18 (6), 1498–1509.

Takano, J., Miwa, K., Fujiwara, T., 2008. Boron transport mechanisms: collaboration of channels and transporters. Trends Plant. Sci. 13 (8), 451–457.

Takano, J., Tanaka, M., Toyoda, A., Miwa, K., Kasai, K., Fuji, K., et al., 2010. Polar localization and degradation of Arabidopsis boron transporters through distinct trafficking pathways. Proc. Natl Acad. Sci. 107 (11), 5220–5225.

Tanaka, M., Wallace, I.S., Takano, J., Roberts, D.M., Fujiwara, T., 2008. NIP6; 1 is a boric acid channel for preferential transport of boron to growing shoot tissues in Arabidopsis. Plant. Cell 20 (10), 2860–2875.

Tanaka, M., Takano, J., Chiba, Y., Lombardo, F., Ogasawara, Y., Onouchi, H., et al., 2011. Boron-dependent degradation of NIP5; 1 mRNA for acclimation to excess boron conditions in Arabidopsis. Plant. Cell 23 (9), 3547–3559.

Tanaka, N., Uraguchi, S., Saito, A., Kajikawa, M., Kasai, K., Sato, Y., et al., 2013. Roles of pollen-specific boron efflux transporter, OsBOR4, in the rice fertilization process. Plant. Cell Physiol. 54 (12), 2011–2019.

Tanaka, M., Sotta, N., Yamazumi, Y., Yamashita, Y., Miwa, K., Murota, K., et al., 2016. The minimum open reading frame, AUG-stop, induces boron-dependent ribosome stalling and mRNA degradation. Plant. Cell 28 (11), 2830–2849.

Thurtle-Schmidt, B.H., Stroud, R.M., 2016. Structure of Bor1 supports an elevator transport mechanism for SLC4 anion exchangers. Proc. Natl Acad. Sci. 113 (38), 10542–10546.
Tombuloglu, G., Tombuloglu, H., Sakcali, M.S., Unver, T., 2015. High-throughput transcriptome analysis of barley (*Hordeum vulgare*) exposed to excessive boron. Gene 557 (1), 71–81.
Torun, A.A., Yazici, A., Erdem, H., Çakmak, İ., 2006. Genotypic variation in tolerance to boron toxicity in 70 durum wheat genotypes. Turkish J. Agricult. Forestry 30 (1), 49–58.
Türker, O.C., Vymazal, J., Türe, C., 2014. Constructed wetlands for boron removal: a review. Ecol. Eng. 64, 350–359.
Voxeur, A., Fry, S.C., 2014. Glycosylinositol phosphorylceramides from Rosa cell cultures are boron-bridged in the plasma membrane and form complexes with rhamnogalacturonan II. Plant. J. 79 (1), 139–149.
Wakuta, S., Mineta, K., Amano, T., Toyoda, A., Fujiwara, T., Naito, S., et al., 2015. Evolutionary divergence of plant borate exporters and critical amino acid residues for the polar localization and boron-dependent vacuolar sorting of AtBOR1. Plant. Cell Physiol. 56 (5), 852–862.
Wang, S., Yoshinari, A., Shimada, T., Hara-Nishimura, I., Mitani-Ueno, N., Ma, J.F., et al., 2017. Polar localization of the NIP5; 1 boric acid channel is maintained by endocytosis and facilitates boron transport in Arabidopsis roots. Plant. Cell 29 (4), 824–842.
Warington, K., 1923. The effect of boric acid and borax on the broad bean and certain other plants. Ann. Botany 37 (148), 629–672.
Winter, D., Vinegar, B., Nahal, H., Ammar, R., Wilson, G.V., Provart, N.J., 2007. An "electronic fluorescent pictograph" browser for exploring and analyzing large-scale biological data sets. PLOS Genet. 8, e718.
Wolkenstein, K., Gross, J.H., Falk, H., 2010. Boron-containing organic pigments from a Jurassic red alga. Proc. Natl Acad. Sci. 107 (45), 19374–19378.
Wolska, J., Bryjak, M., 2013. Methods for boron removal from aqueous solutions—a review. Desalination 310, 18–24.
World Health Organization (WHO), 1996. Trace Elements in Human Nutrition and Health. World Health Organization.
Yamanouchi, M., 1980. The effects of phosphorus, potassium, calcium, magnesium and iron on the absorption and translocation of boron in several crops grown at a high concentration of boron. J. Sci. Soil. Manure (Jpn.) 51 (2), 126–130.
Yau, S.-K., Erskine, W., 2000. Diversity of boron-toxicity tolerance in lentil growth and yield. Genet. Resour. Crop. Evolution 47 (1), 55–62.
Yıldırım, K., 2017. Transcriptomic and hormonal control of boron uptake, accumulation and toxicity tolerance in poplar. Environ. Exp. Botany 141, 60–73.
York, W.S., Darvill, A.G., McNeil, M., Albersheim, P., 1985. 3-Deoxy-D-manno-2-octulosonic acid (KDO) is a component of rhamnogalacturonan II, a pectic polysaccharide in the primary cell walls of plants. Carbohydr. Res. 138 (1), 109–126.
Yoshinari, A., Fujimoto, M., Ueda, T., Inada, N., Naito, S., Takano, J., 2016. DRP1-dependent endocytosis is essential for polar localization and boron-induced degradation of the borate transporter BOR1 in *Arabidopsis thaliana*. Plant. Cell Physiol. 57 (9), 1985–2000.
Zhang, Q., Chen, H., He, M., Zhao, Z., Cai, H., Ding, G., 2017. The boron transporter BnaC4.BOR1;1c is critical for inflorescence development and fertility under boron limitation in *Brassica napus*. Plant, Cell Environ. 40, 1819–1833.

CHAPTER 8

Impact of boron and its toxicity on photosynthetic capacity of plants

Chrysovalantou Antonopoulou and Christos Chatzissavvidis
Department of Agricultural Development, School of Agricultural and Forestry Sciences, Democritus University of Thrace, Orestiada, Greece

8.1 Introduction

Boron (B) is an essential micronutrient for agricultural crops and is required for normal growth and development (Warington, 1923; Gupta, 1993). It is also unique among micronutrients due to the narrow threshold between deficiency and toxicity, while both situations produce symptoms that affect crop yield and production (Camacho-Cristóbal et al., 2008; Yau and Ryan, 2008). The quantities required are small and they vary within certain limits, depending on the plant species, and within species, on the cultivar or variety. Adequate tissue B concentrations for the optimal plant growth of monocots can range from 1 to 6 mg/kg, whereas dicots can range from 20 to 70 mg/kg (Princi et al., 2016; Shah et al., 2017); thus, the available soil boron which is deficient for one crop may exhibit toxic effects on another.

Toxicity in crops can be defined as a situation in which a potentially harmful substance accumulates in the plant tissue to a level affecting its optimal growth and development. Certain essential plant nutrients, as well as non-essential elements, can be absorbed by plants in sufficient quantities to be toxic. Among the essential micronutrients that are more frequent to produce toxicity in plants is B (Bennett, 1993).

Boron toxicity limits crop yield and quality in several agricultural areas worldwide, including Egypt, Turkey, Iraq, Jordan, Syria, Morocco, Chile, USA, Spain, and South Australia (Miwa et al., 2007; Landi et al., 2012; Ayvaz et al., 2013). It is noticeable that nowadays B toxicity has become an important problem, especially in the soils of arid and semiarid regions throughout the world. Under Mediterranean climatic conditions, where

Boron in Plants and Agriculture
DOI: https://doi.org/10.1016/B978-0-323-90857-3.00006-0

rainfall is not sufficient to leach the element from the root zone, B tends to accumulate and, in many cases, to exceed toxic levels (Gupta et al., 1985; Landi et al., 2012; Camacho-Cristóbal et al., 2018). High concentrations of B may occur under the following conditions: (1) in B enriched-soils originating from marine rock materials, (2) in soils overfertilized with minerals high in B, (3) in soils receiving fossil combustion residues, (4) in soils being used as disposal sites of B-containing waste materials, and (5) in soils being irrigated with water high in B content (Gupta et al., 1985; Leyshon and James, 1993; Nable et al., 1997; Chatzissavvidis et al., 2004; Papadakis et al., 2004a, 2004b). Boron frequently exceeds the plant's requirement in arid and semiarid environments due to very poor water drainage and in agricultural soils close to coastal areas (e.g., in Mediterranean countries) due to the intrusion of B-rich seawater in the fresh aquifer, which can contaminate groundwater and increase B concentrations in irrigated water up to 8 mg/L (Kabay et al., 2010; Reid, 2010; Landi et al., 2019). A concentration of B in the irrigation water that exceeds 1 mg/L can negatively affect the yield of most traditional Mediterranean crops such as orange trees (*Citrus sinensis*) and avocados (*Persea americana*) (Gimeno et al., 2012). Referring to Keren and Bingham (1958), the maximum permissible concentrations of B in irrigation water range from 0.3 to 1.0 mg/L for sensitive plants (avocado < apricot < peach < cherry < fig < apple < pear < bean < walnut), and 1.0–2.0 mg/L for semitolerant plants (sweet potato < pumpkin < oat < maize < wheat < barley < olive < tomato < cotton < lettuce < onion < alfalfa < sugar beet < date palm < asparagus). In recent years, B toxicity has attracted increasing interest because of the high demand for desalinated water, in which the B concentration may be too high for healthy irrigation (Parks and Edwards, 2005). B toxicity is more difficult to manage than its deficiency (Takano et al., 2008).

Boron availability in soil is affected by many factors, including soil texture, nature of clay minerals, pH, liming, organic matter, interrelationships with other elements, environmental conditions and management practices (Moraghan and Mascagni, 1991).

Boron is the only element that is normally present in soil solution as a nonionized molecule over the pH range suitable for plant growth. This element is found in soil mainly as boric acid, $B(OH)_3$, and partially as tetra hydroxyl borate anion, $B(OH)_4^-$, which are unevenly distributed in the water and in the organic and mineral deposits (Princi et al., 2016). Boric acid is a weak acid at cytoplasmic pH (about 7.0–7.5) and, under physiological conditions, it can freely pass lipid bilayers, in contrast to other ionic nutrients (Dannel et al., 1998; Tanaka and Fujiwara, 2008; Miwa and Fujiwara, 2010; Landi et al., 2019). However, three main ways through which B enters and exits the cells have been described, depending on its availability: (1) passive, two-way diffusion across the lipid bilayer, (2) two-way passive diffusion through selective or nonselective protein channels, and (3) high-affinity active transport of B (Yoshinari and Takano, 2017). However, active B uptake was observed to occur under suboptimal B availability or even in the condition of B deficiency (Stangoulis et al., 2001), whereas under conditions of adequate or excessive B supply the hypothesis that B is principally passively absorbed by roots is the most widely accepted (Brown et al., 2002; Tanaka and Fujiwara, 2008).

Regarding mobility, B can move within the plants via the apoplastic and symplastic routes through mass flow and diffusion (García-Sánchez et al., 2020). Once absorbed by the roots, B is primarily translocated to mature leaves and leaf tips through nonliving xylem cells (apoplast) driven by transpiration, since mature leaves represent the sites with the highest transpiration rate (Nable et al., 1997; Wimmer et al., 2003; Chatzissavvidis and Therios, 2011). B mobility in the phloem seems to depend on the presence of molecules that can form complexes with B and varies considerably among species. In some plant species, B is phloem-mobile after forming complexes with polyols (e.g., sorbitol, mannitol, dulcitol), which are the primary products of photosynthesis (Brown and Hu, 1996; Brown and Shelp, 1997). Phloem translocation does not follow the transpiration stream and it supplies the major proportion of nutrient requirements to actively growing areas such as young leaves and fruits, organs that do not readily transpire (Brown and Shelp, 1997). Such species belong to *Prunus*, *Malus*, and *Pyrus* genera (prune, pear, apple, cherry, almond, plum, peach, apricot), as well as onion, celery, carrot, olive, bean, pea, cauliflower, cabbage, asparagus, and coffee (Brdar-Jokanović, 2020). On the other hand, B has restricted phloem mobility in species with sucrose as a primary photosynthetic product. These species (e.g., wheat, barley, walnut) represent the majority and its movement mainly occurs through the xylem along the transpiration stream (Nable et al., 1997; Wimmer et al., 2003; Brdar-Jokanović, 2020). Therefore, comparatively higher B concentration in older leaves implies restricted phloem mobility, while a higher concentration in young leaves and fruits indicates significant phloem mobility (Brown and Shelp, 1997; Stangoulis et al., 2001; Stangoulis et al., 2010).

Boron is involved in several metabolic processes (Goldbach et al., 2001; O'Neill et al., 2004). $B(OH)_3$ and $B(OH)_4^-$ easily form complexes with a great number of sugars, proteins and lipids that contain *cis*-hydroxyl groups (Brdar-Jokanović, 2020). The primary function of the element is in the cell wall and plasma membrane synthesis and in maintaining their structure and integrity (Broadley et al., 2012; Brdar-Jokanović, 2020). In the cell wall, it forms rhamnogalacturonan-II-B (RG-II-B) complexes that stabilize the pectin network and regulates cell wall porosity (Fleischer et al., 1999) and tensile strength (Ryden et al., 2003). A recent study reported that glycosylinositol phosphorylceramides (GIPCs), the major components of lipid rafts, participate in the formation of the GIPC–B–RG-II complex, suggesting the structural role of B in membranes (Voxeur and Fry, 2014). The amount of B located in the wall depends on the plant species, organ, and B supply (Goldbach et al., 2000). Many other roles have been proposed for B in plants, such as its influence in reproductive growth and development, stimulation of reproductive tissues, improvement of seed quality, production and quality of the harvest, and its influence on the biosynthesis of some metabolic compounds, that is, antioxidants and polyphenols (Brown et al., 2002; Goldbach and Wimmer, 2007; Camacho-Cristóbal et al., 2008, 2018; Chatzissavvidis and Therios, 2011; Gimeno et al., 2012; Princi et al., 2016). Additionally, involvement in nucleic acid synthesis, phenolic metabolism, carbohydrate biosynthesis and translocation, indole-3-acetic acid oxidase, as well as pollen-tube growth and root elongation, have been recently reported, stressing the B critical role in plant development (González-Fontes et al., 2008; Shireen et al., 2018).

Excess B in soil results in growth impairment, exerting different morpho-anatomical effects to plants, such as progressive necrosis in leaves, stems and malformations of fruits (Reid et al., 2004). Boron does not accumulate uniformly in leaves but typically concentrates in leaf tips of monocotyledons and leaf margins of dicotyledons, where B toxicity symptoms first appear (Gupta, 1993). The effects of B toxicity are also dependent on the ability of a plant species to retranslocate B within the phloem. In species where B is relatively immobile, the symptoms of toxicity are firstly detected in older tissues, whereas in other species, where B is remobilized in the phloem sap, visible symptoms appear principally in actively growing tissues (Brown and Hu, 1996). The most common visible symptom of B toxicity is the presence of burns, which appear as chlorotic and/or necrotic patches, often found on the margins and tips of the mature leaves (García-Sánchez et al., 2020). In species where phloem retranslocation of B occurs, dead apical sprouts, abscission of young shoots and lesions in the shape of dry brown patches next to stems and petioles are observed (García-Sánchez et al., 2020). At stem level, B excess can promote the development of cork and collenchyma cells with increased cell wall thickness in loquat (Papadakis et al., 2018) and thickening in the cell wall of phloem cells in citrus (Huang et al., 2014), that is, storing excessive B in tissues where it could be less harmful. B toxicity has been reported to affect negatively root growth in different crops, such as wheat, barley, maize, apple, orange citrange plants (Choi et al., 2007; Aquea et al., 2012; Esim and Atici, 2013; Oikonomou et al., 2019), inducing alteration of root meristems (Choi et al., 2007), thickening of hypodermis and deposition of suberin in cortical cells (Ghanati et al., 2002), and root lignification (Oikonomou et al., 2019), mainly attributable to a reduction of cell wall extensibility and which can help the plant to reduce the intake of B and limit the effect of toxicity to root system. Many authors mention the absence of visible symptoms in root under B excess, but overall, its length, surface area, and volume decreased significantly (Choi et al., 2007; Reid, 2007; Sheng et al., 2010; Shah et al., 2017). Inhibition of root growth consequently reduces water uptake and aggravates water stress in dry areas (Aquea et al., 2012). High levels of B can even be lethal, especially for sensitive plant species (Miwa et al., 2007).

Toxic B content in plants has different effects on plant physiological and biochemical processes including (1) decreases in the photosynthetic rate and chlorophyll (Chl) contents, (2) negative effects on the efficiency of photosystem II, (3) increase in lipid peroxidation; (4) alterations of the enzymes found in antioxidant pathways, (5) increase in membrane permeability, (6) reduction of the extrusion of protons in the roots, (7) deposition of suberin and lignin, (8) nutritional imbalances, and (9) decrease in stomatal conductance (Nable et al., 1997, Reid et al., 2004; Landi et al., 2013; Simón et al., 2013). Although the physiological basis for B toxicity is not clear, the changes observed in several plant parameters are the results of a cascade of events promoted by three main causes: (1) impairment of cell division and development by binding to ribose, both as the free sugar and as a constituent of RNA, (2) interference with primary metabolism by binding with ribose as a part of ATP, NAD(H) or NADP(H) and nucleotides, and (3) alteration of cell wall structure (Reid et al., 2004).

Conversely, from morpho-anatomical changes, the deleterious effect of B excess to several biochemical and physiological processes which involved the leaves, first among equal, the photosynthetic processes, are more consistently observed.

8.2 Boron toxicity and photosynthesis

The involvement of B in major physiological processes has increasingly been studied in recent years. A lot of research focuses on the influence of B on photosynthesis and several works have shown a positive B effect on photosynthesis of cultivated plants (Zhao and Oosterhuis, 2002; Tavallali, 2017).

Photosynthesis is one of the main metabolic processes impaired by B excess (Sotiropoulos et al., 2002; Papadakis et al., 2004a, b; Ardic et al., 2009; Han et al., 2009; Herrera-Rodríguez et al., 2010; Landi et al., 2013) due to both stomatal or/and nonstomatal limitations (Shah et al., 2017). Under B excess, the reduction of Pn in zucchini (*Cucurbita pepo*) was attributable only to stomatal limitations whereas in cucumber (*C. sativus*) both to stomatal and nonstomatal limitations (Landi et al., 2013). B toxicity can reduce the net photosynthetic rate and downregulate the activities of enzymes involved in photosynthesis (Landi et al., 2012), thus interfering with the growth and development of plants. Pn reflects the ability of leaves to fix CO_2 and accumulate organic matter (Velička et al., 2007) and is positively correlated with crop yield (Lambers, 1987).

Biochemical limitations of excess B on photosynthesis mainly includes the decline of electron transport rate, reduced CO_2 assimilation, impairment of photosystem II (PSII) efficiency and disruption of an antioxidant system (Papadakis et al., 2004a, 2004b; Landi et al., 2013, 2014; Kaya et al., 2018), consequently resulting in visual symptoms such as chlorosis and necrosis of shoots (Han et al., 2009; Reid, 2013). The decline of the carboxylation efficiency is dependent on the effect of B binding to ribulose (Reid et al., 2004), and on the accumulation of soluble sugars (glucose and fructose) which can be responsible for feedback regulation of photosynthesis (Lobo et al., 2015).

Chloroplasts are a primary target for the oxidative load triggered by many environmental stresses (Foyer and Noctor, 2005) and it has been reported that B toxicity induces alterations in photosynthesis (Han et al., 2009; Soylemezoglu et al., 2009; Ruuhola et al., 2011; Landi et al., 2013). Boron excess can cause a decrease in chlorophyll content and structural damage of chloroplasts, which lead to a decrease in the CO_2-assimilatory capacity of the mesophyll cells (Papadakis et al., 2004a, b), ultimately affecting plant growth. In some cases, B excess also resulted in alterations of photosynthetic pigment content and its relative proportion, for example, chlorophyll a, b content and a/b ratio (Huang et al., 2014; Kayihan et al., 2017) and carotenoid (Shah et al., 2017; Sarafi et al., 2018). Photosynthetic pigments, and mainly chlorophylls, are involved in the absorption and transfer of light energy in the process of photosynthesis. When plants were exposed to excess B, lower chlorophyll content in mature leaves and decreased photosynthetic rate have been observed in many studies (Camacho-Cristóbal et al., 2008). Low chlorophyll production in plants might be due to B-induced peroxidation of chloroplast membranes and inhibition of chlorophyll synthesis (Landi et al., 2013; de Souza et al., 2019). Chlorophyll (especially total Chl and Chl b) of apple leaves tended to decrease with added B, probably due to thinning of cell walls (Mouhtaridou et al., 2004; Paparnakis et al., 2013). The increased Chl b/a ratio caused by the added B may be attributed to the decreased thickness of cell walls and, consequently, to enhanced penetration of a wider spectrum of light (Paparnakis et al., 2013).

Boron's ability to form complexes to molecules that are involved in different cellular processes is considered the main basis for the understanding of the role of B in biological systems, independently of its function within them. For instance, the capacity to bind ATP and NADPH (Cervilla et al., 2009) and limit the availability of energy and reduce power, respectively, which are necessary for carbohydrate synthesis in the light-independent reactions of the Calvin–Benson cycle (Cervilla et al., 2009). Therefore, changes in sugar content and partitioning (Roessner et al., 2006; Papadakis et al., 2018), as well as in carbon skeleton devoted to amino acids, which are strictly interlinked (Guo et al., 2014) can be observed as a response to B excess (Sang et al., 2015; Ayvaz et al., 2016).

Due to the B-induced structural damages to chloroplasts and the consequent reduction of CO_2 photoassimilation, the capacity to process incident light energy by photosynthesis is compromised, and excess light energy may exert further stresses by enhancing the production of ROS (Cervilla et al., 2007; Landi et al., 2013). The oxidative stress increases the damage to lipids, proteins and nucleic acids, impairing the cell functionality and duplications, and leading to cell death (Mullineaux and Baker, 2010). This, in addition to a reduction in the photosynthetic area caused by necrosis of leaf margins, can greatly reduce the net carbon assimilation rate (Lovatts and Bates, 1984; Reid et al., 2004; Guidi et al., 2011). Reid et al. (2004) proposed that photooxidative stress is one of the main causes of the inhibition of plant growth under B toxicity, rather than the effect of B on energy supply or proteins biosynthesis. Purple-leafed sweet basil cultivars were more tolerant to B toxicity than other green-leafed cultivars, indicating that anthocyanins serve to mitigate effects of photoinhibitory stress in leaves once the photosynthetic machinery is compromised by B toxicity (Landi et al., 2013). Green genotypes of sweet basil were indeed more prone to photo-inhibition and B-triggered oxidative stress. Anthocyanins might chelate to supernumerary B ions and sequester them in the cell vacuole, and these ions are likely bound with cis−diols and cyanidin derivatives (Landi et al., 2014). Although the chelation of anthocyanins with some metals has been observed, however, there is no experimental evidence for the chelation of anthocyanins with B. However, foliar application of anthocyanins has shown mitigation effects on sweet basil grown in excess B, suggesting that anthocyanins might be a candidate for B toxicity alleviation (Landi et al., 2014). Therefore, the enhancement of the antioxidant apparatus is another typical response of plants when experiencing conditions of B excess (Karabal et al., 2003; Gunes et al., 2006).

In reaction to B excess, apple plants accumulated soluble sugars (among them glucose and fructose) and reduced the level of sucrose, a reaction that can be likely attributed to both the reduction of the photosynthetic level of leaves and the impairment of phloem functionality (Han et al., 2009). The structural damage of the phloem system and/or a reduced demand of sinks may cause an accumulation of sugars in the leaves leading to downregulation of photosynthesis (Tavallali, 2017). This would decrease the demand for CO_2 uptake, which may, in turn, reduce the stomatal aperture. Glucose is a signaling molecule in plants that experience stressful conditions, is a fundamental substrate for cellular respiration and several genes related to photosynthesis are downregulated by glucose-induced phosphorylation (Hua et al., 2020). Moreover, it has been reported that soluble sugars accumulation induced a negative regulation of genes involved in photosynthesis, including the expression of genes involved in the Calvin cycle (Pego et al., 2000). Guo et al. (2014) also found that the effect of B toxicity in citrus not only caused the decrease of CO_2 assimilation but also the expression of genes related to photosynthesis and carbohydrate metabolism (e.g., Rubisco and ADP-glucose pyrophosphorylase), which may lead to the production of ROS, which oxidizes organic molecules as chlorophyll and lipid and can induce cell death.

Some authors have reported that B excess caused a reduction of the transpiration rate, in combination with severe stomata limitations (Papadakis et al., 2018), and increment of the abscisic acid metabolism (Macho-Rivero et al., 2017, 2018), the major player in stomata closure. However, other authors proposed that there is no evidence to support the hypothesis of the generation of B toxicity effects in leaves due to osmotic stress induced by the accumulation of borate or boric acid (Reid et al., 2004). The toxicity to mature tissues is rather due to the accumulated retardation of many cellular processes, enhanced by photooxidative stress (Reid et al., 2004).

Thus, because the capacity to process incident light energy is compromised, plants growing on B-rich soil are more likely to produce a surplus of ROS in the chloroplast, and consequently incur oxidative stress (Cervilla et al., 2007; Landi et al., 2013). Excess B can enhance overaccumulation of ROS such as superoxide radical ($O_2^{\bullet -}$), hydroxyl radical ($OH^{\bullet}$), and hydrogen peroxide (H_2O_2), which may cause oxidative damage to proteins, lipids, pigments, and nucleic acids (Hua et al., 2020). To reduce the harmful effects induced by ROS, plant cells may trigger ROS scavenging antioxidant mechanisms, such as catalase, peroxidase, glutathione reductase and superoxide dismutase (Kayihan et al., 2017). There are several reports of increasing activity of antioxidative enzymes under high B supply (Cervilla et al., 2007; Eraslan et al., 2008; Ardic et al., 2009; Wang et al., 2011). High irradiance appears to increase the harmful effects of B toxicity, probably because elevated B concentrations may impair plant mechanisms to cope with photooxidation stress (Reid et al., 2004). B toxicity affected photosynthesis and the antioxidant machinery of both cucumber and zucchini (Landi et al., 2013). Both species presented an alteration in photosynthetic pigments and a downregulation of PSII photochemical efficiency coupled with an enhancement of the non-photochemical mechanisms involved in the dissipation of the excess excitation energy. These mechanisms are implicated in the attempt to reduce ROS production that occurs when the photosystem electron transport rate is unable to sustain the electrons turn-over based on the intercepted light, which becomes too high. Fv/Fm represents the activity of plant leaves PSII reaction center (Zhang et al., 2016). The reduction of this factor was found in plant species under B toxicity, indicating that the crop is under stress. Reduction of Fv/Fm may result from the accumulation of non-functional PSII reaction centers and partial photoinactivation of PSII (Björkman and Demmig, 1987; Niyogi, 1999). This causes a consequently decrease in actual PSII efficiency and photochemical quenching coefficient (qP) (Schreiber and Bilger, 1993). This may lead to a reorganization of the light-harvesting complexes of PSII into aggregates of polypeptides containing chlorophylls and xanthophylls to maximize thermal energy dissipation as well (Adams et al., 2004). Mechanisms devoted to dissipating the excess of light absorbed by photosystems are necessary under stress conditions to photoprotect the chloroplast against the oxidative load. Boron overdose can change the photochemistry and energy distribution of photosystem II in tomatoes (Guidi et al., 2011). Moreover, an excess of B in the leaf tissue damaged the photochemical machinery of lemon leaves (Guerfel et al., 2009). Fv/Fm was decreased indicating that the loss of the quantum efficiency of PSII was due to damage to the light-harvesting complex. Likewise, in "Navelina" orange, as B concentration in the nutrient solution increased, both chlorophyll fluorescence parameters (Fv/Fm and Fv/F_0) and leaf chlorophyll concentration decreased (Papadakis et al., 2004b). What is more, B toxicity impaired the whole photosynthetic electron transport from PSII donor side up to the reduction of end acceptors of PSI in pomelo leaves (Han et al., 2009).

High B concentration in leaves caused a decrease in photosynthesis of olives combined with a reduction of N concentration in leaves, an element which is a constituent of the chlorophyll molecule, and it is utilized in the photosynthetic process (Chatzissavvidis and Therios, 2010). Boron toxicity may also indirectly affect photosynthesis by decreasing the photosynthetic leaf area and by altering leaf constituents (Lee and Aronoff, 1966). Papadakis et al. (2004a) reported that leaves of orange plants treated with B excess presented a decrease in their thickness and disorganization of mesophyll cells. Similarly, in peach (*Prunus persica* L. Batsch) leaves, shrinkage and deorganization of the cells of the spongy parenchyma were observed under B excess (Kamali and Childers, 1966). Therefore, B toxicity can induce an alteration of leaf structure both in the species where B is phloemimmobile (orange) and in those where B mobility is high as peach (Brown and Shelp, 1997). These anatomical alterations of leaves reduced the electron transport rate and light utilization (Landi et al., 2012).

Under high B conditions, the application of some elements, such as Si, Ca, and Zn are against oxidative stress in plant cells, to increase chlorophyll content leaf stomatal conductance and to facilitate photosynthesis of plants. The use of some elements has been found to grow under B toxicity. For example, Ca has been reported to increase chlorophyll and net photosynthesis in barley (El-Feky et al., 2014) and radish (Siddiqui et al., 2013). Many studies of crops under excess B have shown that Si was able to increase chlorophyll and net photosynthetic rates via increasing light interception due to more erectness in leaves and higher ribulose 1,5-bisphosphate carboxylase (Rubisco) activities (Eraslan et al., 2008; Farooq et al., 2015; de Souza et al., 2019). Chlorophyll content, stomatal conductance and CO_2 assimilation were observed to be increased by Zn addition in pistachio under high levels of B (Tavallali, 2017). Moreover, Zn enhanced the activities of carbonic anhydrase, a Zn-containing enzyme that catalyzes the conversion of CO_2 to HCO_3 and facilitates the supply of CO_2 from the stomatal cavity to the site of CO_2 fixation (Sasaki et al., 1998), which may, in turn, increase the stomatal aperture. Additionally, some plant growth regulators (PGRs) (e.g., melatonin, salicylic acid-SA, jasmonic acid-JA and methyl jasmonate-MeJA), vitamins (e.g., ascorbic acid and thiamin), and biostimulators (e.g., chitosan, 24-epibrassinolide, 28-homobrassinolide and aminolevulinic acid) are mentioned to increase chlorophyll contents in plants grown in high levels of B (Hua et al., 2020). JA and its methyl ester are important plant hormones, that have been reported to be able to improve plant tolerance to high levels of B by regulating antioxidant enzyme activities (Aftab et al., 2011; Sarabandi et al., 2019; Zhao et al., 2019). The reduction of photosynthesis in pepper plants was restored, the reduction of dry weight was prevented, while the B leaf and fruit accumulation was moderated, with the application of both 100 mM resveratrol and 1 mM melatonin (Sarafi et al., 2017). Under excess B, the application of the above compounds might increase chlorophyll in the plant through two mechanisms: (1) by increasing the biosynthesis of chlorophyll, including its precursors such as thiamin and aminolevulinic acid (Genisel et al., 2017; El-Shazoly et al., 2019) and (2) by facilitating the protection against chlorophyll degradation (e.g., melatonin and MeJA) (Aftab et al., 2011; Moussa and Algamal, 2017).

8.3 Conclusion

Boron is an essential micronutrient for higher crops and is involved in several metabolic processes of plants such as cell division, cell wall and membrane synthesis, protein and lipid synthesis, sugars translocation, root elongation, and development of reproductive organs. This element has an extremely narrow range between deficiency and toxicity. Boron accumulation is particularly prominent in soils from arid and semiarid environments. The effects of B toxicity are dependent on the ability of a plant species to retranslocate B within the phloem. In species where B is relatively immobile and is primarily translocated through nonliving xylem cells driven by transpiration, the symptoms of toxicity are firstly detected in older leaves. On the contrary, in other species, where B is remobilized in the phloem sap, visible symptoms appear principally in actively growing young tissues.

The presence of excess B in plant tissues results in alterations of physiological and biochemical functions. Photosynthesis is one of the main metabolic processes impaired, including reduction of photosynthetic rate, electron transport rate, CO_2 assimilation, stomatal conductance, alterations of photosynthetic pigment content and its relative proportion, impairment of photosystem II efficiency, and the production of ROS inducing oxidative stress. The consequences of these events are the visible symptoms of the damage, which are typically located at the leaf margin where B is mainly accumulated, and the decline of plant growth and development.

An approach to alleviating B toxicity in plants is to increase their physiological tolerance, including the enhancement of photosynthesis and the prevention of oxidative stress, by applying nutrient elements, plant growth regulators, vitamins and biostimulators.

Studies regarding B utilization efficiency and toxicity tolerance among plant species and varieties, combined with research about the impact of excess B on the physiology and genetics of plants, can assist the development of efficient and long-term sustainable practices facing the problem of excess B supply.

References

Aftab, T., Khan, M.M.A., Idrees, M., Naeem, M., Hashmi, N., 2011. Methyl jasmonate counteracts boron toxicity by preventing oxidative stress and regulating antioxidant enzyme activities and artemisinin biosynthesis in *Artemisia annua* L. Protoplasma. 248, 601–612.

Aquea, F., Federici, F., Moscoso, C., Vega, A., Jullian, P., Haseloff, J.I.M., et al., 2012. A molecular framework for the inhibition of *Arabidopsis* root growth in response to boron toxicity. Plant. Cell Env. 35, 719–734.

Ardic, M., Sekmen, A.H., Tokur, S., Ozdemir, F., Turkan, I., 2009. Antioxidant responses of chickpea plants subjected to boron toxicity. Plant. Biol. 11, 328–338.

Ayvaz, M., Avci, M.K., Yamaner, C., Koyuncu, M., Guven, A., Fagerstedt, K., 2013. Does excess boron affect malondialdehyde levels of potato cultivars? Eurasia J. Biosci. 7, 47–53.

Ayvaz, M., Guven, A., Blokhina, O., Fagerstedt, K.V., 2016. Boron stress, oxidative damage and antioxidant protection in potato cultivars (*Solanum tuberosum* L.). Acta Agric. Scand, Sect. B-Soil Plant. Sci. 66, 302–316.

Bennett, W.F., 1993. Plant nutrient utilization and diagnostic plant symptoms. Nutr. Deficiencies Toxicities Crop. Plants 1.
Björkman, O., Demmig, B., 1987. Photon yield of O_2 evolution and chlorophyll fluorescence characteristics at 77 K among vascular plants of diverse origins. Planta 170, 489–504.
Brdar-Jokanović, M., 2020. Boron toxicity toxicity and deficiency deficiency in agricultural agricultural plants. Int. J. Mol. Sci. 21, 1424.
Broadley, M., Brown, P., Cakmak, I., Rengel, Z., Zhao, F., 2012. Function of nutrients: micronutrients. Marschner' s Mineral Nutrition of Higher Plants 191–248. Academic Press.
Brown, P.H., Bellaloui, N., Wimmer, M.A., Bassil, E.S., Ruiz, J., Hu, H., et al., 2002. Boron in plant biology. Plant. Biol. 4, 205–223.
Brown, P.H., Hu, H., 1996. Phloem mobility of boron is species dependent: evidence for phloem in sorbitol-rich species. Ann. Bot. 77, 497–505.
Brown, P.H., Shelp, B.J., 1997. Boron mobility in plants. Plant. Soil. 193, 85–101.
Camacho-Cristóbal, J.J., Navarro-Gochicoa, M.T., Rexach, J., González-Fontes, A., Herrera-Rodríguez, M.B., 2018. Plant response to boron deficiency and boron use efficiency in crop plants. Plant Micronutritent Use Efficiency 109–121. Academic Press.
Camacho-Cristóbal, J.J., Rexach, J., González-Fontes, A., 2008. Boron in plants: deficiency and toxicity. J. Integ. Plant. Biol. 50, 1247–1255.
Cervilla, L.M., Blasco, B., Ríos, J.J., Romero, L., Ruiz, J.M., 2007. Oxidative stress and antioxidants in tomato (*Solanum lycopersicum*) plants subjected to boron toxicity. Ann. Bot. 100, 747–756.
Cervilla, L.M., Blasco, B., Ríos, J.J., Rosales, M.A., Rubio-Wilhelmi, M.M., Sánchez-Rodríguez, E., et al., 2009. Response of nitrogen metabolism to boron toxicity in tomato plants. Plant. Biol. 11, 671–677.
Chatzissavvidis, C., Therios, I., 2010. Response of four olive (*Olea europaea* L) cultivars to six B concentrations: growth performance, nutrient status and gas exchange parameters. Sci. Hort. 127, 29–38.
Chatzissavvidis, C.A., Therios, I., Antonopoulou, C., 2004. Seasonal variation of nutrient concentration in two olive (*Olea europaea* L) cultivars irrigated with high boron water. J. Hortic. Sci. Biotechnol. 79, 683–688.
Chatzissavvidis, C., Therios, I., 2011. Boron in higher plants. In: Perkins, G.L. (Ed.), Boron: Compounds, Production and Application. Nova Science Publishers Inc, pp. 147–176.
Choi, E.-Y., Kolesik, P., McNeill, A., Collins, H., Zhang, Q., Huynh, B., et al., 2007. The mechanism of boron tolerance for maintenance of root growth in barley (*Hordeum vulgare* L.). Plant. Cell Env. 30, 984–993.
Dannel, F., Pfeffer, H., Römheld, V., 1998. Compartmentation of boron in roots and leaves of sunflower as affected by boron supply. J. Plant. Physiol. 153, 615–622.
de Souza Jr., P., de Mello Prado, R., dos Santos Sarah, M.M., Felisberto, G., 2019. Silicon mitigates boron deficiency and toxicity in cotton cultivated in nutrient solution. J. Plant. Nutr. Soil. Sci. 182, 805–814.
El-Feky, S.S., El-Shintinawy, F.A., Shaker, E.M., 2014. Role of $CaCl_2$ and salicylic acid on metabolic activities and productivity of boron stressed barley (*Hordeum vulgare* L.). Int. J. Curr. Microbiol. Appl. Sci. 3, 368–380.
El-Shazoly, R.M., Metwally, A.A., Hamada, A.M., 2019. Salicylic acid or thiamin increases tolerance to boron toxicity stress in wheat. J. Plant. Nutr. 42, 702–722.
Eraslan, F., Inal, A., Pilbeam, D.J., Gunes, A., 2008. Interactive effects of salicylic acid and silicon on oxidative damage and antioxidant activity in spinach (*Spinacia oleracea* L. cv. Matador) grown under boron toxicity and salinity. Plant. Growth Regul. 55, 207–219.

Esim, N., Atici, O., 2013. Nitric oxide alleviates boron toxicity by reducing oxidative damage and growth inhibition in maize seedlings (*'Zea mays'* L.). Austr. J. Crop. Sci. 7, 1085–1092.
Farooq, M.A., Saqib, Z.A., Akhtar, J., 2015. Silicon-mediated oxidative stress tolerance and genetic variability in rice (*Oryza sativa* L.) grown under combined stress of salinity and boron toxicity. Turk. J. Agric. For. 39, 718–729.
Fleischer, A., O'Neill, M.A., Ehwald, R., 1999. The pore size of non-graminaceous plant cell walls is rapidly decreased by borate ester cross-linking of the pectic polysaccharide rhamnogalacturonan II. Plant. Physiol. 121, 829–838.
Foyer, C.H., Noctor, G., 2005. Oxidant and antioxidant signalling in plants: a re-evaluation of the concept of oxidative stress in a physiological context. Plant. Cell Env. 28, 1056–1071.
García-Sánchez, F., Simón-Grao, S., Martínez-Nicolás, J.J., Alfosea-Simón, M., Liu, C., Chatzissavvidis, C., et al., 2020. Multiple stresses occurring with boron toxicity and deficiency in plants. J. Hazard. Mater. 397, 122713.
Genisel, M., Turk, H., Dumlupinar, R., 2017. Exogenous aminolevulinic acid protects wheat seedlings against boron-induced oxidative stress. Rom. Biotechnol. Lett. 22, 12741–12750.
Ghanati, F., Morita, A., Yokota, H., 2002. Induction of suberin and increase of lignin content by excess boron in tobacco cells. Soil. Sci. Plant. Nutr. 48, 357–364.
Gimeno, V., Simón, I., Nieves, M., Martínez, V., Cámara-Zapata, J.M., García, A.L., et al., 2012. The physiological and nutritional responses to an excess of boron by Verna lemon trees that were grafted on four contrasting rootstocks. Trees 26, 1513–1526.
Goldbach, H.E., Wimmer, M.A., 2007. Boron in plants and animals: is there a role beyond cell-wall structure? J. Plant. Nutr. Soil. Sci. 170, 39–48.
Goldbach, H.E., Wimmer, M.A., Findeklee, P., 2000. Discussion paper: boron—how can the critical level be defined? J. Plant. Nutr. Soil. Sci. 163, 115–121.
Goldbach, H.E., Yu, Q., Wingender, R., Schulz, M., Wimmer, M., Findeklee, P., et al., 2001. Rapid response reactions of roots to boron deprivation. J. Plant. Nutr. Soil. Sci. 164, 173–181.
González-Fontes, A., Rexach, J., Navarro-Gochicoa, M.T., Herrera-Rodríguez, M.B., Beato, V.M., Maldonado, J.M., et al., 2008. Is boron involved solely in structural roles in vascular plants. Plant. Signal. Behav. 3, 24–26.
Guerfel, M., Baccouri, O., Boujnah, D., Chaïbi, W., Zarrouk, M., 2009. Impacts of water stress on gas exchange, water relations, chlorophyll content and leaf structure in the two main Tunisian olive (*Olea europaea* L) cultivars. Sci. Hort. 119, 257–263.
Guidi, L., Degl'Innocenti, E., Carmassi, G., Massa, D., Pardossi, A., 2011. Effects of boron on leaf chlorophyll fluorescence of greenhouse tomato grown with saline water. Environ. Exp. Bot. 73, 57–63.
Gunes, A., Soylemezoglu, G., Inal, A., Bagci, E.G., Coban, S., Sahin, O., 2006. Antioxidant and stomatal responses of grapevine (*Vitis vinifera* L) to boron toxicity. Sci. Hort. 110, 279–284.
Guo, P., Qi, Y.P., Yang, L.T., Ye, X., Jiang, H.X., Huang, J.H., et al., 2014. cDNA-AFLP analysis reveals the adaptive responses of citrus to long-term boron-toxicity. Plant. Biol. 14, 284.
Gupta, U.C., 1993. Boron and Its Role in Crop Production. CRC Press, Boca Raton, FL, USA.
Gupta, U.C., James, Y.W., Campbell, C.A., Leyshon, A.J., Nicholaichuk, W., 1985. Boron toxicity and deficiency: a review. Can. J. Soil. Sci. 65, 381–409.
Han, S., Tang, N., Jiang, H.X., Yang, L.T., Li, Y., Chen, L.S., 2009. CO_2 assimilation, photosystem II photochemistry, carbohydrate metabolism and antioxidant system of citrus leaves in response to boron stress. Plant. Sci. 176, 143–153.

Herrera-Rodríguez, M.B., González-Fontes, A., Rexach, J., Camacho-Cristóbal, J.J., Maldonado, J.M., Navarro-Gochicoa, M.T., 2010. Role of boron in vascular plants and response mechanisms to boron stresses. Plant. Stress. 4, 115–122.

Huang, J.H., Cai, Z.J., Wen, S.X., Guo, P., Ye, X., Lin, G.Z., et al., 2014. Effects of boron toxicity on root and leaf anatomy in two *Citrus* species differing in boron tolerance. Trees 28, 1653–1666.

Hua, T., Zhang, R., Sun, H., Liu, C., 2020. Alleviation of boron toxicity in plants: mechanisms and approaches. Crit. Rev. Environ. Sci. Technol. 1–41.

Kabay, N., Güler, E., Bryjak, M., 2010. Boron in seawater and methods for its separation-a review. Desalination. 261, 212–217.

Kamali, A., Childers, N., 1966. Effect of boron nutrition on peach anatomy. J. Am. Soc. Hort. Sci. 90, 33–38.

Karabal, E., Yücel, M., Öktem, H.A., 2003. Antioxidant responses of tolerant and sensitive barley cultivars to boron toxicity. Plant. Sci. 164, 925–933.

Kaya, C., Akram, N.A., Ashraf, M., 2018. Kinetin and indole acetic acid promote antioxidant defense system and reduce oxidative stress in maize (*Zea mays* L.) plants grown at boron toxicity. J. Plant. Growth Regul. 37, 1258–1266.

Kayihan, C., Öz, M.T., Eyidoğan, F., Yücel, M., Öktem, H.A., 2017. Physiological, biochemical, and transcriptomic responses to boron toxicity in leaf and root tissues of contrasting wheat cultivars. Plant. Mol. Biol. Rep. 35, 97–109.

Keren, R., Bingham, F.T., 1958. Boron in water, soils, and plants. Advances in Soil Science 229–276. Springer, New York, NY.

Lambers, H., 1987. Does variation in photosynthetic rate explain variation in growth rate and yield? Neth. J. Agric. Sci. 35, 505–519.

Landi, M., Degl'Innocenti, E., Pardossi, A., Guidi, L., 2012. Antioxidant and photosynthetic responses in plants under boron toxicity: a review. Am. J. Agric. Biol. Sci. 7, 255–270.

Landi, M., Guidi, L., Pardossi, A., Tattini, M., Gould, K.S., 2014. Photoprotection by foliar anthocyanins mitigates effects of boron toxicity in sweet basil (*Ocimum basilicum*). Planta. 240, 941–953.

Landi, M., Margaritopoulou, T., Papadakis, I.E., Araniti, F., 2019. Boron toxicity in higher plants: an update. Planta. 250, 1011–1032.

Landi, M., Pardossi, A., Remorini, D., Guidi, L., 2013. Antioxidant and photosynthetic response of a purple-leaved and a green-leaved cultivar of sweet basil (*Ocimum basilicum*) to boron excess. Env. Exp. Bot. 85, 64–75.

Lee, S.G., Aronoff, S., 1966. Investigations on the role of boron in plants. III Anatomical observations. Plant. Physiol. 41, 1570–1577.

Leyshon, J A., James, W.Y., 1993. Boron toxicity and irrigation management. In: Gupta, U. (Ed.), Boron and Its Role in Crop Production. CRC Press Inc., Boca Raton, FL.

Lobo, A.K.M., de Oliveira Martins, M., Neto, M.C.L., Machado, E.C., Ribeiro, R.V., Silveira, J.A.G., 2015. Exogenous sucrose supply changes sugar metabolism and reduces photosynthesis of sugarcane through the down-regulation of Rubisco abundance and activity. J. Plant. Physiol. 179, 113–121.

Lovatts, C.J., Bates, L.M., 1984. Early effects of excess boron on photosynthesis and growth of *Cucurbita pepo*. J. Exp. Bot. 35, 297–305.

Macho-Rivero, M.A., Camacho-Cristóbal, J.J., Herrera-Rodríguez, M.B., Müller, M., Munné-Bosch, S., González-Fontes, A., 2017. Abscisic acid and transpiration rate are involved in the response to boron toxicity in *Arabidopsis* plants. Physiol. Plant. 160, 21–32.

Macho-Rivero, M.A., Herrera-Rodríguez, M.B., Brejcha, R., Schäfner, A.R., Tanaka, N., Fujiwara, T., et al., 2018. Boron toxicity reduces water transport from root to shoot in *Arabidopsis* plants. Evidence for a reduced transpiration rate and expression of major pip aquaporin genes. Plant. Cell Physiol. 59, 841–849.

Miwa, K., Fujiwara, T., 2010. Boron transport in plants: co-ordinated regulation of transporters. Ann. Bot. 105, 1103–1108.
Miwa, K., Takano, J., Omori, H., Seki, M., Shinozaki, K., Fujiwara, T., 2007. Plants tolerant of high boron levels. Science 318, 1417.
Moraghan, J.T., Mascagni Jr, H.J., 1991. Environmental and soil factors affecting micronutrient deficiencies and toxicities. Micronutr. Agriculture 4, 371–425.
Mouhtaridou, G.N., Sotiropoulos, T.E., Dimassi, K.N., Therios, I.N., 2004. Effects of boron on growth, and chlorophyll and mineral contents of shoots of the apple rootstock MM 106 cultured *vitro*. Biol. Plant. 48, 617–619.
Moussa, H.R., Algamal, S.M.A., 2017. Does exogenous application of melatonin ameliorate boron toxicity in spinach plants? Int. J. Veg. Sci. 23, 233–245.
Mullineaux, P.M., Baker, N.R., 2010. Oxidative stress: antagonistic signaling for acclimation or cell death? Plant. Physiol. 154, 521–525.
Nable, R.O., Banuelos, G.S., Paull, J.G., 1997. Boron toxicity. Plant. Soil. 193, 181–198.
Niyogi, K.K., 1999. Photoprotection revisited: genetic and molecular approaches. Annu. Rev. Plant. Physiol. Plant Mol. Biol. 50, 333–359.
O'Neill, M.A., Ishii, T., Albersheim, P., Darvill, A.G., 2004. Rhamnogalacturonan II: structure and function of a borate cross-linked cell wall pectic polysaccharide. Annu. Rev. Plant. Biol. 55, 109–139.
Oikonomou, A., Ladikou, E.V., Chatziperou, G., Margaritopoulou, T., Landi, M., Sotiropoulos, T., et al., 2019. Boron excess imbalances root/shoot allometry, photosynthetic and chlorophyll fluorescence parameters and sugar metabolism in apple plants. Agronomy. 9, 731.
Papadakis, I.E., Dimassi, K.N., Bosabalidis, A.M., Therios, I.N., Giannakoula, A., 2004b. Effects of B excess on some physiological and anatomical parameters of 'Navelina' orange plants grafted on two rootstocks. Env. Exp. Bot. 51, 247–257.
Papadakis, I.E., Dimassi, K.N., Bosabalidis, A.M., Therios, I.N., Patakas, A., Giannakoula, A., 2004a. Boron toxicity in 'Clementine' mandarin plants grafted on two rootstocks. Plant. Sci. 166, 539–547.
Papadakis, I.E., Tsiantas, P.I., Tsaniklidis, G., Landi, M., Psychoyou, M., Fasseas, C., 2018. Changes in sugar metabolism associated to stem bark thickening partially assist young tissues of *Eriobotrya japonica* seedlings under boron stress. J. Plant. Physiol. 231, 337–345.
Paparnakis, A., Chatzissavvidis, C., Antoniadis, V., 2013. How apple responds to boron excess in acidic and limed soil. J. Soil. Sci. Plant. Nutr. 13, 787–796.
Parks, J.L., Edwards, M., 2005. Boron in the environment. Crit. Rev. Env. Sci. Tech. 35, 81–114.
Pego, J.V., Kortstee, A.J., Huijser, C., Smeekens, S.C., 2000. Photosynthesis, sugars and the regulation of gene expression. J. Exp. Bot. 51, 407–416.
Princi, M.P., Lupini, A., Araniti, F., Longo, C., Mauceri, A., Sunseri, F., et al., 2016. Boron toxicity and tolerance in plants: recent advances and future perspectives. Plant. Met. Interact. 115–147. Elsevier: Amsterdam, The Netherlands.
Reid, R.J., Hayes, J.E., Post, A., Stangoulis, J.C.R., Graham, R.D., 2004. A critical analysis of the causes of boron toxicity in plants. Plant. Cell Env. 25, 1405–1414.
Reid, R., 2007. Update on boron toxicity and tolerance in plants. In: Xu, F., et al., (Eds.), Advances in Plant and Animal Boron Nutrition. Springer, pp. 83–90.
Reid, R., 2010. Can we really increase yields by making crop plants tolerant to boron toxicity? Plant. Sci. 178, 9–11.
Reid, R.J., 2013. Boron toxicity and tolerance in crop plants. Crop Improvent Adverse Condition 333–346. Springer, New York, NY.
Roessner, U., Patterson, J.H., Forbes, M.G., Fincher, G.B., Langridge, P., Bacic, A., 2006. An investigation of boron toxicity in barley using metabolomics. Plant. Physiol. 142, 1087–1101.

Ruuhola, T., Keinänen, M., Keski-Saari, S., Lehto, T., 2011. Boron nutrition affects the carbon metabolism of silver birch seedlings. Tree Physiol. 31, 1251–1261.

Ryden, P., Sugimoto-Shirasu, K., Smith, A.C., Findlay, K., Reiter, W.D., McCann, M. C., 2003. Tensile properties of *Arabidopsis* cell walls depend on both a xyloglucan cross-linked microfibrillar network and rhamnogalacturonan II-borate complexes. Plant. Physiol. 132, 1033–1040.

Sang, W., Huang, Z.R., Qi, Y.P., Yang, L.T., Guo, P., Chen, L.S., 2015. An investigation of boron-toxicity in leaves of two citrus species differing in boron-tolerance using comparative proteomics. J. Proteomics. 123, 128–146.

Sarabandi, M., Farokhzad, A., Mandoulakani, B.A., Ghasemzadeh, R., 2019. Biochemical and gene expression responses of two Iranian grape cultivars to foliar application of methyl jasmonate under boron toxicity conditions. Sci. Hortic. 249, 355–363.

Sarafi, E., Chatzissavvidis, C., Therios, I., 2017. Response of two pomegranate (*Punica granatum* L.) cultivars to six boron concentrations: growth performance, nutrient status, gas exchange parameters, chlorophyll fluorescence, proline and carbohydrate content. J. Plant. Nutr. 40, 983–994.

Sarafi, E., Siomos, A., Tsouvaltzis, P., Therios, I., Chatzissavvidis, C., 2018. Boron toxicity effects on the concentration of pigments, carbohydrates and nutrient elements in six non-grafted pepper cultivars (*Capsicum annuum* L.). Indian. J. Plant. Physiol. 23, 474–485.

Sasaki, H., Hirose, T., Watanabe, Y., Ohsugi, R., 1998. Carbonic anhydrase activity and CO_2-transfer resistance in Zn-deficient rice leaves. Plant. Physiol. 118, 929–934.

Schreiber, U., Bilger, W., 1993. Progress in chlorophyll-fluorescence research: major developments during the past years in retrospect. Prog. Bot. 54, 151–173.

Shah, A., Wu, X., Ullah, A., Fahad, S., Muhammad, R., Yan, L., et al., 2017. Deficiency and toxicity of boron: alterations in growth, oxidative damage and uptake by citrange orange plants. Ecotoxic. Env. Safety. 145, 575–582.

Sheng, O., Zhou, G., Wei, Q., Peng, S., Deng, X., 2010. Effects of excess boron on growth, gas exchange, and boron status of four orange scion–rootstock combinations. J. Plant. Nutr. Soil. Sci. 173, 469–476.

Shireen, F., Nawaz, M.A., Chen, C., Zhang, Q., Zheng, Z., Sohail, H., et al., 2018. Boron: functions and approaches to enhance its availability in plants for sustainable agriculture. Int. J. Mol. Sci. 19, 1856.

Siddiqui, M.H., Al-Whaibi, M.H., Sakran, A.M., Ali, H.M., Basalah, M.O., Faisal, M., et al., 2013. Calcium-induced amelioration of boron toxicity in radish. J. Plant. Growth Regul. 32, 61–71.

Simón, I., Díaz-López, L., Gimeno, V., Nieves, M., Pereira, W.E., Martínez, V., et al., 2013. Effects of boron excess in nutrient solution on growth, mineral nutrition, and physiological parameters of *Jatropha curcas* seedlings. J. Plant. Nutr. Soil. Sci. 176, 165–174.

Sotiropoulos, T.E., Therios, I.N., Dimassi, K.N., Bosabalidis, A., Kofidis, G., 2002. Nutritional status, growth, CO_2 assimilation, and leaf anatomical responses in two kiwifruit species under boron toxicity. J. Plant. Nutr. 25, 1249–1261.

Soylemezoglu, G., Demir, K., Inal, A., Gunes, A., 2009. Effect of silicon on antioxidant and stomatal response of two grapevine (*Vitis vinifera* L.) rootstocks grown in boron toxic, saline and boron toxic-saline soil. Sci. Hortic. 123, 240–246.

Stangoulis, J.C.R., Brown, P.H., Bellaloui, N., Reid, R.J., Graham, R.D., 2001. The efficiency of boron utilisation in canola. Aust. J. Plant. Physiol. 28, 1109–1114.

Stangoulis, J., Tate, M., Graham, R., Bucknall, M., Palmer, L., Boughton, B., Reid, R., 2010. The mechanism of boron mobility in wheat and canola phloem. Plant Physiol 153, 876–881.

Takano, J., Miwa, K., Fujiwara, T., 2008. Boron transport mechanisms: collaboration of channels and transporters. Trends Plant. Sci. 13, 451–457.

Tanaka, M., Fujiwara, T., 2008. Physiological roles and transport mechanisms of boron: perspectives from plants. Pflug. Arch. 456, 671–677.

Tavallali, V., 2017. Interactive effects of zinc and boron on growth, photosynthesis, and water relations in pistachio. J. Plant. Nutr. 40, 1588–1603.

Velička, R., Martsinkyavichene, A., Rimkevichené, M., 2007. Photosynthetic characteristics and productivity of spring rape plants as related to crop density. Russ. J. Plant. Physiol. 54, 545–552.

Voxeur, A., Fry, S.C., 2014. Glycosylinositol phosphorylceramides from Rosa cell cultures are boron-bridged in the plasma membrane and form complexes with rhamnogalacturonan II. Plant. J. 79, 139–149.

Wang, J.Z., Tao, S.T., Qi, K.J., Wu, J., Wu, H.Q., Zhang, S.L., 2011. Changes in photosynthetic properties and antioxidative system of pear leaves to boron toxicity. Afr. J. Biotechnol. 10, 19693–19700.

Warington, K., 1923. The effect of boric acid and borax on the broad bean and certain other plants. Ann. Bot. 37, 629–672.

Wimmer, M.A., Muhling, K.H., Lauchli, A., Brown, P.H., Goldbach, H.E., 2003. The interaction between salinity and boron toxicity affects the subcellular distribution of ions and proteins in wheat leaves. Plant. Cell Env. 26, 1267–1274.

Yau, S.K., Ryan, J., 2008. Boron toxicity tolerance in crops: a viable alternative to soil amelioration. Crop. Sci. 48, 854–865.

Yoshinari, A., Takano, J., 2017. Insights into the mechanisms underlying boron homeostasis in plants. Front. Plant. Sci. 8, 1951.

Zhang, H.H., Zhong, H.X., Wang, J.F., Sui, X., 2016. Adaptive changes in chlorophyll content and photosynthetic features to low light in *Physocarpus amurensis* Maxim and *Physocarpus opulifolius* "Diabolo.". Peer J. 23, e2125–e2148.

Zhao, D., Oosterhuis, D.M., 2002. Cotton carbon exchange, nonstructural carbohydrates, and boron distribution in tissues during development of boron deficiency. Field Crop. Res. 78, 75–87.

Zhao, Q., Sun, Q., Dong, P., Ma, C., Sun, H., Liu, C., 2019. Jasmonic acid alleviates boron toxicity in *Puccinellia tenuiflora*, a promising species for boron phytoremediation. Plant. Soil. 445, 397–407.

CHAPTER 9

Comprehensive analyses of gene expression and identification of metabolites for boron stress tolerance

Firat Kurt[1] and Ertugrul Filiz[2]
[1]Faculty of Applied Sciences, Department of Plant Production and Technologies Muş Alparslan University, Muş, Turkey
[2]Cilimli Vocational School, Duzce University, Cilimli, Duzce, Turkey

9.1 Introduction

Boron (B) is known to be an essential nutrient for plant growth and development since 1923 (Warington, 1923). Volatile volcanic emissions and tourmaline are the main sources of B in nature (Nable et al., 1997). It was established that more than 200 minerals contain B. Of these minerals, borax, kernite, colemanite, and ulexite are known to be commercially important hydrated minerals (Herrera-Rodríguez et al., 2010). Hydrated minerals are generally very soluble minerals contrary to tourmaline (Goldberg, 1997).

B has never been present in earth crust in elemental form and found covalently bound to metals or metalloids along with oxygen (O) (Herrera-Rodríguez et al., 2010). B uptake by plants is restricted by pH, soil moisture and texture, temperature, and adsorbing surfaces such as oxides, clay minerals, calcium carbonate, and organic matter (Goldberg, 1997). B is found in soil solution as either boric acid [$B(OH)_3$] or borate anion [$B(OH)_4^-$]. The quantity of boric acid is considerably higher than borate and plants are taken up B as boric acid (Nable et al., 1997). B is mobile in the soil as boric acid and infiltrates lower layers of the soil profile depending on water movement. Therefore, topsoil B concentrations in cool humid climates are low whilst soils in semiarid and arid regions are rich in upper horizontal layers (Kot, 2009). The anthropogenic activities also contribute to the B concentrations of soils. Irrigation water obtained from saline wells, industrial activities, fly ash from electric power plants, and mining increase B contents of soils (Nable et al., 1997).

Boron in Plants and Agriculture
DOI: https://doi.org/10.1016/B978-0-323-90857-3.00004-7

Boric acid is taken up by plants by passive diffusion when available in sufficient concentrations in soil solution. Nonetheless, less quantities of boric acid in growth media necessitates boric acid channels and borate exporters to be involved in the uptake process. Borate exporters are used for the exclusion of excessive B from root cells as well (Yoshinari and Takano, 2017).

B, as borate anion, is used in cell wall synthesis by cross-linking with rhamnogalacturonan II of the cell wall (Matoh, 1997). B deficiency and toxicity damages plant growth and development. The B deficiency impairs various metabolic processes such as cell wall synthesis, membrane integrity, ascorbate and phenol metabolism, RNA metabolism, etc. (Cakmak and Römheld, 1997). The B toxicity in plants damages cell division in roots, photosynthesis by causing a reduction in leaf chlorophyll, stomatal conductance, the accumulation of biopolymers such as lignin and suberin, and plasma membrane (Demiray et al., 2011). Despite the significant roles of B in plant metabolism, B has not been reported so far as a component of any enzymes or there is not any information about the involvement of B in enzymatic processes (Cakmak and Römheld, 1997). However, limited studies with relation to the direct participation of B into metabolite synthesis metabolism are present. In this review, the genes associated with B toxicity or deficiency and the genes linked to the metabolite synthesis under B stresses were discussed (Fig. 9.1).

9.2 BOR1 homologs in plants

Despite its essentiality, the deficiency and the toxicity limits of B in plant metabolism are very narrow and in each mentioned case, plant growth and development are adversely affected (Brdar-Jokanović, 2020). B uptake mechanism of plants remained as a controversial issue for a long time and it was long thought that plants take up B through a passive process by feed-backing changes on membrane permeability and transpiration rates. However, the differences of B concentrations among plant species grown alike conditions (Hu and Brown, 1997), the different fractionation rates of B isotopes among different plant species (Marentes et al., 1997), the differences of B accumulation in symplasm in sunflower roots (Pfeffer et al., 1999), and the transport of boric acid through major intrinsic protein (MIP) PIP1 which is one of aquaporins or other Hg sensitive channels (Dordas et al., 2000) gave evidences suggesting that B uptake occurs through energy-dependent B transporters (Dannel et al., 2002). The first

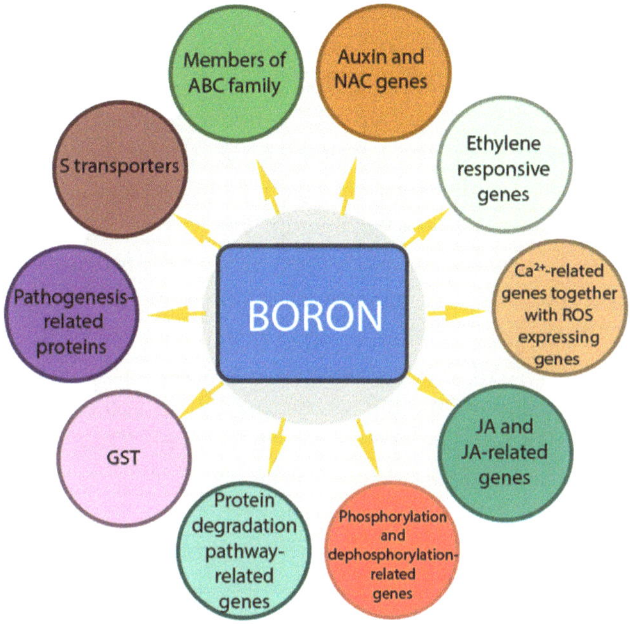

Figure 9.1 The pathways influenced by boron deficiency or toxicity in plants.

identification of B gene was made on high boron requiring *Arabidopsis thaliana* mutant line, called *bor1−1*, in which lower arm of chromosome 2 was ascertained as locus of *BOR1* gene. bor1−1 mutant, showed reduced B uptake and translocation rate compared to wild types and showed different B mobility depending on B-excess or -deficient growth conditions (Noguchi et al., 1997), was used for identification of *BOR1* function and it is suggested that *BOR1* does not participate in the uptake of B (Nakagawa et al., 2007). However, the fact that xylem loading and preferential translocation of B is maintained by the participation of *BOR1* to the carrier- or channel-mediated transport indicated that *BOR1* functions as an efflux transporter (Takano et al., 2001). Proving this suggestion, green fluorescent protein assay and positional cloning procedures showed that *BOR1* is a membrane protein alike bicarbonate transporters in animals.

BOR1 functions as an efflux transporter regulating the export of B from pericycle cells into root stelar apoplasm. It is involved in the xylem loading of B and thereby increases B accumulation in shoot organs at low B supply (Takano et al., 2002). However, when B is supplied at high concentrations in growth media following low B supply, B translocation through BOR1 is reduced, suggesting that BOR1-mediated xylem loading and B accumulation in shoots is regulated against excessive supply of B. Consequently, the participation of BOR1 to B translocation is dynamically regulated through low or high B supply. In this regulation, BOR1 internalization from the cell membrane into endosomes occurs repeatedly under B-deficient media and is degraded in vacuole when B concentrations of growth media become high (Takano et al., 2005). Other than the mentioned posttranscriptional and post-translational mechanisms, *BOR1* expression can also be regulated by a two-step process. It is suggested that if B supply is low, BOR1 undergoes a rapid protein degradation whereas translational suppression occurs under high B conditions (Aibara et al., 2018).

Arabidopsis has seven B transporters (Miwa et al., 2013; Nakagawa et al., 2007) and BOR1-like transporters were characterized and identified in other plant species: 20 BOR1-like transporters in rapeseed (*Brassica napus*) (Chen et al., 2018), six in maize (*Zea mays*) (Chatterjee et al., 2017), one in grapevine (*Vitis vinifera L.)* (Pérez-Castro et al., 2012), four in rice (*Oryza sativa*) (Nakagawa et al., 2007; Tanaka et al., 2013; Yoshinari and Takano, 2017), one in citrus (*Citrus macrophylla*) (Cañon et al., 2013), three in bread wheat (*Triticum aestivum*) (Leaungthitikanchana et al., 2013; Reid, 2007) and one in barley (*Hordeum vulgare*). Using in silico methods, 10 *BOR1*-like genes were predicted in potato (*Solanum tuberosum*) (Kurt and Aydın, 2020) and one *BOR1* gene was identified in *Eucalyptus* (Domingues et al., 2005).

As mentioned previously, four *BOR*-like genes in rice were found as a result of a rice genomic database search by Nakagawa et al. (2007). It was suggested that *OsBOR2, 3* and *4* were more distantly related to *AtBOR1*; however, *OsBOR1* is more similar to *AtBOR1* in terms of sequential and phylogenetic analyses. Nonetheless, *OsBOR1* function differentiates from *AtBOR1* from the view of where it is expressed and B content of growth media. *OsBOR1* is not just involved in the xylem loading of B but also participates in the acquisition of B since it is expressed in endodermis and exodermis under both low and sufficient B containing growth conditions (Nakagawa et al., 2007).

Another homolog of *AtBOR1* was found in maize and called Rotten Ear (RTE) (Chatterjee et al., 2014). ZmRTE1 is established to be expressed in tissues around phloem and xylem, and inflorescence. The inflorescence, the tassel, and the ear sterilities were established as organs in which specific symptoms of mutations in *ZmRTE1* mutant are observable. Moreover, the mutants showed imperfect transport of B even when grown in comparably B-access conditions. Therefore it was suggested that ZmRTE1 is involved in

B transport to above-ground organs and B transport with ZmRTE1 is maintained throughout the plant's life (Chatterjee et al., 2014).

The search of OsBOR1 homologs in wheat (*T. aestivum*) and barley (*H. vulgare*) led to the identification of transporter proteins having more similarity to *OsBOR2*. It is stated that *TaBOR2* and *HvBOR2* genes in B-tolerant wheat and barley cultivars lowered the excessive quantities of root B. Due to this reason, these two genes act in differently than *AtBOR1* and *OsBOR1* which help B be transported to shoots under B-deficient conditions (Reid, 2007; Sutton et al., 2007). In a study, three homologous *BOR1* genes were found to participate in B transport in wheat. The regulations of these three identified *TaBOR1* transcripts are dependent on the B concentrations of growth conditions. *TaBOR1.2* is expressed under high B conditions whilst *TaBOR1.1* and *TaBOR1.3* are generally regulated under B-deficient media. The tissue-specific expressions of *TaBOR1* transcripts are dynamically regulated by B concentrations of growth media (Leaungthitikanchana et al., 2013). Boron tolerance in wheat and barley is suggested to be originated from the divergent evolution of paralogous genes (Pallotta et al., 2014).

CmBOR1 is identified as an efflux transporter in citrus (*Citrus macrophylla*) in which B transport is immobile. Therefore B supply is needed throughout of a plant's life. Like its orthologue in *A. thaliana*, *CmBOR1* is also involved in the transport of B from root cells into xylem loading. The *CmBOR1* expression is also regulated by B contents of growth media. *CmBOR1* is found to be expressed in higher quantities in roots compared to shoots. Flowers are also identified where *CmBOR1* is expressed in large quantities (Cañon et al., 2013).

In grapevine, only one gene (*VvBOR1*) from the predicted six genes is identified as *AtBOR1* homolog. *VvBOR1* is expressed in the early stages of berry development and flowering. In addition, vegetative tissues are sites where *VvBOR1* is highly expressed. Nonetheless, the expression in root tissues is almost two times higher than in flowers. Depending on its higher expression in roots and B transport into the xylem, *VvBOR1* is identified as an efflux transporter involved in loading B into the xylem, although its performance in B export is quite lower than *AtBOR1*. Therefore, it is suggested that it has a possible another role in the grapevine (Pérez-Castro et al., 2012).

Sun et al. (2012) were identified six *BOR1* homologs in *B. napus* using the rapid amplification of cDNA ends method. Interestingly only *BnBOR1;3a* and *BnBOR1;3c* genes are reported to have unique expression patterns from the rest of other transporters. The mentioned two genes have also a similar gene structure to *AtBOR1* in terms of exon numbers. However, instead of *BnBOR1;3a* and *BnBOR1;3c*, it is found that *BnBOR1;1c* and *BnBOR1;2a* are only triggered by B deficiency. In a similar study, a total of 20 BnBOR transporters were identified using a Blast search of seven B transporters in *Arabidopsis*. The organ-specific expressions of these B transporters along with their expressions under B-deficient conditions in a B-efficient (QY10) and B-inefficient (W10) cultivars showed that spatiotemporal expressions of B transporters are dynamically regulated and the responses of BnBORs to B deficiency in growth media show variation. Particularly a homolog of *AtBOR1*, *BnBOR1;1c* (or *BnaC4.BOR1;1c*), was found to be induced by B deficiency and B accumulation in B-efficient cultivar's roots and shoots was higher than the inefficient cultivar (Chen et al., 2018; Zhang et al., 2017).

The transcriptomic studies and data obtained such studies help in the prediction potential of BOR1-like transporters and in the prediction of functional characterization of these genes. In this regard, 80 *BOR1* homologs were identified in several plant species (Ozyigit et al., 2020) and one gene was suggested as a candidate *BOR1* homolog in eucalyptus (Domingues et al., 2005). Similarly, 10 *BOR1*-type genes were identified and characterized in an *in silico* study. Using coexpression and sequence similarity network analyses, it is suggested that BOR1-type transporters interact with genes participating in cell wall fortification and plant immunity. Also, one *BOR1*-type gene in potato (*SlBOR1−8*) was predicted to have acted in different biochemical processes without interacting with the rest of mentioned BOR1-type transporters (Kurt and Aydın, 2020).

Other than BOR1 orthologs and paralogs in plant species, the MIPs participate in B transport as boric acid channels as well (Tanaka and Fujiwara, 2008). BOR1 homologs, paralogs, and MIPs generally affect the uptake, the translocation and the accumulation of B. As a result of the functions of these mentioned genes, the quantities of B in cellular level induce gene expressions of vast and diverse metabolites to alleviate B stress or to tolerate its adverse effects. Therefore, in the following parts of the present article, the identification of B stress-related metabolites and how the gene regulations of these metabolites occur under B toxicity or deficiency was discussed.

9.3 The metabolites for boron stress tolerance

B stress (deficiency and toxicity) impairs plants' growth and development and causes several damages by disrupting several biological processes in plants. Cell wall synthesis, carbohydrate metabolism, ascorbate metabolism, sugar transport, indol acetic acid metabolism, phenol metabolism, generation of reactive oxygen species (ROS) (Cakmak and Römheld, 1997) and reproductive system (Herrera-Rodríguez et al., 2010) are being adversely affected under B deficiency.

B toxicity impairs seed germination, polyphenol oxidase activity, root cell division, the synthesis of photosynthetic pigments, the rate of photosynthesis, the synthesis of cell wall components (Herrera-Rodríguez et al., 2010). In recent years, to enlighten the role of B in the mentioned metabolic pathways, transcriptomic, proteomic, and mapping studies have been conducted. The identified genes under B stress may help improve the tolerance of existed cultivars and develop novel B-tolerant varieties.

9.4 Gene regulation under boron-deficient conditions

The cell recognizes the B deficiency as a result of decreased cell wall strength or redox imbalance, leading to alterations of the cell wall structure. The deficiency signals are transduced to the nucleus by arabinogalactan proteins along with cell wall-plasma membrane-cytoskeleton continuum to initiate B stress machinery (González-fontes et al., 2013). It is also suggested that ABA induces the Ca^{2+} influx pathway, which results in increase of cytosolic Ca^{2+} (hereafter Ca^{2+}) (Allen et al., 1999). ACA10 which is a plasma membrane PIIB-type Ca^{2+}-ATPase and CAX3 which is vacuolar cation/proton exchanger genes in *Arabidopsis*' roots are Ca transporters and their expressions are increased by ABA (Quiles-Pando et al., 2019).

B responsive gene expression is dependent on Ca^{2+} concentrations and it occurs under both low and high Ca^{2+} concentrations. Ca^{2+} is required for a stabilized pectic network under B-deficient conditions and the disruption of this network under low B conditions activates B deficiency signaling pathway. When Ca^{2+} level is very high in a B-deficient growth media, two genes, wound-inducible protein kinase (*WIPK*) and a wound-responsive leucine zipper zinc finger protein (*WIZZ*), which is expressed under wounding conditions, are not transcribed at all (Koshiba et al., 2010). *WIPK* is also involved in the induction of one of the downstream genes in the Ca^{2+}-regulated pathway, *MAPK* (a mitogen-activated protein kinase) in tobacco (Camacho-Cristóbal et al., 2011). Additionally, TFs, including *MYB*, *WRKY*, b*ZIP*, calmodulin-like protein (CMK), and Ca^{2+}-dependent protein kinases (CPK) have roles in the signal transduction of recognition of B deficiency (González-fontes et al., 2013) along with MAPK and ROS. It is suggested that the possible B deficiency signaling under B-deprived media occurs through mechanosensitive Ca^{2+} channels, which are not regulated by the formation of ROS. Following intake of Ca^{2+} into cells, ROS formation is triggered by NAD(P)H oxidase, resulting in secondary Ca^{2+} influx into cells through separate ROS-responsive channels. Thus the B deficiency signaling pathway is activated (Koshiba et al., 2010).

The identification of B deficiency responsive gene expression is a complex task and it is not completely known if the differentially expressed genes under B deficiency, causing numerous changes over diverse biological processes, are directly or indirectly related to B deficiency. However, it was clear that B deficiency affects numerous pathways from cell wall stability to other stress responses (Camacho-Cristóbal et al., 2011). The gene regulation under B deficiency in varieties of the same plant species seems to be different as well. In a study conducted in *C. sinensis* leaves, only 22 miRNAs were expressed in both roots and leaves and even 11 of 22 miRNAs showed differences in their expression levels. In the mentioned study, 14 of 178 miRNAs were identified as novel miRNAs which are either upregulated or downregulated under B deficiency. More specifically, it was found that the alteration of miR393, miR160, and miR3946 transcriptions bestowed tolerance to *C. sinensis* by declining plant growth and development, a plant survival strategy under stress, because of the suppression of auxin response factors (ARFs), and TIR1, a plant development regulator. Moreover, the downregulation of *miR159*, *miR782*, *miR3946*, and *miR7539* was found to trigger the expression of *NACs* by rendering *C. sinensis* more tolerant to B deficiency. The downregulation of a boron exporter, *ATR1*, a member of the major facilitator superfamily, was associated with the upregulation of miR5037. It was stated that in order for stress response machinery and antioxidant system-related genes to be induced miR164, miR6260, miR5929, miR6214, miR3946, and miR3446 should have been transcribed in less quantities. Specifically, miR408 was found to be a downregulated miRNA under B stress in the mentioned study (Lu et al., 2015).

9.5 Gene regulation under conditions of excessive boron

Since B toxicity impedes or damages various vital metabolic pathways, as expected, the B toxicity induces many genes involved in heat, salt, ion, cold, oxidative stress, and drought stresses (Öztürk et al., 2018). The biosynthesis of some stress (Demiray et al., 2011) and pathogen-related (PR) proteins are increased and the production of some hormones as stress-signaling molecules are stimulated under B toxicity (Öz et al., 2009). The expression of B transporters such as *BOR1*, *DUR3*, and *FPS1* is associated with the regulation of B homeostasis (Nozawa et al., 2006). miRNAs participating in stress mechanisms also regulate B homeostasis through translational repression or mRNA cleavage (Ozhuner et al., 2013).

In a study conducted by Öz et al. (2009), the highly upregulated or downregulated genes under boron toxicity in barley were identified using the microarray sequencing method. The results in the mentioned study showed that jasmonic acid (JA), glutathione S-transferase (GST), pathogenesis-related genes (*PR*), and senescence-associated (SA) proteins were upregulated more than twofold under 10 mM $B(OH)_3$ treatment in B-sensitive cultivar Hamidiye. JA-related genes showed variation in their regulation. To be more specific, alcohol dehydrogenase 3, some JA-induced genes, and glutamine-dependent asparagine synthetase were found to be downregulated more than twofolds. Of these genes, thaumatin-like protein (TLP5), B12D protein as SA proteins, glucan *endo*-1,3-beta-glucosidase GII precursor and wheatwin-2 precursor as PR proteins, and allene oxide synthase as JA-related protein were expressed more than threefold. Remarkably, it was stated that two GST proteins were expressed more than fourfolds and all cytochrome *P450* genes were upregulated. Transporter genes such as *MDR*-like *ABC* transporter, *PDR*-like, *ABC* transporter, *ABC* transporter-like, protein-peptide transporter, protein-peptide transporter *PTR2*, amino acid transporter A1, phosphate translocator, anion/sugar transporter, and P-type ATPase expressions were found to be above the twofold change. Of these, the expression of *MDR*-like *ABC* transporter among other transporters was the highest. Moreover, the members of the heat shock protein family such as heat shock cognate 70 kDa protein 2, 16.9 kDa class I HSP, and 17.8 kDa class II HSP were downregulated at 10 mM $B(OH)_3$ treatment. Two transcripts of GAMyb transcription factor particularly were found to be differentially expressed under excessive B treatment. Whilst the expressions of NAC domain, CCT motif family protein, C2H2 zinc finger protein and ethylene-insensitive-3-like protein upregulated, *DRE* binding, *WRKY* binding, and heat shock TFs were downregulated about twofold change. Interestingly, it was reported that some of the mentioned genes were only expressed at the highest B exposure (10 mM B) but not transcribed under less toxic B treatment (5 mM B).

In a proteomic study, when a B-tolerant barley genotype (Anadolu), exhibiting the same level B accumulation like Sahara, was exposed to a toxic level of B, it was found that seven proteins [ribulose 1,5-bisphosphate carboxylase/oxygenase, thaumatin-like protein 5 or TLP5, pathogenesis-related protein 5 or PR5, RNase S-like protein, PSI type III chlorophyll a/b-binding protein, light-harvesting complex I LHC I and vacuolar proton-translocating ATPase subunit E protein (V-ATPase)] were differentially expressed. The authors stated that the overexpression of V-ATPase subunit E gene in heterologous system gave tolerance to yeast cells against toxic B treatment (Atik et al., 2011).

Puccinellia distans, a B hyperaccumulator plant, have ability to restrict the uptake and the translocation of toxic levels of B in shoot and root cells using efflux transporters. Metabolism-related genes in *P. distans* are highly expressed under high B conditions. The genes stimulating generation of ascorbate peroxidase and peroxiredoxin, proline, citrulline, and glycine betaine are induced by high B. The mentioned genes are also reported in *Gypsophila arrostil*, a moderately B-tolerant plant. Many transport-related genes such as ABC transporter C-family member 5 (*MRP5*), H^+-ATPase, ADP and ATP carrier proteins, amino acid transporter, ferritin, and mitochondrial folate transporter are expressed in *P. distans* to avert toxic effects of high B (Padmanabhan et al., 2012).

B toxicity causes high expression of proteins related to stress responses and pathway change in the occurrence of some stress-related proteins. In this regard, it was found that the expression of a key compound for flavonoid biosynthesis and UDP-glycosyltransferase highly increased under high B conditions. On the other hand, the glyoxylate pathway was altered at malate formation which was no longer dependent on the expression of fumarate related proteins (fumarate hydratases). In this altered pathway, malate was generated through glyoxylate and through the gene inducing glyoxylate compounds such as 4-hydroxy-4-methyl-2-oxoglutarate aldolase. Thus, as reported in Al toxicity in wheat, the intracellular pH was regulated by combined efflux of K^+ ions and malate under B toxicity (Öztürk et al., 2018). In terms of B and B-associated transporters, *BOR6* was found as an only upregulated B-efflux transporter in *P. distans*.

Two aquaporin homologs, NOD26-like intrinsic-type (NIP) and TIP-type proteins, increasing tolerance to salt and drought but not cold, were expressed conversely under B-excess conditions in the mentioned study. The authors stated that not all aquaporins are involved in B uptake since the TIP-type aquaporin was downregulated. Moreover, they suggested that vacuole was not the organelle where B was stored. B toxicity in *P. distans* also induced the expression of five members of (B, C, E, F, and I) ABC transporters under high B media in the mentioned study. The upregulation of these ABC transporters was explained by their probable roles in sorbitol transport, including B. Conversely, sulfate transporters are suggested to be affected negatively by the B-excess owing to maintaining and regulating intracellular pH. Öztürk et al. (2018) also found that B toxicity decreased ABA and ABA-related genes like phosphatase whereas increased the expressions of five ethylene response proteins. According to the authors, this regulation is due to an increase of cellular Ca^{2+} concentrations and ROS production and upregulation of Ca^{+2}-related proteins, which stimulates cell wall stability and catabolizes ABA but increases ethylene response under high B.

The gene regulation in leaves of B-tolerant (*C. sinensis*) and B-senstive (*C. grandis*) citrus varieties were tested using cDNA-AFLP method at 10 μM (control) and 400 μM (B-toxic) for 15 weeks by Guo et al. (2014). Authors found that gene expressions did not change in the B-tolerant variety and the expressions of photosynthesis-related genes and ROS scavenging genes were increased more in B-tolerant variety, which is the indication of the protection mechanism of the plant against photo-oxidative damage. More specifically, although photosynthesis-related genes were expressed at higher levels in *C. sinensis*, B toxicity induced sedoheptulose-1,7-bisphosphatase (SBPase), an important enzyme for ribulose 1,5-bisphosphate (RuBP) synthesis, in both tolerant and sensitive citrus varieties. Similarly, protochlorophyllide oxidoreductase, an enzyme triggering Chl synthesizing genes, protecting plants against photo-oxidative damage, was upregulated in both varieties in the mentioned study. The fact that B affects the structural and functional of cell wall stability was suggested to be related to downregulation of caffeic acid 3-*O*-methyltransferase (*COMT*), which is involved in lignin biosynthesis in both varieties. Nonetheless, B tolerance is associated with the preservation of regulation between phosphorylation and dephosphorylation in *C. sinensis*. The authors stated that the upregulation of ubiquitination genes to mark proteins for degradation, of S-adenosylmethionine (AdoMet) for methylation to maintain cells' functions and of inorganic pyrophosphatase 1 (PPase1) genes to assist thermodynamically favorable reactions along with NUDX19, one of the housekeeping genes belonging to Nudix family, also bestow tolerance against toxic B in B-tolerant citrus variety. Autophagy was damaged through reduction of transcript forming autophagosomes, a transducin/WD40 domain-containing protein (ATG18a), under high B conditions in the mentioned study. Pseudo-response regulator 5, negatively regulating Chl, Car, ABA, and α-tocopherol pathways were reported to be upregulated and an ABA-induced gene, ERD6-like 5, a glucose transporter protein in vacuole membranes, was downregulated in *C. sinensis*. Sugar transport into and out of cells in B-tolerant variety was found to increase by the upregulation of *SWEET7* gene (Guo et al., 2014). On the other hand, the gene responses and regulations to B toxicity in sensitive variety were different. A member of thioredoxin superfamily protein, the late embryogenesis abundant protein 5 (LEA5), thaumatin-like protein 1 (TLP1), LIM domain-containing protein which is responsible for actin cytoskeleton regulation were upregulated and they enhanced B tolerance in B-sensitive variety. On the contrary, the decrease of the expressions of stress-related protein (14−3-3 protein) and heterotrimeric GTP-binding protein subunit beta 1, a gene bestowing heavy metal tolerance (*MTP*), are suggested to decrease B tolerance of sensitive cultivar (Guo et al., 2014).

B toxicity in roots and leaves induces the expression of hormone signaling genes, kinases, *MYB*, *AP2*, and *bHLH* transcription factors along with some important enzymes involved in ROS scavenging mechanism in tolerant (Bolal-2873) and sensitive wheat (Atay-85) cultivars when exposed to 10 μM and 5 mM boric acid for five days (Kayıhan et al., 2017). It is reported that B toxicity did not affect the growth parameters of two cultivars, although 5 mM boric acid caused some visual B toxicity symptoms. Photosynthetic activities of both cultivars are also reported not changed even though three genes (2Fe-2S iron-sulfur cluster binding domain-containing protein, oxygen-evolving enhancer 3 PsbQ-family protein, a chloroplast gene encoding subunit IV of the cytochrome b6/f complex) are differentially expressed. Likewise, the concentration of proline, involved in osmoregulation, and of glycine betaine, an osmoprotectant, was found not changed in roots and leaf tissues of both cultivars. Some genes were commonly expressed in tolerant and sensitive cultivars. In this regard, peptidase and endopeptidase inhibitor activity genes and peroxidases in root tissues of both cultivars and protein degradation-related gene members; belonging to ubiquitin/proteasome machinery, *GST*, and *NIP4;1* (nodulin-26-like intrinsic proteins) were found to be highly expressed. As mentioned previously, despite the fact that some TFs were highly upregulated under B toxicity, *WRKY* transcription factors were found to be downregulated. On the other hand, *C2H2* TFs were detected to be upregulated only in the sensitive cultivar. In terms of plant hormones, B toxicity seems to induce JA and JA-related genes in B-sensitive wheat and barley varieties. In addition to this, gibberellic acid and ethylene biosynthesis were upregulated in B-sensitive wheat cultivar. Authors also found that expansin-like genes, involved in cell wall modification, were downregulated in the sensitive cultivar (Kayıhan et al., 2017).

It is known that ROS are scavenged by intermediates like GSH in the ascorbate and GSH pathway and GST acts in the removal of ROS by linking GSH with ROS. In this regard, Kayihan et al. (2019) conducted a study to identify members of the GST superfamily and miRNAs participating in the GSH-associated detoxification mechanism. In this regard, the expressions of some members of the GST superfamily, miRNA169 which targets GST, and miRNA156 which is involved in *GST5* regulation, were tested under 1 mM (1B) and 3 mM (3B) boric acid treatments in *A. thaliana*. It was reported that one member from each tau (AtGSTU19) and zeta (AtGST1) subfamilies of GST along with miRNA169 were highly expressed under both B-toxic concentrations. Moreover, due to the insignificant expression of PCS1 and glutathione synthetase (GSH2), the phytochelatin and GSH levels did not change under both toxic B levels. However, glutamate cysteine ligase (*GSH1*) expression remarkably increased under 3B treatment.

miRNAs are important for posttranscriptional regulation of gene expressions. Ozhuner et al. (2013) were identified four novel miRNAs together with 32 studied miRNAs in barley under B toxicity of which targets are conserved TFs such as *MYB*, *auxin*, *NAC*, *NFY*, *AP2*, etc. Of these, the most significantly downregulated miR408 (16-folds) in barley leaves targets *heterotrimeric G protein alpha (a) subunit* and *ATPase family gene 1* (AFG1) which are involved in signal transduction. miR172, miR319 (ethylene related), and miR398 in wheat were found to be associated with not just B toxicity responsive genes but also with genes involved in Fe, P, S, and Cu deficiencies and toxicities. Additionally, B toxicity induced leaf senescence genes with the combined effect of JA, miR172, and miR319 in the mentioned study.

Some housekeeping proteins are highly expressed under conditions of excessive B. A proteomic study conducted in carrot using SDS page analysis showed that major allergen Dau c1 (CR16), isocitrate dehydrogenase ($NADP^+$), adenosyl homocysteinase and EDPG precursor proteins are expressed both in control and fivefolds B containing media. However, it was found that *glutathione peroxidase* and *glyoxylase I* expressions were only detected under B stress. Glutathione peroxidase and glyoxylase I were reported to be involved in the reduction of oxidative stress and plant defense mechanism, respectively (Demiray et al., 2011).

As previously mentioned, ABC transporters are important components of the B stress response mechanism. De Abreu Neto et al. (2017) found that some ABC proteins in B-sensitive rice are differentially expressed under high B conditions. The other proteins induced by B-excess in B-sensitive rice genotype were identified as PHD-finger family protein, MYB and WAD40-like domain-containing proteins as TFs, a bifunctional monodehydroascorbate reductase which is an intermediate of ascorbate glutathione cycle, and OsORP7, a JA regulating hormone. Nonetheless, lysine ketoglutarate reductase trans-splicing related 1 which is a transcriptional regulator protein of amino acid metabolism under abiotic stress, *OsHIPP35* which takes place in a heavy metal detoxification pathway, BURP domain-containing protein which acts in diverse stress responses, and a starch synthase which is involved in starch accumulation mechanism in leaves were found by authors to be highly induced by the toxic level of B in tolerant rice genotype (De Abreu Neto et al., 2017).

9.6 Conclusion

B deficiency and toxicity cause the alteration of numerous pathways in plants. The genes involved in plants' adaptive responses to B toxicity are more studied compared to B deficiency. The responses of plant species to B toxicity or deficiency are both different and species-specific. Nevertheless, there are some common plant responses at the molecular level to B toxicity or deficiency. In this regard, the following generalization can be remarked:

- Stress-related transcription factors have not only different expression patterns between tolerant and sensitive genotypes of the same species but also among species.
- Members of the ABC family are involved in the B tolerance mechanism and these genes should be taken into consideration in plant breeding studies.
- Phosphorylation and dephosphorylation-related genes play important roles in gaining B tolerance.
- The involvement of S transporters in the B tolerance mechanism is very complex.

- Pathogenesis-related protein and GST are expressed under both boron deficiency and toxicity
- Protein degradation pathway-related genes are generally upregulated under B tolerance. This process is affected by a plant hormone JA
- JA and JA-related genes are generally upregulated in B-sensitive wheat and barley varieties
- There are very a few studies, as opposed to what is expected, reporting the expression of arabinogalactan proteins which are suggested to be involved in B stress
- B deficiency and toxicity is recognized by cellular Ca^{2+} concentrations and ROS. Therefore in this mechanism, Ca^{2+}-related genes together with ROS expressing genes are activated under both B-deficient and B-excess conditions.
- Auxin and *NAC* genes are activated under both B deficiency and toxicity
- ABA triggers the Ca^{2+} influx pathway, which leads to an increase of cytosolic Ca^{2+} under B deficiency and cellular Ca^{2+} concentrations along with ROS are important components activating the B deprivation signaling pathway. However, ABA is downregulated whilst some ethylene-responsive genes are activated in B-tolerant plant varieties under high B conditions. Interestingly, cellular Ca^{2+} concentrations, Ca^{+2-}related proteins, and ROS are again key components of the B toxicity response mechanism although it is still not known how stress response mechanisms are regulated or altered under both B-deficient and toxic conditions.

References

Aibara, I., Hirai, T., Kasai, K., Takano, J., Onouchi, H., Naito, S., et al., 2018. Boron-dependent translational suppression of the borate exporter BOR1 contributes to the avoidance of boron toxicity. Plant. Physiol. 177, 759–774. Available from: https://doi.org/10.1104/pp.18.00119.

Allen, G.J., Kuchitsu, K., Chu, S.P., Murata, Y., Schroeder, J.I., 1999. Arabidopsis abi1-1 and abi2-1 phosphatase mutations reduce abscisic acid-induced cytoplasmic calcium rises in guard cells. Plant. Cell 11, 1785–1798. Available from: https://doi.org/10.1105/tpc.11.9.1785.

Atik, A.E., Bozdağ, G.O., Akinci, E., Kaya, A., Koç, A., Yalçin, T., et al., 2011. Proteomic changes during boron tolerance in barley (*Hordeum vulgare*) and the role of vacuolar proton-translocating ATPase subunit E. Turk. J. Botany 35, 379–388. Available from: https://doi.org/10.3906/bot-1007-29.

Brdar-Jokanović, M., 2020. Boron toxicity and deficiency in agricultural plants. Int. J. Mol. Sci. 21. Available from: https://doi.org/10.3390/ijms21041424.

Cakmak, I., Römheld, V., 1997. Boron deficiency-induced impairments of cellular functions in plants. Plant. Soil. 193, 71–83. Available from: https://doi.org/10.1007/978-94-011-5580-9_6.

Camacho-Cristóbal, J.J., Rexach, J., Herrera-Rodríguez, M.B., Navarro-Gochicoa, M.T., González-Fontes, A., 2011. Boron deficiency and transcript level changes. Plant. Sci. 181, 85–89. Available from: https://doi.org/10.1016/j.plantsci.2011.05.001.

Cañon, P., Aquea, F., Rodríguez-Hoces de la Guardia, A., Arce-Johnson, P., 2013. Functional characterization of *Citrus macrophylla* BOR1 as a boron transporter. Physiol. Plant. 149, 329–339. Available from: https://doi.org/10.1111/ppl.12037.

Chatterjee, M., Liu, Q., Menello, C., Galli, M., Gallavotti, A., 2017. The combined action of duplicated boron transporters is required for maize growth in boron-deficient conditions. Genetics 206, 2041–2051. Available from: https://doi.org/10.1534/genetics.116.198275/-/DC1.1.

Chatterjee, M., Tabi, Z., Galli, M., Malcomber, S., Buck, A., Muszynski, M., et al., 2014. The boron efflux transporter ROTTEN EAR is required for maize inflorescence development and fertility. Plant. Cell 26, 2962–2977. Available from: https://doi.org/10.1105/tpc.114.125963.

Chen, H., Zhang, Q., He, M., Wang, S., Shi, L., Xu, F., 2018. Molecular characterization of the genome-wide BOR transporter gene family and genetic analysis of BnaC04.BOR1;1c in *Brassica napus*. BMC Plant. Biol. 18, 1–14. Available from: https://doi.org/10.1186/s12870-018-1407-1.

Dannel, F., Pfeffer, H., Römheld, V., 2002. Update on boron in higher plants—uptake, primary translocation and compartmentation. Plant. Biol. 4, 193–204. Available from: https://doi.org/10.1055/s-2002-25730.

De Abreu Neto, J.B., Hurtado-Perez, M.C., Wimmer, M.A., Frei, M., 2017. Genetic factors underlying boron toxicity tolerance in rice: genome-wide association study and transcriptomic analysis. J. Exp. Bot. 68, 687–700. Available from: https://doi.org/10.1093/jxb/erw423.

Demiray, H., Dereboylu, A.E., Altan, F., Zeytünlüoğlu, A., 2011. Identification of proteins involved in excess boron stress in roots of carrot (*Daucus carota* L.) and role of niacin in the protein profiles. African J. Biotechnol. 10, 15545–15551. Available from: https://doi.org/10.5897/AJB11.1117.

Domingues, D.S., Leite, S.M.M., Farro, A.P.C., Coscrato, V.E., Mori, E.S., Furtado, E.L., et al., 2005. Boron transport in Eucalyptus. 2. Identification in silico of a putative boron transporter for xylem loading in eucalypt. Genet. Mol. Biol. 28, 625–629. Available from: https://doi.org/10.1590/s1415-47572005000400017.

Dordas, C., Chrispeels, M.J., Brown, P.H., 2000. Permeability and channel-mediated transport of boric acid across membrane vesicles isolated from squash roots. Plant. Physiol. 124, 1349–1361. Available from: https://doi.org/10.1104/pp.124.3.1349.

Goldberg, S., 1997. Reactions of boron with soils. Plant Soil 35–48. Available from: https://doi.org/10.1007/978-94-011-5580-9_3.

González-fontes, A., Rexach, J., Quiles-pando, C., Herrera-rodríguez, M.B., Camacho-cristóbal, J.J., Navarro-gochicoa, M.T., 2013. Transcription factors as potential participants in the signal transduction pathway of boron deficiency. Plant. Signal. Behav. 8 (11), 10–12. e261.

Guo, P., Qi, Y.P., Yang, L.T., Ye, X., Jiang, H.X., Huang, J.H., et al., 2014. cDNA-AFLP analysis reveals the adaptive responses of citrus to long-term boron-toxicity. BMC Plant. Biol. 14, 1–22. Available from: https://doi.org/10.1186/s12870-014-0284-5.

Herrera-Rodríguez, M., González-Fontes, A., Rexach, J., Camacho-Cristóbal, J., Maldonado, J., Navarro-Gochicoa, M., 2010. Role of boron in vascular plants and response mechanisms to boron stresses. Plant. Stress. 4, 115–122.

Hu, H., Brown, P.H., 1997. Absorption of boron by plant roots. Plant. Soil. 193, 49–58. Available from: https://doi.org/10.1007/978-94-011-5580-9_4.

Kayihan, D.S., Kayihan, C., Çiftçi, Y.Ö., 2019. Regulation of boron toxicity responses via glutathione-dependent detoxification pathways at biochemical and molecular levels in *Arabidopsis thaliana*. Turk. J. Botany 43, 749–757. Available from: https://doi.org/10.3906/bot-1905-7.

Kayıhan, C., Öz, M.T., Eyidoğan, F., Yücel, M., Öktem, H.A., 2017. Physiological, biochemical, and transcriptomic responses to boron toxicity in leaf and root tissues of contrasting wheat cultivars. Plant. Mol. Biol. Report. 35, 97–109. Available from: https://doi.org/10.1007/s11105-016-1008-9.

Koshiba, T., Kobayashi, M., Ishihara, A., Matoh, T., 2010. Boron nutrition of cultured tobacco BY-2 Cells. VI. calcium is involved in early responses to boron deprivation. Plant. Cell Physiol. 51, 323–327. Available from: https://doi.org/10.1093/pcp/pcp179.

Kot, F.S., 2009. Boron sources, speciation and its potential impact on health. Rev. Environ. Sci. Biotechnol. 8, 3–28. Available from: https://doi.org/10.1007/s11157-008-9140-0.

Kurt, F., Aydın, A., 2020. An in-silico study: interaction of bor1-type boron (b) transporters with a small group of functionally unidentified proteins under various stresses in potato (*Solanum tuberosum*) commagene. J. Biol. 4, 134–139. Available from: https://doi.org/10.31594/commagene.798805.

Leaungthitikanchana, S., Fujibe, T., Tanaka, M., Wang, S., Sotta, N., Takano, J., et al., 2013. Differential expression of three BOR1 genes corresponding to different genomes in response to boron conditions in hexaploid wheat (*Triticum aestivum* L.). Plant. Cell Physiol. 54, 1056–1063. Available from: https://doi.org/10.1093/pcp/pct059.

Lu, Y.-P., Qi, Y.P., Yang, L.T., Guo, P., Li, Y., Chen, L.S., 2015. Boron-deficiency-responsive microRNAs and their targets in *Citrus sinensis* leaves. BMC Plant. Biol. 15, 1–15. Available from: https://doi.org/10.1186/s12870-015-0642-y.

Marentes, E., Vanderpool, R.A., Shelp, B.J., 1997. Boron-isotope fractionation in plants. Can. J. Plant. Sci. 77, 627–629.

Matoh, T., 1997. Boron in plant cell walls. Plant. Soil. 193, 59–70. Available from: https://doi.org/10.1007/978-94-011-5580-9_5.

Miwa, K., Wakuta, S., Takada, S., Ide, K., Takano, J., Naito, S., et al., 2013. Roles of BOR2, a boron exporter, in cross linking of rhamnogalacturonan II and root elongation under boron limitation in arabidopsis. Plant. Physiol. 163, 1699–1709. Available from: https://doi.org/10.1104/pp.113.225995.

Nable, R.O., Ba, G.S., Paull, J.G., 1997. Boron toxicity. Plant. Soil. 193, 181–198.

Nakagawa, Y., Hanaoka, H., Kobayashi, M., Miyoshi, K., Miwa, K., Fujiwara, T., 2007. Cell-type specificity of the expression of Os BOR1, a rice efflux boron transporter gene, is regulated in response to boron availability for efficient boron uptake and xylem loading. Plant. Cell 19, 2634–2635. Available from: https://doi.org/10.1105/tpc.106.049015.

Noguchi, K., Yasumori, M., Imai, T., Naito, S., Oda, H., Hayashi, H., et al., 1997. bor1-1, an *Arabidopsis thaliana* mutant that requires a high level of boron. Plant Physiol. 115, 901–906.

Nozawa, A., Takano, J., Kobayashi, M., Von Wirén, N., Fujiwara, T., 2006. Roles of BOR1, DUR3, and FPS1 in boron transport and tolerance in *Saccharomyces cerevisiae*. FEMS Microbiol. Lett. 262, 216–222. Available from: https://doi.org/10.1111/j.1574-6968.2006.00395.x.

Öz, M.T., Yilmaz, R., Eyidoğan, F., de Graaff, L., Yücel, M., Öktem, H.A., 2009. Microarray analysis of late response to boron toxicity in barley (*Hordeum vulgare* L.) leaves. Turkish J. Agric. For. 33, 191–202. Available from: https://doi.org/10.3906/tar-0806-22.

Ozhuner, E., Eldem, V., Ipek, A., Okay, S., Sakcali, S., Zhang, B., et al., 2013. Boron stress responsive microRNAs and their targets in barley. PLoS One 8. Available from: https://doi.org/10.1371/journal.pone.0059543.

Öztürk, S.E., Göktay, M., Has, C., Babaoğlu, M., Allmer, J., Doğanlar, S., et al., 2018. Transcriptomic analysis of boron hyperaccumulation mechanisms in *Puccinellia distans*. Chemosphere 199, 390–401. Available from: https://doi.org/10.1016/j.chemosphere.2018.02.070.

Ozyigit, I.I., Filiz, E., Saracoglu, I.A., Karadeniz, S., 2020. Exploration of two major boron transport genes BOR1 and NIP5;1 in the genomes of different plants. Biotechnol. Biotechnol. Equip. 34, 455–468. Available from: https://doi.org/10.1080/13102818.2020.1773311.

Padmanabhan, P., Babaoğlu, M., Terry, N., 2012. A comparative transcriptomic analysis of the extremely boron tolerant plant *Puccinellia distans* with the moderately boron tolerant *Gypsophila arrostil*. Plant. Cell Rep. 31, 1407–1413. Available from: https://doi.org/10.1007/s00299-012-1256-6.

Pallotta, M., Schnurbusch, T., Hayes, J., Hay, A., Baumann, U., Paull, J., et al., 2014. Molecular basis of adaptation to high soil boron in wheat landraces and elite cultivars. Nature 514, 88–91. Available from: https://doi.org/10.1038/nature13538.

Pfeffer, H., Dannel, F., Römheld, V., 1999. Are there two mechanisms for boron uptake in sunflower? J. Plant. Physiol. 155, 34–40. Available from: https://doi.org/10.1016/S0176-1617(99)80137-0.

Pérez-Castro, R., Kasai, K., Gainza-Cortés, F., Ruiz-Lara, S., Casaretto, J.A., Peña-Cortés, H., et al., 2012. VvBOR1, the grapevine ortholog of AtBOR1, encodes an efflux boron transporter that is differentially expressed throughout reproductive development of vitis vinifera L. Plant. Cell Physiol. 53, 485–494. Available from: https://doi.org/10.1093/pcp/pcs001.

Quiles-Pando, C., Navarro-Gochicoa, M.T., Herrera-Rodríguez, M.B., Camacho-Cristóbal, J.J., González-Fontes, A., Rexach, J., 2019. Boron deficiency increases cytosolic Ca2 + levels mainly via Ca2 + influx from the apoplast in *Arabidopsis thaliana* roots. Int. J. Mol. Sci. 20, 5–7. Available from: https://doi.org/10.3390/ijms20092297.

Reid, R., 2007. Identification of boron transporter genes likely to be responsible for tolerance to boron toxicity in wheat and barley. Plant. Cell Physiol. 48, 1673–1678. Available from: https://doi.org/10.1093/pcp/pcm159.

Sun, J., Shi, L., Zhang, C., Xu, F., 2012. Cloning and characterization of boron transporters in *Brassica napus*. Mol. Biol. Rep. 39, 1963–1973. Available from: https://doi.org/10.1007/s11033-011-0930-z.

Sutton, T., Baumann, U., Hayes, J., Collins, N.C., Shi, B.J., Schnurbusch, T., et al., 2007. Boron-toxicity tolerance in barley arising from efflux transporter amplification. Sci. (80-) 318, 1446–1449. Available from: https://doi.org/10.1126/science.1146853.

Takano, J., Miwa, K., Yuan, L., Von Wirén, N., Fujiwara, T., 2005. Endocytosis and degradation of BOR1, a boron transporter of *Arabidopsis thaliana*, regulated by boron availability. Proc. Natl. Acad. Sci. U. S. A. 102, 12276–12281. Available from: https://doi.org/10.1073/pnas.0502060102.

Takano, J., Noguchi, K., Yasumori, M., 2002. Arabidopsis boron transporter for xylem loading. Nature 420, 337–340.

Takano, J., Yamagami, M., Noguchi, K., Hayashi, H., Fujiwara, T., 2001. Preferential translocation of boron to young leaves in *Arabidopsis thaliana* regulated by the BOR1 gene. Soil. Sci. Plant. Nutr. 47, 345–357. Available from: https://doi.org/10.1080/00380768.2001.10408398.

Tanaka, M., Fujiwara, T., 2008. Physiological roles and transport mechanisms of boron: perspectives from plants. Pflug. Arch. Eur. J. Physiol. 456, 671–677. Available from: https://doi.org/10.1007/s00424-007-0370-8.

Tanaka, N., Uraguchi, S., Saito, A., Kajikawa, M., Kasai, K., Sato, Y., et al., 2013. Roles of pollen-specific boron efflux transporter, OsBOR4, in the rice fertilization process. Plant. Cell Physiol. 54, 2011–2019. Available from: https://doi.org/10.1093/pcp/pct136.

Warington, K., 1923. The effect of boric acid and borax on the broad bean and certain other plants. Ann. Bot. 37, 629–672.

Yoshinari, A., Takano, J., 2017. Insights into the mechanisms underlying boron homeostasis in plants. Front. Plant. Sci. 8, 1–8. Available from: https://doi.org/10.3389/fpls.2017.01951.

Zhang, Q., Chen, H., He, M., Zhao, Z., Cai, H., Ding, G., et al., 2017. The boron transporter BnaC4.BOR1;1c is critical for inflorescence development and fertility under boron limitation in *Brassica napus*. Plant. Cell Env. 40, 1819–1833. Available from: https://doi.org/10.1111/pce.12987.

CHAPTER 10

Transcription factors and target genes involved in plant responses to high boron adaptation

Emre Aksoy[1] and Ceyhun Kayıhan[2]
[1]Faculty of Arts and Sciences, Department of Biological Sciences, Middle East Technical University, Ankara, Turkey
[2]Faculty of Sciences and Letters, Department of Molecular Biology and Genetics, Başkent University, Ankara, Turkey

10.1 Introduction

The increased vulnerability of crops to a wide variety of biotic and abiotic stresses may have severe effects on the growth and yield of crops. As sessile organisms, plants must rapidly transmit extracellular stimuli/stresses to endogenous signals via signal transduction cascades and gain physiological adaptation. Therefore plants have evolved a variety of complex defense mechanisms based on signal networks in response to various environmental conditions. Transcription factors (TFs) are activated by different pathways of signal transduction and may associate directly or indirectly with cis-acting elements or TF binding sites (TFBSs) to modulate the transcription efficiency of target genes (Javed et al., 2020).

With some exceptions, a plant TF has a DNA-binding domain, a nuclear localization signal, a transcription-regulation domain, and an oligomerization site. Most plant TFs have only one type of DNA-binding and oligomerization site, sometimes in multiple copies, while some contain two different types. DNA-binding regions are generally overlapped with or adjacent to oligomerization sites and their combined complex tertiary structure regulates TF activity. Many plant TFs have pairs of nuclear localization signals and basic amino acid residues have a key role in their function. Multigene families express plant TFs, with members either clustered on the same chromosome or distributed throughout the genome. Sequence and distribution analyses show that plant TF families have evolved through mutation, exon capture, gene duplication, and translocation.

Boron in Plants and Agriculture
DOI: https://doi.org/10.1016/B978-0-323-90857-3.00012-6

TFs are categorized into several families or groups (Phillips, 2008). MYB, AP2/ERF, WRKY, bHLH are NAC (NAM/ATAF/CUC) are the major plant TF families which regulate various biological processes (Seo and Choi, 2015). TFBSs are generally highly conserved and used to classify TFs, including MADS, APETALA2/ethylene-responsive factors (AP2/ERF) and WRKY. Furthermore, TFs may also be classified based on their three-dimensional protein structures and compositions, including zinc finger proteins, basic helix-loop-helix (bHLH), and helix-turn-helix (Pulverer, 2005).

Boron (B) is one of the key micronutrients for crop growth and development (Marschner, 2012) because it is engaged in several important plant physiological and biochemical processes (Blevins and Lukaszewski, 1998). As an essential micronutrient, B plays a key role in plant growth, yield, crops quality, sugar transport, carbohydrate, auxin metabolism, cell wall structure, phenols, lignification, tissue development, membrane-associated reactions, petal and bud formation, ribose nucleic acid (RNA) metabolism, sugar and hydrocarbon metabolism, pollen germination and seed formation, cytokinin production, phenols and indole acetic acid metabolism (Goldbach and Monika, 2007).

B is taken by the plant roots in the form of boric acid ($H_3(BO)_3$). Although B is an essential micronutrient for plants, a high concentration of B is toxic to plants (Aquea et al., 2012). B toxicity is one of the important environmental factors affecting plant growth and production. Its symptoms in plants are more difficult to distinguish than other toxic ions, but normally it can be seen on older leaves as drying, yellowing, and spotting of the tips and edges of leaf tissues (Brdar-Jokanović, 2020). The common noticeable B toxicity symptoms are the weak plant vigor, reduced size and number of fruits, discoloration, death of leaf margins, leaf burn, necrotic patches on older leaves, black spots on older foliage, brown margin, and the tip of leaves, leaf area reduction, restricted cell wall expansion, reduction in photosynthetic rate and chlorophyll contents (Brdar-Jokanović, 2020). Similarly, other associated effects and symptoms of B toxicity are the reductions in root growth and dry weight (Turan et al., 2009), accumulation of reactive oxygen species (ROS) in barley (Karabal et al., 2003), oxidative damage and hydrogen peroxide (H_2O_2) accumulation in grapevine (Gunes et al., 2006), lignin and suberin deposition (Esim et al., 2013). Similarly, in dicots, the B toxicity can be seen around the leaf margins while the black patches on leaf tips are developed in grasses like wheat and barley (Roessner et al., 2006).

B toxicity causes several changes in the transcription of downstream genes, which leads to biochemical and physiological consequences observed as the symptoms. These transcriptional changes are regulated by several TF families in plants, and they have been identified in different monocot and dicots under B toxicity (Öz et al., 2009; Aquea et al., 2012; Tombuloglu et al., 2013; de Abreu Neto et al., 2017; Kayihan et al., 2017; Yıldırım and Uylaş, 2016, 2017; Feng et al. 2021). Here, we describe the identification and potential functions of some of these TFs identified under B toxicity in plants.

10.2 Transcription factors identified in *Arabidopsis thaliana* under boron toxicity

To investigate the molecular basis of B-induced inhibition of root growth, Aquea et al. (2012) carried out global gene expression analysis under 5 mM $H_3(BO)_3$ treatment in *Arabisopsis thaliana* roots. B caused a reduction in meristem size due to decreased cell division. The number of mitotic cells was reduced significantly and the expression level of key nuclear cell cycle regulators was modulated. Interestingly, the modulation of the cell cycle does not appear to act via cytokinin and auxin signaling. TFs that positively regulate the cell cycle (*E2Fa, E2Fb*) were not affected by B toxicity while the TF that negatively regulates the cell cycle (*DEL1*) was suppressed after 12 hours of boric acid treatment. Moreover, B toxicity induced the expression of genes involved in abscisic acid (ABA) signaling and ABA-responsive cell wall modifications while the genes encoding for water transporters were repressed. These results suggest that B toxicity produces a reduction in water and boric acid uptake, triggering a stress response that produces inhibition of root growth (Aquea et al., 2012). In addition to some general stress-responsive genes, some TFs are also differentially expressed under B toxicity. For instance, *HOMEOBOX7 (HB7), HOMEOBOX12 (HB12), ABI FIVE BINDING PROTEIN1 (AFP1), INDUCER OF CBF EXPRESSION1 (ICE1)* were significantly upregulated under B toxicity.

Feng et al. (2021) found a B toxicity-responsive TF WRKY47 in *A. thaliana*. In comparison to the wild-type Col-0 plants, the T-DNA insertion mutant *wrky47* caused increased tolerance to B toxicity and better growth parameters under high B conditions. Under B toxicity conditions, *WRKY47* mRNA was downregulated, suggesting it could function as a negative regulator under B toxicity. In contrast to the phenotype of *wrky47* mutants, overexpression of *WRKY47* in Col-0 caused increased sensitivity to B toxicity since less chlorophyll content and lower biomass were obtained in the overexpression lines. Moreover, the B concentration in shoots was lower in the *wrky47* mutants and higher in the overexpression lines than in Col-0 plants. These results show that WRKY47 has a key role in regulating plant tolerance under B toxicity.

In a yeast complementation experiment, *Arabidopsis* MYB13 and MYB63 conferred tolerance to B toxicity when overexpressed in B efflux transporter *ScBOR1* mutant (YNL275W), suggesting that they may function as positive regulators downstream genes involved in B toxicity tolerance (Nozawa et al., 2006).

Recently, it has been suggested that B forms a complex with anthocyanin. B-anthocyanin complexes in vacuoles have been suggested as a tolerance mechanism against B toxicity (Landi et al., 2015). Anthocyanin−GSH or −GST complexes can transiently bind to metal or metalloid ions thereby forming GST-anthocyanin-metal complexes and/or glutathionylanthocyanin-metal complexes, which can be sequestered into the vacuoles. To understand the expression levels of genes encoding for anthocyanin biosynthesis and transport, *A. thaliana* Col-0 ecotype was grown under 1 and 3 mM boric acid and qRT-PCR was performed (Kayihan et al., 2021). Accordingly, the TFs regulating the anthocyanin biosynthesis and/or transporter genes such as *MYB75, MYB114*, and *TT8* were significantly induced under B toxicity. The induction rate was higher in response to high B concentrations, suggesting the involvement of anthocyanin accumulation as a novel tolerance response in plants under B toxicity.

10.3 Transcription factors identified in barley under boron toxicity

Öz et al. (2009) aimed to identify differentially expressed genes of barley (*Hordeum vulgare* L.) under B toxicity and elucidate the molecular mechanism of B tolerance or toxicity. This was the first report on global expression analysis of barley seedlings under B toxicity. In this study, barley seedlings were treated with 5 or 10 mM boric acid ($B(OH)_3$) for 5 days, and the transcriptome changes in the leaves of a sensitive barley cultivar were investigated by the Barley1 GeneChip. In comparison to the control, 5 mM $B(OH)_3$ treatment caused differential expression of 168 genes at least by two folds. Furthermore, 10 mM $B(OH)_3$ treatment caused at least twofold reduction or induction in the expression of 312 transcripts.

37 and 61 of them exhibited significantly altered levels of expression under 5 and 10 mM $B(OH)_3$ treatments, respectively. Differential expression profiles showed that the networks of signaling or molecular responses change in B toxicity. Moreover, jasmonic acid (JA)-related genes were found to be an important late response to B toxicity (Öz et al., 2009). Some TFs such as *SQUAMOSA PROMOTER BINDING PROTEIN-LIKE7 (SPL7)*, *MYB DOMAIN PROTEIN33 (MYB33)* and *RELATED TO AP2.7/TARGET OF EARLY ACTIVATION TAGGED (EAT)1* (*RAP2.7/TOE1*) were significantly induced under 5 mM $B(OH)_3$ treatment whereas *PSEUDO-RESPONSE REGULATOR3 (PRR3)*, *NAC DOMAIN-CONTAINING PROTEIN29/NAC-LIKE, ACTIVATED BY AP3/PI (NAC29/NAP)* and *C-REPEAT BINDING FACTOR3/DEHYDRATION RESPONSE ELEMENT B1A (CBF3/DREB1A)* were significantly upregulated under 10 mM $B(OH)_3$ treatment. *WOUND INDUCED DEDIFFERENTIATION1/RELATED TO AP2 4 (WIND1/RAP2.4)* and *WRKY41* were downregulated under 5 mM $B(OH)_3$ treatment whereas *C-REPEAT/DRE BINDING FACTOR2/DRE/CRT-BINDING PROTEIN 1C/FREEZING TOLERANCE QTL4 (CBF2/DREB1C/FTQ4)* and *HEAT SHOCK TRANSCRIPTION FACTOR A6B (HSFA6B)* were significantly repressed under 10 mM $B(OH)_3$ treatment.

MYB TF family is one of the largest TFs in plants. It contains highly conserved repeats (1R, R2R3, 3R, and 4R) in the N-terminus and has diverse sequences in the C-terminal, which help the protein gain distinct functions, including response to biotic and abiotic stress, secondary metabolism production, controlling development and hormonal regulations. To identify R2R3-MYB TFs, and understand their interaction partners under B toxicity, Tombuloglu et al. (2013) carried out a genome-wide transcriptomic study in barley tissues under control and toxic B conditions. Using protein motif and conserved domain analyses, a total of 320 protein sequences were found as MYB TFs while 51 of them were identified as R2R3-type MYB TFs. Phylogenetic relations with *A. thaliana* MYBs categorized the barley R2R3 MYB TFs into 34 subgroups. Conserved domains of the MYB proteins were compared with the DNA-binding domains of *A. thaliana*, *Zea mays*, and *Brachypodium distachyon*. The results showed that *Brachypodium*, maize, and *Arabisopsis* DNA-binding domains had high conservation with barley. By comparing *MYB* genes with their *Brachypodium* homologs, chromosome distribution and probable location of the *MYB* genes were predicted. Moreover, barley MYB TFs were quantitatively evaluated and transcript accumulations were compared between B toxicity and control in leaf and root tissues, separately, and important R2R3 MYB TFs induced under B toxicity were identified (Tombuloglu et al., 2013). Among them, *HvMYB20, HvMYB32, HvMYB33,* and *HvMYB40* were upregulated in the leaves; *HvMYB01, HvMYB07, HvMYB09, HvMYB11, HvMYB12, HvMYB13, HvMYB15, HvMYB18, HvMYB19, HvMYB21, HvMYB23, HvMYB25, HvMYB26, HvMYB27, HvMYB28, HvMYB29, HvMYB30, HvMYB35, HvMYB36, HvMYB37, HvMYB39, HvMYB41, HvHvMYB42, HvMYB45, HvMYB46, HvMYB47,* and *HvMYB49* were upregulated in the roots whereas *HvMYB01, HvMYB14, HvMYB16, HvMYB22, HvMYB31, HvMYB43, HvMYB44, HvMYB48,* and *HvMYB51* were upregulated in both tissues under B toxicity. *HvMYB01, HvMYB07, HvMYB09, HvMYB11, HvMYB12, HvMYB13, HvMYB18, HvMYB19, HvMYB21, HvMYB23, HvMYB25, HvMYB26, HvMYB27, HvMYB28, HvMYB35, HvMYB36, HvMYB37, HvMYB41, HvMYB42, HvMYB45, HvMYB46,* and *HvMYB49* were downregulated in the leaves; *HvMYB20, HvMYB32, HvMYB34* and *HvMYB40* were downregulated in the roots whereas *HvMYB10, HvMYB17, HvMYB32,* and *HvMYB50* were downregulated in both tissues under B toxicity.

10.4 Transcription factors identified in poplar under boron toxicity

Poplar (genus *Populus*) is one of the species tolerant to B toxicity and accumulation. To identify the transcriptomic, physiological, and hormonal regulations under B toxicity in poplar, Yıldırım and Uylaş (2016) performed a microarray-based transcriptome experiment with elevated B toxicity treated leaves and roots of black poplar clones. The results showed that black poplar is quite excellent for phytoremediation of B pollution. It could withstand 15 ppm soil B content and >1500 ppm B accumulation in leaves, which are extremely toxic concentrations to practically all agricultural plants. Transcriptomic results revealed a total of 1419 and 1625 altered probe sets under 15 ppm B toxicity in root and leaf tissues, respectively. The highest induction was documented for the probe sets annotated to glutathione S transferases, ATP binding cassette transporters, metallochaperone proteins, and tyrosine aminotransferase. Upregulation of these genes was attributed to the existence of B detoxification mechanism and B sequestration in the vacuole. Furthermore, several other candidate genes were found to be differentially expressed related to transcriptional regulation, B uptake and transport, and ROS scavenging. Most of the genes identified in this study have homologs in many other plants. Among differentially expressed genes, some TF families such as NAC, ERF, and bZIP were highly induced and some such as MYB, NAC, and ERF are highly downregulated under B toxicity.

To investigate the TF families involved in B toxicity in poplar, Yıldırım (2017) studied the physiological and transcriptomic responses of *Populus alba* (P.a) and *Populus nigra* (P.n) clones under B toxicity. Of the poplar clones collected for this study, *Populus alba* (Pa) (PN.92.237) and *Populus nigra* (Pa) (PA.29.126) were recorded as the lowest and highest boron accumulator plants, respectively. Then, these clones were used in a microarray-based transcriptome study to find out genes, transcriptional networks, and phytohormone-related molecular pathways that may be responsible for boron accumulation and toxicity tolerance in differential tissues. There were 1006 and 1902 differentially regulated transcripts for the roots and leaves of P.a under B toxicity, respectively. In P.a tissues, several transcripts are responsible for salicylic acid (SA)-dependent gene regulation (PR proteins, WRKYs, chitinases, proteases, lipases, and protease inhibitors), and SA production (such as salicylic acid-binding protein 2) were significantly upregulated under B toxicity. Many different TFs were significantly differentially expressed in roots and leaves of both species under B toxicity. Some TF families differentially expressed include ERF, MYB, NAC, WRKY, bZIP, and zinc finger.

10.5 Transcription factors identified in rice under boron toxicity

To investigate the genetic factors underlying B toxicity tolerance in rice, de Abreu Neto et al. (2017) performed a genome-wide association study (GWAS) coupled to transcriptomic analysis. A panel of 137 *indica* genotypes from a wide range of backgrounds were phenotyped in toxic B conditions according to root length, leaf bronzing, and root and shoot dry weights. GWAS of these genotypes identified candidate loci containing 75 gene models, among which 9 were coding for TFs. These TFs are categorized in the families of MYB, BHLH, zinc finger and RING/U-box. MYB12 and RING/U-box proteins were highly associated with root length, BHLH TF (PYE) was associated with leaf bronzing score, and zinc finger RHL41/ZAT12 was highly associated with both shoot and root dry weights. Then, two contrasting genotypes were selected and differentially expressed genes were determined under B toxicity in transcriptomic analyses. Characteristic gene expression patterns related to tolerance or sensitivity were observed in contrasting rice genotypes. Among the TFs identified in GWAS, only the RING/U-box protein was significantly expressed in tolerant cultivar than the sensitive one under B toxicity.

In another study, a NAC-like TF boron excess tolerant 1 (BET-1) was identified as a major locus associated with B tolerance in rice recombinant inbreeding lines by eco-TILLING analysis (Ochiai et al., 2011). When the gene is silenced, the transgenic plants showed enhanced B tolerance, indicating a critical role of these TFs in rice.

10.6 Transcription factors identified in wheat under boron toxicity

Kayihan et al. (2017) investigated transcriptomic, biochemical, and physiological responses in toxic B-treated roots and leaves of wheat cultivars. A high level of B did not cause a delayed development or reduced vigor in the roots and leaves of both cultivars. In comparison to the control conditions, 1248 and 957 differentially expressed genes were responsive to B toxicity in the roots while 995 and 892 genes were differentially expressed in the leaves of cv. Atay-85 (B sensitive) and cv. Bolal-2973 (B tolerant). B toxicity differentially affected the genes associated with kinases, hormone signaling, TFs such as MYB and WRKY, and major enzymes in ROS scavenging mechanisms in both cultivars. A total of 12 TF genes were differentially expressed in both cultivars under B toxicity. C2H2, MYB, and WRKY were the families represented the most among differentially expressed TFs.

10.7 Transcription factors identified in *Puccinellia distans* under boron toxicity

Genome-wide transcriptomic studies have been also done in other plant families under B toxicity. Some plant species are highly adapted to high levels of B. One of these species is common alkali grass (*Puccinellia distans*) which can accumulate up to 6 g/kg B in the shoots without any obvious side effects (Bar, 2015). Its relative *Puccinellia frigida* is also known for its high B accumulation in the shoots (Rámila et al., 2016). Therefore *Puccinellia* genus can be investigated for the identification of new TFs involved in B hyperaccumulation. In a recent study, the transcriptomic changes of *P. distans* collected from a boron mining site in Turkey were investigated under high B conditions in hydroponics (Öztürk et al., 2018). 1% of all differentially expressed transcripts were found to be related to TF activity gene ontology (GO). Some of the differentially expressed TFs in *P. distans* include FAR1-RELATED SEQUENCE4 (FRS4), FRS5, and YIN YANG1 (YY1).

10.8 miRNAs involved in post-transcriptional control under boron toxicity

miRNAs are capable of targeting single/multiple mRNA targets. Some miRNA targets are TFs. Studies between miRNAs and TFs from different plant species help to understand the mechanism of action and interaction between these regulators in plant growth and development under normal and environmental stress conditions. Ozhuner et al. (2013) characterized miRNAs related to B toxicity in barley leaves and roots and found 31 known and three new miRNAs. In silico analysis of miRNA target genes showed that several of them are conserved TFs found in the MYB, the auxin response factor (ARF), the NAC and the MADS-box protein families. For example, hvu-miR159 and hvu-miR319a target a MYB TF, hvu-miR160 targets an ARF TF, hvu-miR164 targets a NAC TF, and hvu-miR444 targets a MADS-box TF.

Huang et al. (2016) determined the miRNAs differentially expressed under B toxicity in *Citrus sinensis* (tolerant) and *C. grandis* (sensitive) leaves via next-generation sequencing. 51 miRNAs in *C. grandis* and 20 miRNAs in *C. sinensis* were differentially expressed under B toxicity. The targets of these miRNAs were predicted by software and their gene expression levels were also determined in B-treated plant leaves. Among the candidate targets, TF genes are of interest. For instance, it was predicted that miR858b can target different MYB family of proteins while miR473a can target a GRAS family of TF, known as *RGA-LIKE1* (*RGL1*), which is a negative regulator of gibberellic acid responses and is involved in reducing ROS accumulation in response to stress by upregulating the transcription of *SUPEROXIDE DISMUTASE* genes (Park et al., 2013). miR2622.2 is predicted to target *TCP FAMILY TRANSCRIPTION FACTOR4* (*TCP4*), which is involved in heterochronic regulation of leaf differentiation as regulated by miR319 (Vadde et al., 2019). ci-miRN31 was predicted to target *ALTERED SEED GERMINATION4/REVEILLE3* (*ASG4/REV3*) that plays an insignificant role in clock regulation (Gray et al., 2017).

miR347 was predicted to target *APETALA2/FLORAL MUTANT2/FLOWER1* (*AP2/FL1/FLO2*), which is a member of the AP2/EREBP family of TFs, and is involved in the specification of floral organ identity, the establishment of floral meristem identity, suppression of floral meristem indeterminacy, and development of the ovule and seed coat (Zumajo-Cardona et al., 2021). miR847 was predicted to target *MYB, WRKY9*, and *RAP2.7/TOE1.* Finally, miR164a and miR319a were predicted to target *NAC98* and *MYB33*, respectively.

In another study from the same group, differentially expressed miRNAs were again identified in *C. grandis* and *C. sinensis* roots by next-generation sequencing (Huang et al., 2019). The results demonstrated that 11 miRNAs in *C. sinensis* and 37 miRNAs in *C. grandis* were differentially expressed. miR319/miR159 was found to target *MYB65* while miR171 targeted a GRAS family of TF, *HAIRY MERISTEM3/LOST MERISTEMS3* (*HAM3/LOM3*). miR319 also targeted another TF in the family of TCP, namely *TEOSINTE BRANCHED 1, CYCLOIDEA AND PCF TRANSCRIPTION FACTOR2* (*TCP2*), which can control multiple aspects of growth and development in plants (He et al., 2021).

In another study, Jyothi et al. (2018) confirmed that seven miRNAs were highly responsive to toxic B levels via RT-qPCR in *Phaseolus vulgaris.* Target prediction and their functional analysis showed that some of the miRNA targets represented TFs regulating the expression of stress-related genes. miR398 targets a GATA-type zinc finger TF and miR845 targets *MYB20.*

Kayihan et al. (2019) determined the expression levels of miRNAs targeting the TFs involved in JA and ethylene biosynthesis and/or signaling including miR159, miR172, miR319 and miR394 under toxic B conditions in *A. thaliana.* The expression levels of these three miRNAs were induced remarkably in moderate B toxicity (1 mM H_3BO_3) but not in severe B toxicity (3 mM H_3BO_3). miR159 is predicted to target *UAS-TAGGED ROOT PATTERNING3 (UPR3/OBP2)*, which is a member of the DOF TF family. It is prominently expressed in the phloem of leaves and other organs, and the expression is induced by wounding, methyljasmonate and insect feeding. It positively regulates glucosinolate biosynthesis while negatively regulating boundary formation at the zone of cell division (Skirycz et al., 2006; Miyashima et al., 2019; Smit et al., 2020). It can also target MYB TFs such as *MYB1/MYB101, MYB104* and *DUO POLLEN1* (*DUO1*) that is required for male gamete formation, specifically for entry of the generative cell into mitosis (Borg et al., 2011; Zheng et al., 2017).

miR172 is predicted to target an AP2 family of proteins *TOE1, TOE2, TOE3*, and *AP2/FL1*/FLO2 which are involved in the regulation of flowering and innate immunity as well as floral organ identity (Du et al., 2020; Zou et al., 2020; Ó'Maoiléidigh et al., 2021; Zumajo-Cardona et al., 2021). miR319 is predicted to target *MYB097* whereas miR394 can target a putative F-box protein *LEAF CURLING RESPONSIVENESS* (*LCR*) that is involved in the regulation of leaf morphology (Song et al., 2013). The same group later investigated the expression patterns of *miR172, miR319*, and *miR398* and their targets under toxic B levels in two wheat cultivars, Atay-85 (B-sensitive) and Bolal-2973 (B-tolerant) (Kayihan et al., 2021). *miR172* targeting TF *TOE1* was only induced under B toxicity in Atay while *MYB3* as the target of miR319 was significantly upregulated under B toxicity in both cultivars.

10.9 Long noncoding RNAs involved in post-transcriptional control under B toxicity

In a recent study, Unver, Tombuloglu (2020) performed the genome-wide discovery of approximately 8000 long noncoding RNAs (lncRNAs) in barley under excessive B conditions and assessed their expression patterns. According to the tissue-specific comparison, leaves showed more B-responsive differentially expressed lncRNAs than the roots. Furthermore, the coding transcripts, which are coexpressed with lncRNAs were functionally annotated. Around 32 barley lncRNAs were found as eTMs (endogenous target mimics) that possibly negatively regulate the transcriptional repression activity of 18 miRNAs. Some lncRNAs (i.e., *TCONS_00056415* and *TCONS_00045190*) were only induced in B-tolerant variants under excessive B conditions. Therefore this study demonstrates that the B toxicity response may also be regulated by lncRNA expression via miRNA-eTM-encoder interaction (Unver, Tombuloglu, 2020).

10.10 Known functions of identified transcription factor families under boron toxicity

Overall, a total of 126 TF genes were determined as differentially expressed or predicted as miRNA targets in different plant species under B toxicity (Table 10.1). Some of them have been characterized under B toxicity itself. However, we still do not know the functions of many of them in tolerance to B toxicity. GO enrichment analyses of TFs identified

Table 10.1 Arabidopsis orthologs of differentially expressed transcription factors under boron toxicity in different plant species.

Plant species	Arabidopsis ortholog ID	Gene abbreviation	Gene name	References
Arabidopsis	AT1G06180	*MYB13*	*MYB DOMAIN PROTEIN13*	Nozawa et al. (2006)
	AT5G65790	*MYB68*	*MYB DOMAIN PROTEIN68*	
	AT2G46680	*HB7*	*HOMEOBOX7*	Aquea et al. (2012)
	AT3G61890	*HB12*	*HOMEOBOX12*	
	AT1G69260	*AFP1*	*ABI FIVE BINDING PROTEIN1*	
	AT3G26744	*ICE1*	*INDUCER OF CBF EXPRESSION1*	
	AT5G14000	*NAC084*	*NAC DOMAIN-CONTAINING PROTEIN84*	
	AT1G07640	*UPR3/OBP2*	*UAS-TAGGED ROOT PATTERNING3*	Kayihan et al. (2019)
	AT2G32460	*MYB1/MYB101/ABS7*	*MYB DOMAIN PROTEIN1/MYB DOMAIN PROTEIN101/ABNORMAL SHOOT7*	
	AT3G60460	*DUO1*	*DUO POLLEN 1*	
	AT2G26950	*MYB104*	*MYB DOMAIN PROTEIN104*	
	AT2G28550	*RAP2.7/TOE1*	*RELATED TO AP2.7/TARGET OF EARLY ACTIVATION TAGGED (EAT)1*	
	AT5G60120	*TOE2*	*TARGET OF EARLY ACTIVATION TAGGED (EAT)2*	
	AT5G67180	*TOE3*	*TARGET OF EARLY ACTIVATION TAGGED (EAT)3*	
	AT4G36920	*AP2/FL1/FLO2*	*APETALA2/FLOWER1/FLORAL MUTANT2*	
	AT4G26930	*MYB097*	*MYB DOMAIN PROTEIN097*	
	AT1G27340	*LCR*	*LEAF CURLING RESPONSIVENESS*	
	AT1G66380	*MYB114*	*MYB DOMAIN PROTEIN114*	Kayihan (2021)
	AT1G56650	*MYB75/PAP1/SIAA1*	*MYB DOMAIN PROTEIN75 /PRODUCTION OF ANTHOCYANIN PIGMENT1 / SUC-INDUCED ANTHOCYANIN ACCUMULATION1*	
	AT4G09820	*TT8*	*TRANSPARENT TESTA8*	
	AT4G01720	*WRKY47*	*WRKY47*	Feng et al. (2021)

Barley	AT5G18830	*SPL7*	*SQUAMOSA PROMOTER BINDING PROTEIN-LIKE7*	Öz et al. (2009)
	AT2G28710	–	C2H2-type zinc finger family protein	
	AT5G06100	*MYB33*	*MYB DOMAIN PROTEIN33*	
	AT2G28550	*RAP2.7/TOE1*	*RELATED TO AP2.7/TARGET OF EARLY ACTIVATION TAGGED (EAT)1*	
	AT2G27050	*EIL1*	*ETHYLENE-INSENSITIVE3-LIKE1*	
	AT5G60100	*PRR3*	*PSEUDO-RESPONSE REGULATOR3*	
	AT1G69490	*NAC29/NAP*	*NAC DOMAIN-CONTAINING PROTEIN29/NAC-LIKE, ACTIVATED BY AP3/PI*	
	AT4G25480	*CBF3/DREB1A*	*C-REPEAT BINDING FACTOR3/DEHYDRATION RESPONSE ELEMENT B1A*	
	AT1G78080	*WIND1/RAP2.4*	*WOUND INDUCED DEDIFFERENTIATION1/RELATED TO AP2 4*	
	AT4G11070	*WRKY41*	*WRKY41*	
	AT4G25470	*CBF2/DREB1C/FTQ4*	*C-REPEAT/DRE BINDING FACTOR2/DRE/CRT-BINDING PROTEIN 1C/FREEZING TOLERANCE QTL4*	
	AT3G22830	*HSFA6B*	*HEAT SHOCK TRANSCRIPTION FACTOR A6B*	
	AT4G38620	*MYB04*	*MYB DOMAIN PROTEIN04*	Tombuloglu et al. (2013)
	AT2G16720	*MYB07*	*MYB DOMAIN PROTEIN07*	
	AT3G23250	*MYB15*	*MYB DOMAIN PROTEIN15*	
	AT5G15310	*MYB16*	*MYB DOMAIN PROTEIN16*	
	AT4G25560	*MYB18*	*MYB DOMAIN PROTEIN18*	
	AT5G52260	*MYB19*	*MYB DOMAIN PROTEIN19*	
	AT3G27810	*MYB21*	*MYB DOMAIN PROTEIN21*	
	AT5G40350	*MYB24*	*MYB DOMAIN PROTEIN24*	
	AT3G28910	*MYB30*	*MYB DOMAIN PROTEIN30*	
	AT5G57620	*MYB36*	*MYB DOMAIN PROTEIN36*	
	AT1G57560	*MYB50*	*MYB DOMAIN PROTEIN50*	
	AT1G17950	*MYB52*	*MYB DOMAIN PROTEIN52*	

(*Continued*)

Table 10.1 (Continued)

Plant species	Arabidopsis ortholog ID	Gene abbreviation	Gene name	References
	AT1G73410	*MYB54*	*MYB DOMAIN PROTEIN54*	
	AT1G08810	*MYB60*	*MYB DOMAIN PROTEIN60*	
	AT1G09540	*MYB61*	*MYB DOMAIN PROTEIN61*	
	AT3G24310	*MYB71*	*MYB DOMAIN PROTEIN71*	
	AT4G13480	*MYB79*	*MYB DOMAIN PROTEIN79*	
	AT3G49690	*MYB84*	*MYB DOMAIN PROTEIN84*	
	AT5G26660	*MYB86*	*MYB DOMAIN PROTEIN86*	
	AT2G02820	*MYB88*	*MYB DOMAIN PROTEIN88*	
	AT4G21440	*MYB102*	*MYB DOMAIN PROTEIN102*	
	AT3G01140	*MYB106*	*MYB DOMAIN PROTEIN106*	
	AT3G55730	*MYB109*	*MYB DOMAIN PROTEIN109*	
Poplar	AT2G38340	*DREB19*	*DEHYDRATION RESPONSE ELEMENT-BINDING PROTEIN19*	Yıldırım and Uylaş (2016)
	AT5G24590	*NAC091/TIP*	*NAC DOMAIN-CONTAINING PROTEIN91/TCV-INTERACTING PROTEIN*	
	AT4G28500	*NAC073/SND2*	*NAC DOMAIN-CONTAINING PROTEIN73/ SECONDARY WALL-ASSOCIATED NAC DOMAIN PROTEIN 2*	
	AT5G59820	*RHL41/ZAT12*	*RESPONSIVE TO HIGH LIGHT41*	
	AT5G13330	*RAP2.6L*	*RELATED TO AP2 6L*	
	AT5G50080	*ERF110*	*ETHYLENE RESPONSE FACTOR110*	
	AT4G27950	*CRF4*	*CYTOKININ RESPONSE FACTOR4*	
	AT2G37430	*ZAT11*	*ZINC FINGER OF ARABIDOPSIS THALIANA11*	
	AT5G49450	*BZIP1*	*BASIC LEUCINE-ZIPPER1*	
	AT3G62420	*BZIP53*	*BASIC REGION/LEUCINE ZIPPER MOTIF53*	
	AT5G13080	*WRKY75*	*WRKY DNA-BINDING PROTEIN75*	
	AT5G61890	*ERF114*	*ETHYLENE RESPONSE FACTOR114*	
	AT5G61430	*NAC100*	*NAC DOMAIN-CONTAINING PROTEIN 100*	

	AT1G14600	–	Homeodomain-like superfamily protein	
	AT5G25810	*ERF40/TNY*	*ETHYLENE RESPONSE FACTOR40/TINY*	
	AT5G01380	–	Homeodomain-like superfamily protein	
	AT2G38340	*DREB19*	*DEHYDRATION RESPONSE ELEMENT-BINDING PROTEIN19*	Yıldırım (2017)
	AT5G13330	*RAP2.6L*	*RELATED TO AP2 6L*	
	AT5G50080	*ERF110*	*ETHYLENE RESPONSE FACTOR110*	
	AT1G25340	*MYB116*	*MYB DOMAIN PROTEIN116*	
	AT1G68320	*MYB62*	*MYB DOMAIN PROTEIN62*	
	AT5G24590	*NAC091/TIP*	*NAC DOMAIN-CONTAINING PROTEIN91/TCV-INTERACTING PROTEIN*	
	AT5G18270	*NAC087*	*NAC DOMAIN-CONTAINING PROTEIN87*	
	AT4G27410	*NAC72/RD26*	*NAC DOMAIN-CONTAINING PROTEIN72/RESPONSIVE TO DESICCATION 26*	
	AT5G13080	*WRKY75*	*WRKY DNA-BINDING PROTEIN75*	
	AT1G29860	*WRKY71/EXB1*	*WRKY DNA-BINDING PROTEIN71/EXCESSIVE BRANCHES1*	
	AT1G80840	*WRKY40*	*WRKY DNA-BINDING PROTEIN40*	
	AT3G56400	*WRKY70*	*WRKY DNA-BINDING PROTEIN70*	
	AT5G49450	*BZIP1*	*BASIC LEUCINE-ZIPPER1*	
	AT3G62420	*BZIP53*	*BASIC REGION/LEUCINE ZIPPER MOTIF53*	
	AT2G37430	*ZAT11*	*ZINC FINGER OF ARABIDOPSIS THALIANA11*	
	AT5G59820	*RHL41/ZAT12*	*RESPONSIVE TO HIGH LIGHT41*	
	AT1G25580	*SOG/NAC008*	*NAC DOMAIN-CONTAINING PROTEIN8/SUPPRESSOR OF GAMMA RADIATION1*	Ochiai et al. (2011)
Rice	AT2G47460	*MYB12*	*MYB DOMAIN PROTEIN12*	de Abreu Neto et al. (2017)
	AT3G05670	–	RING/U-box protein	
	AT3G47640	*PYE*	*POPEYE*	
	AT5G59820	*RHL41/ZAT12*	*RESPONSIVE TO HIGH LIGHT41*	

(*Continued*)

Table 10.1 (Continued)

Plant species	Arabidopsis ortholog ID	Gene abbreviation	Gene name	References
Wheat	AT3G17980	*C2/ATGAP1*	*C2 DOMAIN/HOMOLOG OF OSGAP1*	Kayihan et al. (2017)
	AT2G45190	*YAB1/AFO/FIL*	*YABBY1/ABNORMAL FLORAL ORGANS/ FILAMENTOUS FLOWER,*	
	AT5G43170	*ZFP3*	*ZINC-FINGER PROTEIN3*	
	AT3G09600	*LCL5/RVE8*	*LHY-CCA1-LIKE5/REVEILLE 8*	
	AT5G26170	*WRKY50*	*WRKY DNA-BINDING PROTEIN50*	
	AT1G68840	*RAV2/EDF2/TEM2*	*RELATED TO ABI3/VP1 2/ETHYLENE RESPONSE DNA BINDING FACTOR2/ TEMPRANILLO 2*	
	AT4G22680	*MYB85*	*MYB DOMAIN PROTEIN85*	
	AT4G14805	*WRKY45*	*WRKY DNA-BINDING PROTEIN45*	
	AT4G31550	*WRKY11*	*WRKY DNA-BINDING PROTEIN11*	
	AT1G75490	–	member of the DREB subfamily A-2 of ERF/AP2 transcription factor family	
	AT5G58010	*BHLH82/DROP3/LRL3*	*BASIC HELIX-LOOP-HELIX82/DEFECTIVE REGION OF POLLEN3/LJRHL1-LIKE 3*	
	AT5G55020	*MYB120*	*MYB DOMAIN PROTEIN120*	Kayihan et al. (2019)
	AT4G27330	*NZZ/SPL*	*NOZZLE/SPOROCYTELESS*	
	AT3G28910	*MYB30*	*MYB DOMAIN PROTEIN30*	
	AT2G31220	*BHLH10*	*BASIC HELIX LOOP HELIX PROTEIN10*	
	AT2G28550	*RAP2.7/TOE1*	*RELATED TO AP2.7/TARGET OF EARLY ACTIVATION TAGGED (EAT)1*	Kayihan et al. (2021)
	AT4G36920	*AP2/FL1/FLO2*	*APETALA2/FLOWER1/FLORAL MUTANT2*	
	AT5G67180	*TOE3*	*TARGET OF EARLY ACTIVATION TAGGED (EAT)3*	
	AT5G06100	*MYB33*	*MYB DOMAIN PROTEIN33*	
	AT2G32460	*MYB1/MYB101/ABS7*	*MYB DOMAIN PROTEIN1/MYB DOMAIN PROTEIN101/ABNORMAL SHOOT7*	
	AT3G11440	*MYB65*	*MYB DOMAIN PROTEIN65*	

	AT3G16770	*ERF72/RAP2.3*	*ETHYLENE RESPONSE FACTOR72/RELATED TO AP2 3*	
	AT1G72360	*ERF73/HRE1*	*ETHYLENE RESPONSE FACTOR73/HYPOXIA RESPONSIVE ERF1*	
	AT2G47520	*ERF71/HRE2*	*ETHYLENE RESPONSE FACTOR71/HYPOXIA RESPONSIVE ERF2*	
	AT2G37740	*ZFP10*	*ZINC-FINGER PROTEIN10*	
	AT2G42410	*ZFP11*	*ZINC-FINGER PROTEIN11*	
	AT3G23130	*FLO10/FON1/SUP*	*FLORAL DEFECTIVE10/FLORAL ORGAN NUMBER1/SUPERMAN*	
Citrus cinensis and *C. grandis*	AT3G15030	*MEE3/TCP4*	*MATERNAL EFFECT EMBRYO ARREST35/TCP FAMILY TRANSCRIPTION FACTOR4*	Huang et al. (2016)
	AT1G01520	*ASG4/REV3*	*ALTERED SEED GERMINATION4/REVEILLE3*	
	AT4G36920	*AP2/FL1/FLO2*	*APETALA2/FLORAL MUTANT2/FLOWER1*	
	AT4G38620	*MYB4*	*MYB DOMAIN PROTEIN4*	
	AT1G68150	*WRKY9*	*WRKY DNA-BINDING PROTEIN9*	
	AT2G28550	*RAP2.7/TOE1*	*RELATED TO AP2.7/TARGET OF EARLY ACTIVATION TAGGED (EAT)1*	
	AT5G49330	*MYB111/PFG3*	*MYB DOMAIN PROTEIN111/PRODUCTION OF FLAVONOL GLYCOSIDES3*	
	AT2G47460	*MYB12/PFG1*	*MYB DOMAIN PROTEIN12/PRODUCTION OF FLAVONOL GLYCOSIDES1*	
	AT3G23250	*MYB15*	*MYB DOMAIN PROTEIN15*	
	AT2G31180	*MYB14*	*MYB DOMAIN PROTEIN14*	
	AT1G66350	*RGL1*	*RGA-LIKE1*	
	AT5G53950	*NAC98/CUC2*	*NAC DOMAIN-CONTAINING PROTEIN98/CUP-SHAPED COTYLEDON2*	
	AT5G06100	*MYB33*	*MYB DOMAIN PROTEIN33*	
	AT3G11440	*MYB65*	*MYB DOMAIN PROTEIN65*	
	AT3G11440	*MYB65*	*MYB DOMAIN PROTEIN65*	Huang et al. (2019)
	AT4G00150	*HAM3/LOM3*	*HAIRY MERISTEM3/LOST MERISTEMS3*	
	AT4G18390	*TCP2*	*TEOSINTE BRANCHED 1, CYCLOIDEA AND PCF TRANSCRIPTION FACTOR2*	

(*Continued*)

Table 10.1 (Continued)

Plant species	Arabidopsis ortholog ID	Gene abbreviation	Gene name	References
Puccinellia distans	AT4G38180	*FRS5*	*FAR1-RELATED SEQUENCE5*	Öztürk et al. (2018)
	AT1G76320	*FRS4/CPD25*	*FAR1-RELATED SEQUENCE4/CHLOROPLAST DIVISION MUTANT25*	
	AT1G74520	*HVA22A*	*HVA22 HOMOLOG A*	
	AT1G78300	*GF14 OMEGA/GRF2*	*14–3-3 PROTEIN G-BOX FACTOR14 OMEGA/GENERAL REGULATORY FACTOR2*	
	AT4G06634	*YY1*	*YIN YANG1*	
	AT1G62370	–	RING/U-box superfamily protein	
	AT5G08720	*HCF145/PPP1*	*HIGH CHLOROPHYLL FLUORESCENCE145/PIN2 PROMOTER BINDING PROTEIN1*	

in previous transcriptomic studies showed that they were highly related to the regulation of flavonol and flavonoid biosynthesis as well as signaling pathways mediated by gibberellic acid, JA, and ethylene (Table 10.2). Some of the TFs identified in previous studies as highly differentially expressed under B toxicity in different plant species were actually linked with the regulation of flavonol, flavonoid and anthocyanin biosynthesis (Kayihan, 2021), suggesting the internal B tolerance mechanism. Some of these TFs have also been discussed for their essential roles in hormonal regulation of responses and tolerance under B toxicity in different plant species.

Moreover, a common set of TFs was identified as B toxicity-responsive in different plant species (Table 10.3). These TFs include several proteins in MYB and AP2/ERF families, and some others are found in NAC, WRKY, bZIP, and zinc finger families. This suggests that a common cross-species response signaling pathway may be triggered under B toxicity in plants. Indeed, a cross-species comparison between B-sensitive *A. thaliana* and its N-tolerant extremophyte relative *Schrenkiella parvula* was performed recently (Wang et al., 2021). It was shown that the total number of transcripts affected in *Arabidopsis thaliana* was much higher than *S. parvula*, and the RNA metabolism and translation were interrupted more in *Arabidopsis thaliana*. The main reason for this was attributed to *S. parvula*'s ability to exclude excess B more efficiently than *Arabidopsis thaliana*. Among the

Table 10.2 Gene ontology (biological process) enrichment analysis of boron-responsive transcription factors in plants.[a]

GO biological process complete	Fold enrichment	*P* value
Protein storage vacuole organization	>100	1.00E-02
Regulation of flavonol biosynthetic process	>100	1.00E-02
Positive regulation of flavonoid biosynthetic process	98.96	6.90E-04
Pollen sperm cell differentiation	40.12	5.73E-05
Gibberellic acid-mediated signaling pathway	32.99	1.65E-04
Cellular response to gibberellin stimulus	27.49	4.45E-04
Floral organ morphogenesis	27.49	5.06E-03
Microgametogenesis	25.16	7.23E-04
Jasmonic acid-mediated signaling pathway	24.39	8.44E-05
Ethylene-activated signaling pathway	21.33	1.61E-02

[a]GO enrichment is performed according to Mi et al. (2019) by using Fisher's Exact testing followed by the Bonferroni correction for multiple testing ($p < .05$).

Table 10.3 Common set of boron-responsive transcription factors in plants.

Arabidopsis ortholog ID	Gene abbreviation	Gene name
AT2G32460	*MYB1/MYB101/ ABS7*	*MYB DOMAIN PROTEIN1/MYB DOMAIN PROTEIN101/ ABNORMAL SHOOT7*
AT2G28550	*RAP2.7/TOE1*	*RELATED TO AP2.7/TARGET OF EARLY ACTIVATION TAGGED (EAT)1*
AT5G67180	*TOE3*	*TARGET OF EARLY ACTIVATION TAGGED (EAT)3*
AT4G36920	*AP2/FL1/FLO2*	*APETALA2/FLOWER1/FLORAL MUTANT2*
AT5G06100	*MYB33*	*MYB DOMAIN PROTEIN33*
AT4G38620	*MYB04*	*MYB DOMAIN PROTEIN04*
AT3G23250	*MYB15*	*MYB DOMAIN PROTEIN15*
AT3G28910	*MYB30*	*MYB DOMAIN PROTEIN30*
AT2G38340	*DREB19*	*DEHYDRATION RESPONSE ELEMENT-BINDING PROTEIN19*
AT5G24590	*NAC091/TIP*	*NAC DOMAIN-CONTAINING PROTEIN91/TCV-INTERACTING PROTEIN*
AT5G59820	*RHL41/ZAT12*	*RESPONSIVE TO HIGH LIGHT41*
AT5G13330	*RAP2.6L*	*RELATED TO AP2 6L*
AT5G50080	*ERF110*	*ETHYLENE RESPONSE FACTOR110*
AT2G37430	*ZAT11*	*ZINC FINGER OF ARABIDOPSIS THALIANA11*
AT5G49450	*BZIP1*	*BASIC LEUCINE-ZIPPER1*
AT3G62420	*BZIP53*	*BASIC REGION/LEUCINE ZIPPER MOTIF53*
AT5G13080	*WRKY75*	*WRKY DNA-BINDING PROTEIN75*
AT3G11440	*MYB65*	*MYB DOMAIN PROTEIN65*

common set of TFs, *MYB15, ZAT11*, and *WRKY75* were induced while *MYB30* and *BZIP1* were repressed in Arabidopsis shoots under B toxicity. Also, *DREB19, RAP2.6L*, and *RHL41/ZAT12* were highly induced in both roots and shoots of *Arabidopsis thaliana* under B toxicity. Among these TFs, *BZIP1* was downregulated uniquely in *Arabidopsis thaliana* whereas *RAP2.6L* was upregulated in both species. bZIP1 is known to be a positive regulator of plant tolerance to salt, osmotic, and drought stresses. Its overexpression enhances lateral root formation and biomass growth under drought stress in poplar (Dash et al., 2017), and it coordinates the salt-induced primary C- and N-metabolism reprogramming in Arabidopsis roots together with bZIP53 (Hartmann et al., 2015). Interestingly, bZIP1 was shown to regulate the short-term responses to N deficiency (Para et al., 2014). Some of its targets were bound with bZIP1 but not regulated by it. They were enriched in genes involved in response to biotic/abiotic stimuli and transport of divalent ions. Most probably these targets require some other nonnutrient based signaling for bZIP1-based activation. On the other hand, some target genes were bound by bZIP1 and regulated by it. These target genes were not enriched in any specific GOs. Some target genes were categorized in a transient binding, suggesting that bZIP1 transiently binds to their promoters under N deprivation. The genes in this third group were defined as the "First Responders" in rapid N signaling, and they include *NITRATE TRANSPORTER2.1* (*NRT2.1*), *NIN-LIKE TRANSCRIPTION FACTOR3* (*NLP3*) and *LOB DOMAIN-CONTAINING PROTEIN39* (*LBD39*). They are enriched in GO terms, including N-related biological processes, amino acid metabolism, and N assimilation. Because of its known functions in N signaling, bZIP1 can be a candidate TF that should be studied further for its roles in B tolerance in plants.

10.11 Conclusion

B toxicity is an emerging problem that crop species can face in the field. By inhibiting many essential biological mechanisms, it decreases growth and development, thereby restraining crop yield and production. Plants respond to B toxicity in different ways depending on their tolerance levels. However, the main tolerance mechanisms are common in many plant species, suggesting the response mechanisms are regulated by some collective families of TFs under B toxicity. GWAS in different plant species has identified mainly MYB, WRKY, NAC, BHLH, and ERF/AP2 families in common differentially expressed under B toxicity. These TFs are potential targets for developing B toxicity-tolerant crops by conventional transgenic or genome editing approaches. Moreover, further quantitative trait locus mapping or GWAS are required to identify new TFs linked with the B toxicity tolerance in crop species to develop molecular markers that can be incorporated in marker-assisted breeding studies in future.

References

Aquea, F., Federici, F., Moscoso, C., Vega, A., Jullian, P., Haseloff, J.I.M., et al., 2012. A molecular framework for the inhibition of Arabidopsis root growth in response to boron toxicity. Plant. Cell Env. 35 (4), 719–734.

Bar, C. 2015. Elucidation of boron hyperaccumulation and tolerance mechanisms in *Puccinellia Distans* (Jacq.) Parl. using proteomics approach. Ph.D. Dissertation. Izmir Institude of Technology, Izmir, Turkey.

Blevins, D.G., Lukaszewski, K.M., 1998. Boron in plant structure and function. Annu. Rev. Plant. Physiol. Plant. Mol. Biol. 49, 481–500. Available from: http://dx.doi.org/10.1146/annurev.arplant.49.1.481. 15012243.

Borg, M., Brownfield, L., Khatab, H., Sidorova, A., Lingaya, M., Twell, D., 2011. The R2R3 MYB transcription factor DUO1 activates a male germline-specific regulon essential for sperm cell differentiation in Arabidopsis. Plant. Cell 23 (2), 534–549.

Brdar-Jokanović, M., 2020. Boron toxicity and deficiency in agricultural plants. Int. J. Mol. Sci. 21 (4). Available from: http://dx.doi.org/10.3390/ijms21041424. 32093172.

Dash, M., Yordanov, Y.S., Georgieva, T., Tschaplinski, T.J., Yordanova, E., Busov, V., 2017. Poplar Ptab ZIP 1-like enhances lateral root formation and biomass growth under drought stress. Plant. J. 89 (4), 692–705.

de Abreu Neto, J.B., Hurtado-Perez, M.C., Wimmer, M.A., Frei, M., 2017. Genetic factors underlying boron toxicity tolerance in rice: genome-wide association study and transcriptomic analysis. J. Exp. Bot. 68 (3), 687–700.

Du, S.S., Li, L., Li, L., Wei, X., Xu, F., Xu, P., et al., 2020. Photoexcited cryptochrome2 interacts directly with TOE1 and TOE2 in flowering regulation. Plant. Physiol. 184 (1), 487–505.

Esim, N., Tiryaki, D., Karadagoglu, O., Atici, O., 2013. Toxic effects of boron on growth and antioxidant system parameters of maize (*Zea mays* L.) roots. Toxicol. Ind. Health. 29 (9), 800–805. Available from: http://dx.doi.org/10.1177/0748233712442729. 22491723.

Feng, Y., Cui, R., Huang, Y., Shi, L., Wang, S., Xu, F., 2021. Repression of transcription factor AtWRKY47 confers tolerance to boron toxicity in *Arabidopsis thaliana*. Ecotoxicol. Environ. Saf. 220, 112406. Available from: http://dx.doi.org/10.1016/j.ecoenv.2021.112406. 34119927.

Goldbach, H., Monika, W., 2007. Boron in plants and animals: is there a role beyond cell-wall structure? J. Plant. Nutr. Soil Sci. 170 (1), 39–48. Available from: http://dx.doi.org/10.1002/jpln.200625161.

Gray, J.A., Shalit-Kaneh, A., Chu, D.N., Hsu, P.Y., Harmer, S.L., 2017. The REVEILLE clock genes inhibit growth of juvenile and adult plants by control of cell size. Plant. Physiol. 173 (4), 2308–2322.

Gunes, A., Soylemezoglu, G., Inal, A., Bagci, E.G., Coban, S., Sahin, O., 2006. Antioxidant and stomatal responses of grapevine (*Vitis vinifera* L.) to boron toxicity. Scientia Horticulturae 110 (3), 279–284.

Hartmann, L., Pedrotti, L., Weiste, C., Fekete, A., Schierstaedt, J., Göttler, J., et al., 2015. Crosstalk between two bZIP signaling pathways orchestrates salt-induced metabolic reprogramming in Arabidopsis roots. Plant. Cell 27 (8), 2244–2260.

He, Z., Zhou, X., Chen, J., Yin, L., Zeng, Z., Xiang, J., et al., 2021. Identification of a consensus DNA-binding site for the TCP domain transcription factor TCP2 and its important roles in the growth and development of Arabidopsis. Mol. Biol. Rep. 48 (3), 2223–2233.

Huang, J.H., Lin, X.J., Zhang, L.Y., Wang, X.D., Fan, G.C., Chen, L.S., 2019. MicroRNA sequencing revealed Citrus adaptation to long-term boron toxicity through modulation of root development by miR319 and miR171. Int. J. Mol. Sci. 20 (6), 1422.

Huang, J.H., Qi, Y.P., Wen, S.X., Guo, P., Chen, X.M., Chen, L.S., 2016. Illumina microRNA profiles reveal the involvement of miR397a in Citrus adaptation to long-term boron toxicity via modulating secondary cell-wall biosynthesis. Sci. Rep. 6 (1), 1–14.

Javed, T., Shabbir, R., Ali, A., Afzal, I., Zaheer, U., Gao, S.J., 2020. Transcription factors in plant stress responses: challenges and potential for sugarcane improvement. Plants 9 (4), 491.

Jyothi, M.N., Usha, S., Suchithra, B., Ulfath, T.S., Devaraj, V.R., Babu, R.N., 2018. Boron toxicity induces altered expression of miRNAs in French bean (*Phaseolus vulgaris* L.). J. Appl. biol. 6 (06), 001–010.

Karabal, E., Yücel, M., Öktem, H.A., 2003. Antioxidant responses of tolerant and sensitive barley cultivars to boron toxicity. Plant Sci. 164 (6), 925–933.

Kayihan, C., 2021. The involvement of the induction of anthocyanin biosynthesis and transport in toxic boron responsive regulation in *Arabidopsis thaliana*. TURK. J. BOT. 45 (3), 181–191.

Kayihan, C., Öz, M.T., Eyidoğan, F., Yücel, M., Öktem, H.A., 2017. Physiological, biochemical, and transcriptomic responses to boron toxicity in leaf and root tissues of contrasting wheat cultivars. Plant. Mol. Biol. Rep. 35 (1), 97–109.

Kayihan, D.S., Aksoy, E., Kayihan, C., 2021. Identification and expression profiling of toxic boron-responsive microRNAs and their targets in sensitive and tolerant wheat cultivars. Turk. J. Agric. For. 45 (4), 411–433.

Kayihan, D.S., Kayihan, C., Çiftçi, Y.Ö., 2019. Moderate level of toxic boron causes differential regulation of microRNAs related to jasmonate and ethylene metabolisms in *Arabidopsis thaliana*. TURK. J. BOT. 43 (2), 167–172.

Landi, M., Tattini, M., Gould, K.S., 2015. Multiple functional roles of anthocyanins in plant-environment interactions. Environ. Exp. Bot. 119, 4–17.

Marschner, P., 2012. In: Marschner, Petra (Ed.), Marschner's Mineral Nutrition of Higher Plants, 3rd Academic Press.

Mi, H., Muruganujan, A., Ebert, D., Huang, X., Thomas, P.D., 2019. PANTHER version 14: more genomes, a new PANTHER GO-slim and improvements in enrichment analysis tools. Nucleic Acids Res. 47 (D1), D419–D426.

Miyashima, S., Roszak, P., Sevilem, I., Toyokura, K., Blob, B., Heo, J.O., et al., 2019. Mobile PEAR transcription factors integrate positional cues to prime cambial growth. Nature 565 (7740), 490–494.

Nozawa, A., Miwa, K., Kobayashi, M., Fujiwara, T., 2006. Isolation of *Arabidopsis thaliana* cDNAs that confer yeast boric acid tolerance. Biosci. Biotechnol. Biochem. 70 (7), 1724–1730. Available from: http://dx.doi.org/10.1271/bbb.60065. 16861809.

Ó'Maoiléidigh, D.S., van Driel, A.D., Singh, A., Sang, Q., Le Bec, N., Vincent, C., et al., 2021. Systematic analyses of the MIR172 family members of Arabidopsis define their distinct roles in regulation of APETALA2 during floral transition. PLoS Biol. 19 (2), e3001043.

Ochiai, K., Shimizu, A., Okumoto, Y., Fujiwara, T., Matoh, T., 2011. Suppression of a NAC-like transcription factor gene improves boron-toxicity tolerance in rice. Plant. Physiol. 156 (3), 1457–1463.

Öz, M.T., Yılmaz, R., Eyidoğan, F., Graaff, L.D., 2009. Microarray analysis of late response to boron toxicity in barley (*Hordeum vulgare* L.) leaves. Turk. J. Agric. For. 33 (2), 191–202.

Ozhuner, E., Eldem, V., Ipek, A., Okay, S., Sakcali, S., Zhang, B., et al., 2013. Boron stress responsive microRNAs and their targets in barley. PLoS one 8 (3), e59543.

Öztürk, S.E., Göktay, M., Has, C., Babaoğlu, M., Allmer, J., Doğanlar, S., et al., 2018. Transcriptomic analysis of boron hyperaccumulation mechanisms in *Puccinellia distans*. Chemosphere 199, 390–401.

Para, A., Li, Y., Marshall-Colón, A., Varala, K., Francoeur, N.J., Moran, T.M., et al., 2014. Hit-and-run transcriptional control by bZIP1 mediates rapid nutrient signaling in Arabidopsis. PNAS. 111 (28), 10371–10376.

Park, J., Nguyen, K.T., Park, E., Jeon, J.S., Choi, G., 2013. DELLA proteins and their interacting RING Finger proteins repress gibberellin responses by binding to the promoters of a subset of gibberellin-responsive genes in Arabidopsis. Plant. Cell 25 (3), 927–943.

Phillips, T., 2008. Regulation of transcription and gene expression in eukaryotes. Nat. Educ. 1, 199.

Pulverer, B., 2005. Sequence-specific DNA-binding transcription factors. Nat. Cell Biol. 13, 1–2.

Rámila, C.D.P., Contreras, S.A., Di Domenico, C., Molina-Montenegro, M.A., Vega, A., Handford, M., Bonilla, C.A., Pizarro, G.E., 2016. Boron stress response and accumulation potential of the extremely tolerant species *Puccinellia frigida*. J. Hazard. Mater. 317, 476–484. Available from: http://dx.doi.org/10.1016/j.jhazmat.2016.05.086. 27322905.

Roessner, U., Patterson, J.H., Forbes, M.G., Fincher, G.B., Langridge, P., Bacic, A., 2006. An investigation of boron toxicity in barley using metabolomics. Plant. Physiol. 142 (3), 1087–1091. Available from: http://dx.doi.org/10.1104/pp.106.084053. 16998089.

Seo, E., Choi, D., 2015. Functional studies of transcription factors involved in plant defenses in the genomics era. Brief. Funct. Genomics 14 (4), 260–267.

Skirycz, A., Reichelt, M., Burow, M., Birkemeyer, C., Rolcik, J., Kopka, J., et al., 2006. DOF transcription factor AtDof1. 1 (OBP2) is part of a regulatory network controlling glucosinolate biosynthesis in Arabidopsis. Plant. J. 47 (1), 10–24.

Smit, M.E., McGregor, S.R., Sun, H., Gough, C., Bågman, A.M., Soyars, C.L., et al., 2020. A PXY-mediated transcriptional network integrates signaling mechanisms to control vascular development in Arabidopsis. Plant. Cell 32 (2), 319–335.

Song, J.B., Gao, S., Sun, D., Li, H., Shu, X.X., Yang, Z.M., 2013. miR394 and LCR are involved in Arabidopsis salt and drought stress responses in an abscisic acid-dependent manner. BMC Plant. Biol. 13 (1), 1–16.

Tombuloglu, H., Kekec, G., Sakcali, M.S., Unver, T., 2013. Transcriptome-wide identification of R2R3-MYB transcription factors in barley with their boron responsive expression analysis. Mol. Genet. Genom. 288 (3), 141–155.

Turan, M.A., Taban, N., Taban, S., 2009. Effect of calcium on the alleviation of boron toxicity and localization of boron and calcium in cell wall of wheat. Notulae. Botanicae. Horti. Agrobotanici. Cluj-Napoca. 37 (2), 99–103.

Unver, T., Tombuloglu, H., 2020. Barley long non-coding RNAs (lncRNA) responsive to excess boron. Genomics 112 (2), 1947–1955.

Vadde, B.V.L., Challa, K.R., Sunkara, P., Hegde, A.S., Nath, U., 2019. The TCP4 transcription factor directly activates TRICHOMELESS1 and 2 and suppresses trichome initiation. Plant. Physiol. 181 (4), 1587–1599.

Wang, G., DiTusa, S.F., Oh, D.H., Herrmann, A.D., Mendoza-Cozatl, D.G., O'Neill, M.A., et al., 2021. Cross species multi-omics reveals cell wall sequestration and elevated global transcript abundance as mechanisms of boron tolerance in plants. N. Phytol. 230 (5), 1985–2000.

Yıldırım, K., Uylaş, S., 2016. Genome-wide transcriptome profiling of black poplar (Populus nigra L.) under boron toxicity revealed candidate genes responsible in boron uptake, transport and detoxification. Plant. Physiol. Biochem. 109, 146–155.

Yıldırım, K., 2017. Transcriptomic and hormonal control of boron uptake, accumulation and toxicity tolerance in poplar. Environ. Exp. Bot. 141, 60–73.

Zheng, Z., Reichel, M., Deveson, I., Wong, G., Li, J., Millar, A.A., 2017. Target RNA secondary structure is a major determinant of miR159 efficacy. Plant. Physiol. 174 (3), 1764–1778.

Zou, Y., Wang, S., Lu, D., 2020. MiR172b-TOE1/2 module regulates plant innate immunity in an age-dependent manner. Biochem. Biophys. Res. Commun. 531 (4), 503–507.

Zumajo-Cardona, C., Pabón-Mora, N., Ambrose, B.A., 2021. The evolution of euAPETALA2 genes in vascular plants: from plesiomorphic roles in sporangia to acquired functions in ovules and fruits. Mol. Biol. Evol. 38 (6), 2319–2336.

CHAPTER 11

Alleviation of boron toxicity in plants by silicon: mechanisms and approaches

Muhammad Riaz[1,2], Muhammad Kamran[3], Omar Aziz[4] and Xiurong Wang[1]

[1]State Key Laboratory for Conservation and Utilization of Subtropical Agro-bioresources, Root Biology Center, College of Natural Resources and Environment, South China Agricultural University, Guangzhou, China
[2]College of Resources and Environment, Zhongkai University of Agriculture and Engineering, Guangzhou, China
[3]School of Agriculture, Food and Wine, The University of Adelaide, Adelaide, SA, Australia
[4]Department of Soil & Environmental Sciences, University of Agriculture Faisalabad Sub Campus Depalpur, Okara, Pakistan

11.1 Introduction

Boron (B) is a vital plant nutrient and is a very specular micronutrient for crop growth and productivity worldwide (Warington, 1923; Brown et al., 2002). B is present in the hydrosphere and lithosphere (Shorrocks, 1997; Camacho-Cristóbal et al., 2008). B deficiency and toxicity restrict crop productivity and quality in areas with moderate rainfall as well as dry or heavily mineralized soils. The relationship between B deficiency and toxicity in plants is complex (Loomis and Durst, 1992; Riaz et al., 2018a; Riaz et al., 2019). Many soils naturally contain high levels of B, including those in South Australia, West Asia, and North Africa. There may be many reasons for the excessive accumulation of B in the soil (Shah et al., 2017; Wu et al., 2019; Hua et al., 2020). Deficiency of either B or its toxicity can cause inhibition of plant development. The deficit of B can be replenished by supplying B enriched fertilizers (Shah et al., 2017). However, B poisoning is a major issue for the normal growth of the plant and is very challenging, especially in arid and semiarid regions where this problem exists worldwide (Landi et al., 2019; Sharma et al., 2020). Water sources (due to irrigation) and soils that are higher in B than normal soils are thought to be the two primary factors involved in the accumulation of high B in soil (Wu et al., 2019). At neutral pH, boric acid is present undissociated. B exists in various forms in the soil, except for a small

Boron in Plants and Agriculture
DOI: https://doi.org/10.1016/B978-0-323-90857-3.00007-2

number of boric acid molecules (H_3BO_3) and ions (HBO_4^-) in the soil solution, most of which exist in the soil solid phase in inorganic or organic forms. According to different forms, it can be divided into four forms: insoluble B, slow-acting B, soluble B, and organic B. Different forms of B can be converted into each other (Shorrocks, 1997; Landi et al., 2019; Wu et al., 2019). Plants can directly absorb and use the available B in the soil. The available B in the soil includes the water-soluble B in the soil solution and the B adsorbed by organic matter. Plants are very sensitive to B. Generally, when the soil available B is 0.5–1.0 mg/kg, it is beneficial to the growth of plants. If the amount of B in the soil is insufficient, it will cause plant nutrient deficiency, but if it is present in excess, it may be toxic to plants, causing a variety of biological changes and physiological functions, which in turn affect plant quality and reduce crop yields (Reid, 2007; Brdar-Jokanović, 2020). Plants with too much B can cause disturbances in cell walls, metabolism, inhibit cell division and differentiation, and cause oxidative damage. B poisoning can cause yellowing of plant leaves, as well as dry and curled leaf edges, spots, and even withering, therefore, lowering crop yields (Camacho-Cristóbal et al., 2008; Shireen et al., 2018). Adjusting environmental factors to mitigate the toxic effects of B on plants has become a new research direction in recent years to increase plant viability in high-B soil. The problem of B toxicity has attracted attention worldwide. Changes in soil nitrogen, phosphorus, zinc, calcium, and silicon (Si) nutrient conditions have been shown to reduce B toxicity in plants to some degree.

In addition to oxygen, the most abundant element in the earth's crust is Si. The lack of Si in plants depends largely on the concentration of available Si in the soil, rather than the total concentration of Si. The bioavailability of Si is closely related to the biogeochemistry of soil Si, including the biogeochemical cycle of Si, the type of Si in the soil, and its solubility (Farooq and Dietz, 2015; Farooq et al., 2015). The Si that enters the plant has a variety of physiological functions, such as improving plant growth and biomass accumulation, agricultural product yield and quality, and improving plant resistance to nonbiological and abiotic stress. Si is a potentially beneficial factor for plant development. The amount of available Si in the soil is considered to be one of the most effective ways to mitigate B toxicity (Gunes et al., 2007; Farooq and Dietz, 2015). The majority of Si in the soil is in the form of oxidized or silicate, making it difficult for plants to use directly. Increased soil Si nutrition has been shown in studies to help improve plant stress resistance (Rizwan et al., 2016; Riaz et al., 2021a), including B tolerance. For example, studies have shown that applying Si to crops like tomatoes, canola, and wheat can improve their B tolerance (Gunes et al., 2007; Kaya et al., 2011; Liu et al., 2017; Metwally et al., 2018). High soil salinity is often followed by high B content in arid and semiarid regions. In plants, high salt levels can lead to water loss, ion imbalance, and ion (B) poisoning. The balance of reactive oxygen species (ROS) in the metabolic system in plants is disrupted by high B stress, allowing more ROS to accumulate in the body and causing oxidative damage (Soylemezoglu et al., 2009; Farooq et al., 2015; Kaya et al., 2020b, 2020c). Furthermore, previous studies on using Si to cure B toxicity in plants concentrated on plant growth and antioxidant system and related mechanisms. Therefore, Si has broad application prospects in agriculture. Si has attracted increasing attention due to its unique forms and its effects on plant physiology. The impact of Si on plant physiology and ecology provides a scientific and effective theoretical basis for exploring the application prospects of Si in agriculture, forestry, and horticulture and whether they can be used as environmentally friendly materials.

11.2 Silicon-induced alleviation of boron toxicity

Si has a positive effect on plant growth. Therefore, Si may enhance the resistance by different mechanisms. Due to Si's good impacts on plant development and stress resistance, Si materials can be used to improve plant tolerance and agronomic practices. A few mechanisms reported in the alleviation of B toxicity are by restricting B uptake, enhancing growth parameters, regulating antioxidative systems and photosynthesis. A detailed explanation of different mechanisms has been discussed in the subsections.

11.2.1 Plant growth traits

The plant's response to abiotic stressors is an important criterion for assessing its metal/metalloid tolerance. Several studies showed that B toxicity had a predominant effect on the inhibition of growth parameters, such as citrus (Riaz et al., 2019), rice (Riaz et al., 2021a), orange (Wu et al., 2019), and wheat (Gunes et al., 2007). Kaya et al. (2011) suggested that the application of Si in conjunction with B toxicity in tomato plants improved shoot, root, and total dry biomass accumulation, as compared to plants treated with B toxicity alone. Another study reported by

(Liu et al., 2017) showed through a greenhouse pot experiment that Si application (50 mg/kg) had a positive response on the alleviation of B stress in wheat, which is an important crop and often faces the problem of different stresses, and B toxicity is one of them. Si improved the shoot biomass by 39% and the roots by 17% when B toxicity at 300 mg/kg was used, which showed that B had more effect on the shoots than the roots. B and Si did not affect net photosynthetic rates; however, Si had more effect on the increase of stomatal conductance and transpiration rate. The source of Si, doses of Si, and application methods may all have different effects on the efficacy of B toxicity alleviation in different plants. For example, in one experiment when Si was applied through the foliar application, it alleviated the symptoms of B deprivation as well as B toxicity in cotton and promoted chlorophyll *a*, *b*, and total chlorophyll. Furthermore, during the root application, the results were not as promising as during foliar application. Root and shoot dry mass accumulation were also increased with Si supply under B-stressed cotton plants (Pereira de Souza Junior et al., 2019). In another study, canola (*Brassica napus*) plants exhibited optimistic effects by alleviating undesirable influences of B toxicity by Si treatments. The overall increase in different parameters was observed in growth, water contents, seed index, siliqua weight, and siliqua number (Metwally et al., 2018). Toxic leaves are one of the visual indicators of toxicity. The study of (Celikkol Akcay and Erkan, 2016) showed that after 3 days of exposure to B toxicity, barley plants showed necrosis at the tips of leaves, and the observed effects of these parameters typically did not show up for those studied under standard growth conditions. There were no major changes in the lengths of the barley tissues or the ion leakage rate by the application of either B or Si. Under the influence of the current levels of treatment, the overall water content in the root cells generally remained stable under B—Si treated cells, except for the significant reductions in root tissue upon B exposure. Besides B toxicity, salt stress is also one of the limiting factors in the development of plants and productivity. (Eraslan et al., 2008) reported that both B toxicity and salt stress in spinach plants had lower fresh weight than the treatment of Si. Adding Si also enhanced chlorophyll meter values. There were strong B—Si interactions that increased chlorophyll meter readings. The results of several investigations show that B toxicity has a deleterious effect on plant growth and development in both short and long-term tests and Si is one of the promising elements in the alleviation of B toxicity symptoms and consequences on plant growth.

11.2.2 Reduction of boron transport from roots to shoots

Tolerance mechanisms in plants are the reduced B transport from roots to shoots by holding and storing B in the roots (Shireen et al., 2018; Hua et al., 2020). The tolerance to B toxicity in plants is generally believed to reduce the uptake of B from the soil solution by roots or efflux of excess from the plant cells via different mechanisms, because high B in the plant tissues has been suggested as the major cause of the appearance of B toxicity symptoms in different plant tissues, except for the hyperaccumulator plants, which can accumulate high concentrations of B without showing symptoms of B toxicity due to possessing different mechanisms of tolerance, that is, physiological, metabolic, and molecular (Reid, 2007; Landi et al., 2019). Most of the research findings suggested that the process of B uptake by roots is similar to other ions in the soil solution which may cause a competitive effect in the uptake of B and therefore reduce B uptake and toxicity to the plants and these ions are sulfate, Si, nitrate, and phosphate, Al. Recent studies have shown that Si has a positive effect on reducing B uptake in plants (Gunes et al., 2007). Thus, the reports explained that the beneficial effects of Si on B toxicity were a result of Si competitive function in transport through B transporters at high B concentrations. They discovered that Si is involved in three transport-related genes (known as *BOR2, PIP1*, and *PIP1;1*) that are active in tissues and organs in which B and Si can be transported or transported out of tissues (Celikkol Akcay and Erkan, 2016). The studies showed that Si may form a complex with B in the roots of plants which is regarded as a mechanism of tolerance (Inal et al., 2009). Another mechanism induced by Si is delaying the entry of B into the plant root through thick lignified/suberized exodermis. On the other hand, some researchers asserted that no competitive activity between B and Si was observed in various barley genotypes. Additionally, the competitiveness of B with metal ions was investigated. For example, copper (Cu) can greatly decrease B uptake by plant species by preserving the membrane's permeability, but the mechanism is unknown. Aluminum (Al), another metal element, has been found to have antagonistic effects on plants that contain B. However, there was no evidence of uptake competition between Al and B in citrus seedlings (Yan et al., 2021). (Liu et al., 2017) reported that during the application of Si to wheat plants, B concentration in shoots was significantly decreased under high B treatments, indicating a mechanism of tolerance. Shoot B accumulation is thought to be important for emergence and sensitivity to

B toxicity. Therefore, Si accumulated more B in the roots, and less was transported to the shoots (Stiles et al., 2010; Liu et al., 2017). One of the previous studies also confirmed that in spinach, higher root B concentrations were associated with Si supply, which resulted in lower shoot B concentrations, suggesting a greater impact of Si to withstand high concentrations of B in the roots (Gunes et al., 2007). Soil B–Si compound formation reduces B availability and so tissue B buildup (Inal et al., 2009). When compared to B-grown plants at a high concentration of B, that is, 100 mg/kg/ soil, Metwally et al. (2018) found that Si reduced total B export to the plant body by 34.13% and to shoots by 41.48%. The study also found that B content, as well as B distribution and total B accumulation in whole plant, were lower in Si-treated plants, confirming that the use of Si limits the availability of B, leading to reduction in B uptake.

According to the findings of Fleck et al. (2011), Si supplementation reduced root oxidation power and enhanced Casparian bands and lignin depositions in rice plants. Rhizophoracious exodermis, which can directly hinder the entry of the roots into metals, can be due to higher metal tolerances of mangrove species (Pb, Zn, and Cu). At neutral pH, both Si and B are metalloids that form uncharged acidic compounds and are absorbed in nonionic form into plant tissues. It is somewhat unusual for the same problems to exist when dealing with other nutrients (Reid, 2010). Celikkol Akcay and Erkan (2016) discovered that Si suppresses root and shoot levels of B when it is under B stress. They also indicated that rivalry through a channel protein and transporter protein preference for Si intake may be another possible explanation. According to these findings, Inal et al. (2009) proposed that Si reduces B uptake by the formation of B-silicate complexes in the soil. Gunes et al. (2007) discovered that Si decreased B translocation in spinach plants, due to Si in the cell walls and lumens irreversibly precipitated as ($SiO_2.nH_2O$). One of the most important causes of B toxicity is high soil B. This high concentration of B in soil has a detrimental effect on the growth of plant cells and tissue as it concentrates in the leaves, reaching its highest levels at the leaf tips and resulting in regionally caused cell death in the form of necrosis in the leaves' margins, causing leaf necrosis to spread backward along all the leaf area. It was recently discovered that plants are naturally very variable in their level of sensitivity to B, and one tolerance mechanism found in plants is B expulsion from the root system. It appears that root efflux expression of the gene for B transport (encoding for the molecule) is stronger in more resistant crops. It appears that the same gene also increases in leaves and it is similar to that in the root, but contrasting in the leaves, B efflux transporter did not change the tissue B concentration. Alternatively, (Reid, 2010) proposed that in tolerant genotypes, these transporters redistribute B where it is not nearly as susceptible to damage. For example, B is detrimental in the intracellular phase, leading to even greater toxicity in the cell death of the tissue, and even more damaging to the tissue's cells, and these transporters redistribute less toxic apoplast phase. There is a high correlation between cultivar toxicity and leaf concentrations of the same B concentration, providing an explanation for why different cultivars may have different symptoms. It has also been shown by Reid (2010) that rain can remove large quantities of B, leading to substantial increases in leave and root growth. B reaches the leaf through the xylem and keeps building up as the leaf develops. With increasingly higher levels of B in leaves, the older sections of the leaf first become necrotic while younger basal tissues begin to expand as plants at high concentrations of B. Under the intolerance conditions, B is absorbed in the cytoplasm where it can be toxic to the cell. Many tolerant cultivars rely on expelling B from the cytoplasm where it is toxic while preserving its concentration in the cell wall; this means less cellular abnormalities and reducing toxic effects on plant growth. Sensitive cultivars have a very low B efflux capability, retaining far higher levels in the cell than in resistant cultivars. Huge quantities of B can be removed from the leaves by rain, which reduces the toxicity. The same B flux transporters in leaves are used for pumping B of tolerant cultivars from the cells into the external medium. This decreases root toxicity and restricts the amount of B that enters and reaches the leaves.

11.2.3 Oxidative stress and plant defense system

B toxicity may cause negative effects on the membranes, which is also one of the major consequences of biotic stress. The current studies showed that a high concentration of B in the medium resulted in the peroxidation of the membrane in the form of high contents of MDA in plant tissues (Riaz et al., 2021a). During the process of metabolism, ROS are produced as signaling molecules. However, their concentration is also regulated by different mechanisms. Once ROS are generated in a high quantity, they may cause toxic effects on the plants, resulting in oxidative stress (Mittler, 2002; Wu et al., 2019) and the content of MDA indicates the extent of damage to the membranes, which is a good indicator of oxidative stress (Riaz et al., 2019). Previous investigations showed that B toxicity caused the accumulation of higher oxidative stress and damaged the leaf membranes of barley (Karabal et al., 2003; Molassiotis et al., 2006). Other studies also reported that B toxicity induced lipid peroxidation of membranes and high accumulations of ROS in terms of H_2O_2 (Gunes et al., 2007). The photosynthetic mechanisms may have been adversely affected by lipid peroxidation and hydrogen peroxide (H_2O_2) toxicity. B-toxicity stresses commonly exhibit an oxidative mechanism that can be reduced by the presence of antioxidant enzymes (Eraslan et al., 2008). B-toxicity or abiotic stressors, such as UV light or lack of water, can lead to ROS and be harmful to plants unless the antioxidant system works well. According to research, enzymes that delay or counteract the toxic effects of B on plants have the capability of increasing the lifetime of stress resistance to B stress. These enzymes are antioxidants, which help the plant to avoid or get rid of oxidative stress as well as keep it safe from it. It is the primary line of defense enzyme that acts as an antioxidant for all aerobic organisms that have an O_2-oxidase, and its job is to convert O_2 to H_2O_2. It is also known as a superoxide dismutase (SOD). Recent evidence suggests that SOD is increased in many plant species after they are subjected to nonbiologically stressful conditions such as B toxicity (Riaz et al., 2019; Riaz et al., 2021a). Oliveira et al. (2020) analysis showed that increased SOD activity in the field pea plants was observed under high B, indicating that the SOD activity is enhanced under B stress in field peas. SOD activity was higher with the application of exogenous Si treatment, as shown by (Eraslan et al., 2008), while that work showed O_2^- was more efficiently converted into H_2O_2. However, SOD activity was stated by (Eraslan et al., 2008) as being significantly higher with exogenous Si, showing the greater performance (Singh et al., 2015; Rahman et al., 2017; Ahmad et al., 2019). The H_2O_2 that is formed in response to SOD or other metabolism-activating enzymes can be subsequently reduced to H_2O and O_2 in the presence of catalase (CAT), glutathione (GSH), ascorbic peroxidase (APX), and other reductive enzymes (Riaz et al., 2018b; Riaz et al., 2021b; Yan et al., 2021). The changes in the activity of antioxidant enzymes as a result of metal/metalloid stress often result in oxidative stress and the formation of ROS in plant tissues. In fighting stress, antioxidant enzymes such as SOD, CAT, guaiacol peroxidase (POD), and APX play an important role, which reduces ROS in cells. The activity level of these enzymes can be an indicator of the severity of stress. Many scientists have studied the effect of Si application under stress caused by excess cadmium on changes in the activity of antioxidant enzymes (Rizwan et al., 2016). Farooq and Dietz (2015) showed a significant increase in the activity of antioxidant enzymes in plants poisoned with Cd as compared to plants grown on a medium without Cd. The addition of Si to the Cd medium caused a further increase in SOD, POD, CAT, and APX in cotton leaves and roots. The authors state that this indicates an increase in the activity of these enzymes due to the application of Si, and thus increases the plant's ability to fight stress.

Changes in the malondialdehyde (MDA) content are a marker of oxidative stress. An increase in its content indicates oxidative damage to lipids (lipid peroxidation). Decreasing its content in the tissues proves the reduction of stress. Oliveira et al. (2020) reported that in-field pea B toxicity caused a reduction in growth parameters and oxidative stress in terms of the high accumulation of ROS. On the other hand, Si in a combination of SA (salicylic acid) showed a positive consequence by reducing the content of MDA and H_2O_2, indicating regulation of ROS generation and their degradation. The addition of Si + SA increased the CAT activity in the cultivars of field peas and APX, and GSH-Px showed a difference between cultivars. Similarly, the activity of GR also had a cultivar-dependent response under Si + SA + B treatment. Moreover, the study of (Metwally et al., 2018) showed that canola plants had high B content in the ground and aboveground parts and a high accumulation of total B accumulation in all plants as compared to control. However, Ca, Si, and SA application resulted in a decrease in these parameters. B toxicity increased ascorbic acid, ascorbate, α-tocopherol, dehydroascorbate, and phenolics while Si exhibited a significant reduction in all these traits, excluding α-tocopherol, that was higher at low B levels. The glutathione content, which is another important component in the plant defense system, was not changed by B stress. However, under B stress, Ca, Si, and SA additions reduced its accumulation. Liu et al. (2017) found SOD to be higher in plants treated with high soil B, and Si had a slight effect on the increase of SOD activity. It was also observed that CAT did not change significantly at medium B level but increased in Si treatments at high B concentration and CAT activity was increased by Si application under B toxicity. Peroxidase (POD) activity was observed to increase with B and decrease with the use of Si. The findings of a previous study were not consistent with (Gunes et al., 2007), where the shoot SOD and CAT were decreased by the addition of Si under B treatments. Data from (Liu et al., 2017) has shown that Si addition reduced the MDA content at

300 mg B/kg. The effects of Si are controversial on antioxidant enzymes and lipid peroxidation (Gunes et al., 2007; Inal et al., 2009; Kaya et al., 2011). Findings by Liu et al. (2017) indicate that although some improvements were not significant in antioxidant enzymes and MDA, Si had only a minor impact on oxidative damage relief in the wheat plant. In general, nearly all preceding studies believed that B stress would cause oxidative damages, and the application of Si reduces its damage.

11.2.4 Silicon induced improvement in the photosynthesis under boron toxicity

Photosynthesis is an important trait that participates in the formation of food. However, environmental stress such as B toxicity severely affects photosynthetic traits (Shah et al., 2017). Kaya et al. (2020a) showed that photosynthetic-related attributes were deleteriously reduced by B stress. The possible explanation for why chlorophyll and Fv/Fm have been reduced by B toxicity may probably be linked to the overaccumulation of H_2O_2 in plants' cells (Genisel et al., 2017). In addition, a Si-induced increase in the chlorophyll and Fv/Fm content of the B-stressed plants and a reduction in H_2O_2 content was observed by (Kaya et al., 2020a), indicating that Si helps to reduce the harmful consequences for chlorophyll and Fv/Fm caused by B toxicity, probably by reducing the amount of H_2O_2. The photosynthesis rate data showed that the photosynthesis effectiveness was decreased by salinity and/or B toxicity. The photosynthetic system damage of the two varieties was most severe (up to 65%). The photosynthesis efficiency of genotype KS-282 improved significantly in both control and saline environments. Si also enhanced photosynthetic efficiency in both genotypes, with improvements ranging from 2 to 3.6 factors. Under both control and saline conditions, Si addition increased RWC. Salinity and/or B toxicity due to higher salt content in the rooting medium severely influenced water potential gradient, as revealed by the transpiration rate data. Under any form of applied stress, this resulted in the reduced upward movement of water within the plant body, lowering the transpiration rate. However, the greatest decrease in transpiration was observed compared to the control due to the combined salinity stress and B toxicity. Previously, it has been shown that a decrease in growth and an increase in B concentration in plant tissues as a result of B toxicity have been shown in tomatoes (Güneş et al., 1999), sunflower (Ruiz et al., 2003), barley (Inal et al., 2009) and wheat (Sonmez et al., 2009). Supplementary Si restored dry weights to control levels but did not affect

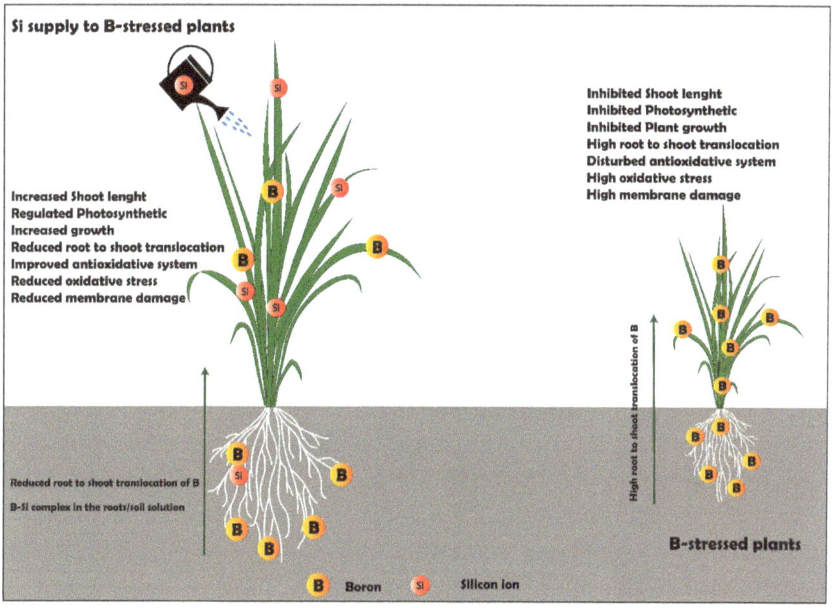

Figure 11.1 Schematic diagram explaining silicon (Si) induced alleviation of boron (B) toxicity. Si forms a complex with B in the soil solution and plant body, retaining most of the B in the roots and less B is absorbed and translocated to aboveground parts. Reducing upward movement of B results in the resistance to B toxicity and alleviation of B toxicity symptoms on leaves. Si enhances plant growth and improved the antioxidant system which ultimately reduced B indued oxidative damages on membranes and cell organelles.

RWC. Si addition increases barley plant growth under B toxicity (Gunes et al., 2007; Inal et al., 2009). The WUE rises with increased stress, the claim being that it causes an improvement in reduced transpiration leading to a decline in water loss (Smith et al., 2013). WUE has been raised with B and many plants have been shown to change WUE in response to stresses, due to Si. For example, Si causes an improvement in overall water use efficiency through mechanisms such as (1) the reduces loss of water by reducing transpiration or (2) by the silicate crystals deposition beneath the epidermal cells of leaves and stems (Romero-Aranda et al., 2006). Fig. 11.1 shows the schematic diagram of Si-induced alleviation of B toxicity in plants.

11.3 Prospects and challenges

High B soils are found in arid and semiarid regions of the world. In high B areas, excessive B in the soil has seriously threatened the development of agricultural production and indirectly affected human health. At present, the following improvement measures are generally taken for high B soil: (1) Irrigation and water leaching. The disadvantage of this method is that it consumes a lot of water and is not suitable for arid areas. (2) Use of chelators. The disadvantage of this method is that it is not economical enough to form a chelate with boric acid to reduce the effective B in the soil. (3) Applying lime can increase the pH value of the soil and increase the adsorption capacity of the soil to B, thereby reducing the B toxicity, but this method is easy to causes the soil pH to be too high and affect the growth of plants, so it is not suitable for the improvement of alkaline soil. Another way is from the perspective of ecological restoration, using hyperaccumulating plants to restore high B soil. In recent years, most scholars have focused on the impact of B poisoning on crops, but relatively few reports on phytoremediation research have been reported.

In view of the limitations of high B soil improvement technology, researchers are trying to overcome this difficulty by economical and feasible methods like the use of Si. The research on soil B is mostly based on nutrition. The research on B toxicity of plants has not been paid attention to, and there are rare reports on the use of plants to remediate high B soils. Field application of Si should be investigated to gain real results; therefore, detailed and planned experiments must be carried out with food crops to get high productivity while reducing B toxicity.

References

Ahmad, P., Tripathi, D.K., Deshmukh, R., Pratap Singh, V., Corpas, F.J., 2019. Revisiting the role of ROS and RNS in plants under changing environment.

Brdar-Jokanović, M., 2020. Boron toxicity and deficiency in agricultural plants. Int. J. Mol. Sci. 21 (4), 1424.

Brown, P.H., Bellaloui, N., Wimmer, M., Bassil, E.S., Ruiz, J., Hu, H., et al., 2002. Boron in plant biology. Plant. Biol. 4 (02), 205–223.

Camacho-Cristóbal, J.J., Rexach, J., González-Fontes, A., 2008. Boron in plants: deficiency and toxicity. J. Integr. plant. Biol. 50 (10), 1247–1255.

Celikkol Akcay, U., Erkan, I., 2016. Silicon induced antioxidative responses and expression of BOR2 and two PIP family aquaporin genes in barley grown under boron toxicity. Plant. Mol. Biol. Report. 34 (1).

Eraslan, F., Inal, A., Pilbeam, D.J., Gunes, A., 2008. Interactive effects of salicylic acid and silicon on oxidative damage and antioxidant activity in spinach (*Spinacia oleracea* L. cv. Matador) grown under boron toxicity and salinity. Plant. Growth Regul. 55 (3), 207–219.

Farooq, M.A., Dietz, K.-J., 2015. Silicon as versatile player in plant and human biology: overlooked and poorly understood. Front. Plant. Sci. 6, 994.

Farooq, M.A., Saqib, Z.A., Akhtar, J., 2015. Silicon-mediated oxidative stress tolerance and genetic variability in rice (*Oryza sativa* L.) grown under combined stress of salinity and boron toxicity. Turkish J. Agricult Forestry 39 (5), 718–729.

Fleck, A.T., Nye, T., Repenning, C., Stahl, F., Zahn, M., Schenk, M.K., 2011. Silicon enhances suberization and lignification in roots of rice (*Oryza sativa*). J. Exp. Botany 62 (6), 2001–2011.

Genisel, M., Turk, H., Dumlupinar, R., 2017. Exogenous aminolevulinic acid protects wheat seedlings against boron-induced oxidative stress. Romanian Biotechnol. Lett. 22 (4), 12741–12750.

Güneş, A., Alpaslan, M., Çikili, Y., Özcan, H., 1999. Effect of zinc on the alleviation of boron toxicity in tomato. J. Plant. Nutr. 22 (7), 1061–1068.

Gunes, A., Inal, A., Bagci, E., Coban, S., Sahin, O., 2007. Silicon increases boron tolerance and reduces oxidative damage of wheat grown in soil with excess boron. Biologia Plant. 51 (3), 571–574.

Hua, T., Zhang, R., Sun, H., Liu, C., 2020. Alleviation of boron toxicity in plants: mechanisms and approaches. Crit. Rev. Environ. Sci. Technol. 1–41.

Inal, A., Pilbeam, D.J., Gunes, A., 2009. Silicon increases tolerance to boron toxicity and reduces oxidative damage in barley. J. Plant. Nutr. 32 (1), 112–128.

Karabal, E., Yücel, M., Öktem, H.A., 2003. Antioxidant responses of tolerant and sensitive barley cultivars to boron toxicity. Plant. Sci. 164 (6), 925–933.

Kaya, C., Tuna, A.L., Guneri, M., Ashraf, M., 2011. Mitigation effects of silicon on tomato plants bearing fruit grown at high boron levels. J. Plant. Nutr. 34 (13), 1985–1994.

Kaya, C., Ashraf, M., Al-Huqail, A.A., Alqahtani, M.A., Ahmad, P., 2020a. Silicon is dependent on hydrogen sulphide to improve boron toxicity tolerance in pepper plants by regulating the AsA-GSH cycle and glyoxalase system. Chemosphere 257, 127241.

Kaya, C., Ashraf, M., Alyemeni, M.N., Ahmad, P., 2020b. Responses of nitric oxide and hydrogen sulfide in regulating oxidative defence system in wheat plants grown under cadmium stress. Physiologia Plant. 168 (2), 345–360.

Kaya, C., Ashraf, M., Alyemeni, M.N., Ahmad, P., 2020c. The role of nitrate reductase in brassinosteroid-induced endogenous nitric oxide generation to improve cadmium stress tolerance of pepper plants by upregulating the ascorbate-glutathione cycle. Ecotoxicol. Environ. Saf. 196, 110483.

Landi, M., Margaritopoulou, T., Papadakis, I.E., Araniti, F., 2019. Boron toxicity in higher plants: an update. Planta 250 (4), 1011–1032.

Liu, C., Lu, W., Ma, Q., Ma, C., 2017. Effect of silicon on the alleviation of boron toxicity in wheat growth, boron accumulation, photosynthesis activities, and oxidative responses. J. Plant. Nutr. 40 (17), 2458–2467.

Loomis, W.D., Durst, R., 1992. Chemistry and biology of boron. BioFactors (Oxford, Engl.) 3 (4), 229–239.

Metwally, A.M., Radi, A.A., El-Shazoly, R.M., Hamada, A.M., 2018. The role of calcium, silicon and salicylic acid treatment in protection of canola plants against boron toxicity stress. J. plant. Res. 131 (6), 1015–1028.

Mittler, R., 2002. Oxidative stress, antioxidants and stress tolerance. Trends plant. Sci. 7 (9), 405–410.

Molassiotis, A., Sotiropoulos, T., Tanou, G., Diamantidis, G., Therios, I., 2006. Boron-induced oxidative damage and antioxidant and nucleolytic responses in shoot tips culture of the apple rootstock EM 9 (*Malus domestica* Borkh). Environ. Exp. Botany 56 (1), 54–62.

Oliveira, K.R., Junior, J.P.S., Bennett, S.J., Checchio, M.V., de Cássia Alves, R., Felisberto, G., et al., 2020. Exogenous silicon and salicylic acid applications improve tolerance to boron toxicity in field pea cultivars by intensifying antioxidant defence systems. Ecotoxicol. Environ. Saf. 201, 110778.

Pereira de Souza Junior, J., de Mello Prado, R., Machado dos Santos Sarah, M., Felisberto, G., 2019. Silicon mitigates boron deficiency and toxicity in cotton cultivated in nutrient solution. J. Plant. Nutr. Soil. Sci. 182 (5), 805–814.
Rahman, M.F., Ghosal, A., Alam, M.F., Kabir, A.H., 2017. Remediation of cadmium toxicity in field peas (*Pisum sativum* L.) through exogenous silicon. Ecotoxicol. Environ. Saf. 135, 165–172.
Reid, R., 2007. Update on boron toxicity and tolerance in plants. Adv. plant. Anim. boron Nutr. 83–90.
Reid, R., 2010. Can we really increase yields by making crop plants tolerant to boron toxicity? Plant. Sci. 178 (1), 9–11.
Riaz, M., Yan, L., Wu, X., Hussain, S., Aziz, O., Jiang, C., 2018a. Boron deprivation induced inhibition of root elongation is provoked by oxidative damage, root injuries and changes in cell wall structure. Environ. Exp. Botany 156, 74–85.
Riaz, M., Yan, L., Wu, X., Hussain, S., Aziz, O., Wang, Y., et al., 2018b. Boron alleviates the aluminum toxicity in trifoliate orange by regulating antioxidant defense system and reducing root cell injury. J. Environ. Manag. 208, 149–158.
Riaz, M., Yan, L., Wu, X., Hussain, S., Aziz, O., El-Desouki, Z., et al., 2019. Excess boron inhibited the trifoliate orange growth by inducing oxidative stress, alterations in cell wall structure, and accumulation of free boron. Plant. Physiol. Biochem. 141, 105–113.
Riaz, M., Kamran, M., El-Esawi, M.A., Hussain, S., Wang, X., 2021a. Boron-toxicity induced changes in cell wall components, boron forms, and antioxidant defense system in rice seedlings. Ecotoxicol. Environ. Saf. 216, 112192.
Riaz, M., Kamran, M., Rizwan, M., Ali, S., Parveen, A., Malik, Z., et al., 2021b. Cadmium uptake and translocation: synergetic roles of selenium and silicon in Cd detoxification for the production of low Cd crops: a critical review. Chemosphere 129690.
Rizwan, M., Ali, S., Adrees, M., Rizvi, H., Zia-ur-Rehman, M., Hannan, F., et al., 2016. Cadmium stress in rice: toxic effects, tolerance mechanisms, and management: a critical review. Environ. Sci. Pollut. Res. 23 (18), 17859–17879.
Romero-Aranda, M.R., Jurado, O., Cuartero, J., 2006. Silicon alleviates the deleterious salt effect on tomato plant growth by improving plant water status. J. plant. Physiol. 163 (8), 847–855.
Ruiz, J.M., Rivero, R.M., Romero, L., 2003. Preliminary studies on the involvement of biosynthesis of cysteine and glutathione concentration in the resistance to B toxicity in sunflower plants. Plant. Sci. 165 (4), 811–817.
Shah, A., Wu, X., Ullah, A., Fahad, S., Muhammad, R., Yan, L., et al., 2017. Deficiency and toxicity of boron: alterations in growth, oxidative damage and uptake by citrange orange plants. Ecotoxicol. Environ. Saf. 145, 575–582.
Sharma, S.S., Kumar, V., Dietz, K.-J., 2020. Emerging trends in metalloid-dependent signaling in plants. Trends plant. Sci.
Shireen, F., Nawaz, M.A., Chen, C., Zhang, Q., Zheng, Z., Sohail, H., et al., 2018. Boron: functions and approaches to enhance its availability in plants for sustainable agriculture. Int. J. Mol. Sci. 19 (7), 1856.
Shorrocks, V.M., 1997. The occurrence and correction of boron deficiency. Plant. Soil. 193 (1), 121–148.
Singh, R., Hemantaranjan, A., Patel, P.K., 2015. Salicylic acid improves salinity tolerance in field pea (*Pisum sativum* L.) by intensifying antioxidant defense system and preventing salt-induced nitrate reductase (NR) activity loss. Legume Research-An Int. J. 38 (2), 202–208.
Smith, T., Grattan, S., Grieve, C., Poss, J., Läuchli, A., Suarez, D., 2013.) pH dependent salinity-boron interactions impact yield, biomass, evapotranspiration and boron uptake in broccoli (*Brassica oleracea* L.). Plant. Soil. 370 (1), 541–554.

Sonmez, O., Aydemir, S., Kaya, C., 2009. Mitigation effects of mycorrhiza on boron toxicity in wheat (*Triticum durum*) plants. N. ZealJ. Crop. Horticult Sci. 37 (2), 99–104.
Soylemezoglu, G., Demir, K., Inal, A., Gunes, A., 2009. Effect of silicon on antioxidant and stomatal response of two grapevine (*Vitis vinifera* L.) rootstocks grown in boron toxic, saline and boron toxic-saline soil. Sci. Horticulturae 123 (2), 240–246.
Stiles, A.R., Bautista, D., Atalay, E., Babaoğlu, M., Terry, N., 2010. Mechanisms of boron tolerance and accumulation in plants: a physiological comparison of the extremely boron-tolerant plant species, *Puccinellia distans*, with the moderately boron-tolerant *Gypsophila arrostil*. Environ. Sci. Technol. 44 (18), 7089–7095.
Warington, K., 1923. The effect of boric acid and borax on the broad bean and certain other plants. Ann. Botany 37 (148), 629–672.
Wu, X., Lu, X., Riaz, M., Yan, L., Jiang, C., 2019. Boron toxicity induced specific changes of cell ultrastructure and architecture of components in leaf center and tip of trifoliate orange [*Poncirus trifoliata* (L.) Raf.]. J. Environ. Manag. 246, 426–433.
Yan, L., Riaz, M., Liu, J., Liu, Y., Zeng, Y., Jiang, C., 2021. Boron reduces aluminum deposition in alkali-soluble pectin and cytoplasm to release aluminum toxicity. J. Hazard. Mater. 401, 123388.

CHAPTER 12

Agronomic aspects of boron: fertilizers, agronomical strategy, and interaction with other nutrients

Seyed Majid Mousavi[1], Seyed Ali Ghaffari Nejad[1], Fereydun Nourgholipour[1] and Shirzad Abbaszadeh Zoshkey[2]
[1]Agricultural Research Education and Extension Organization (AREEO), Soil and Water Research Institute (SWRI), Department of Soil Fertility and Plant Nutrition, Karaj, Iran
[2]Department of Agronomy and Plant Breeding, College of Agricultural and Natural Resources, The University of Tehran, Tehran, Iran

12.1 Introduction

Boron (B) is an element from the metalloid group, with intermediate properties between metals and nonmetals. Its essential role for the optimum and normal growth of plants was reported more than 80 years ago (Marschner, 2011). Its chemical properties and roles in biology have caused B to be known as a unique nutrient in agriculture. When B was introduced as one of the essential nutrients for plants, its position as an agricultural chemical has significantly promoted. Nowadays, numerous studies have shown that Boron availability in soil and irrigation water is vital in agricultural production. Boron has several physiological and biochemical functions during plants' life cycle and directly and indirectly affects the growth and quality of productions.

On the other hand, Boron deficiency and toxicity disturb the optimum growth and yield of plants. Direct effects of B refer to its specific functions on the physiological and biochemical properties of the plant. In contrast, its indirect effects are more related to its interactions with other nutrients. Same as other nutrients, B has different interactions with soil particles, soil properties, as well as other nutrients, which determine its final fate in the soil solution in respect to bioavailability for plants. Different strategies for the management of B in the agricultural soils been suggested by researchers both under deficiency and toxicity conditions

Boron in Plants and Agriculture
DOI: https://doi.org/10.1016/B978-0-323-90857-3.00011-4

have that emphasize the importance of B in agriculture. In this chapter, the authors, by using a different perspective, try to show the position and importance of B in agriculture, its vital functions in plants, its interactions with soil properties and nutrients, and the applicable strategies for management of B under deficiency and toxicity conditions.

12.2 Status of boron in soils

Boron is present in various forms in the soil with the main source of Boron in soil including minerals and rocks (Singh et al., 2020). In nature, B is mainly presenting in oxygen compounds mostly as borate. A small percent of natural B is found in boric acid and rare cases in combination with fluoride ions (Goldberg, 1997). Kumar et al. (Kumar et al., 2020) reported that the main forms of B in soils include primary minerals such as tourmaline as well as B rich micas, along with secondary minerals absorbed on clays, hydrated ferrous and aluminum oxides, water-soluble boric acid or borates and organic matters. The distribution and forms of B in the soil can change based on soil management practices. Plants are able to access only a small percent of the total B present in the soil (~5%–10%) (Farooq et al., 2018). Available forms of B for plant usage include hot water-soluble (HWS) B, as well as Boron extractable using dilute calcium chloride (0.01 or 0.02 M) and dilute mannitol (0.10 M). Although none of these extraction methods are 100% efficient, HWS extraction of boron is the main extraction method for available B (Rasheed, 2009). Total B concentration of soils can vary between 2 and 100 mg/kg, with soluble B making up less than 0.1–10 mg/kg, and the concentration of hot water extractable B varying between 0.5 and 30 mg/kg (Goldberg, 1997). Boron in the soil is mainly found in the form of an insoluble mineral known as tourmaline. B enters the soil solution as boric acid during the weathering process, which is a weak acid and forms the borate anion by accepting a hydroxyl ion, thus being available for plants (Nable et al., 1997).

Boron content can be used to divide soils into two categories: low B content soils (B content of less than 10 mg/kg) and s high B content soils (B content of 10–100 mg/kg). The soils with an available boron content of less than 0.5 mg/kg are considered boron deficient soils (Brdar-Jokanović, 2020). The actual critical limit of B depends on the type of plant, but for general purposes, available soil boron content of 0.0–0.02 mg/kg is considered as low, 0.21–0.60 is considered as a medium, 0.61–1.1 mg/kg is considered as high, 1.1–3.0 mg/kg is considered very high, and higher B contents are considered as toxic. As a result, B content between 0.5 and 2.0 is often considered as optimum (Brdar-Jokanović, 2020).

12.3 Importance and functions of boron in plants

Despite the fact that B has been known for a long time as an essential micronutrient for plants (Warington, 1923), its widespread importance for crops has only become known recently. B deficiency in plants has been known as one of the most usual economically important micronutrient stresses and is the second most common deficiency worldwide (Graham, 2008; Rerkasem and Jamjod, 2004). A specified function of B in cell walls of plants has been reported recently (Matoh, 1997; O'Neill et al., 2001). Based on the report of Hu and Brown (1994), about 95%–98% of B is concentrated in the cell walls of leaves in tobacco and squash. B is also involved in the control of the tensile strength and permeability of the cell wall through its cross-linking with pectin assembly, glycosylinositol phosphorylceramides, and rhamnoglacturonan-II (RG-II) (O'Neill et al., 2004; Ryden et al., 2003; Voxeur and Fry, 2014), and under B deficiency, RG-II dimer formation is reduced resulting in abnormal (course and thick) cell wall formation (Ishii et al., 2001). B plays other important roles in protein and enzymatic functioning of the cell membrane, resulting in increased membrane integrity (Goldbach et al., 2001). Goldbach and Wimmer (2007) reported that the optimal level of B increases the plasma membrane hyperpolarization, whereas B deficiency impresses the membrane potential and decreases the activity of H^+-ATPase. Lawrence et al. (1995), in a study on chickpea, reported that activity of ATPase in plasmalemma-enriched vesicles of chickpea roots is decreased under low availabilities of B compared to the control treatment.

The young growing tissues of plants require higher B concentrations due to boron's vital role in cell division and elongation (Dell and Huang, 1997). B deficiency has adverse effects on root elongation and causes flower and fruit deformation through impairing cell division in the meristematic region (Gupta and Solanki, 2013). The role of B in phenolic metabolism has been established in previous studies (Marschner, 2011) which have reported that accumulation of phenol is a common disorder in plants growing under B-deficiency conditions. Enzymatic and nonenzymatic Oxidation is stimulated by B deficiency because phenol is used as substrate under B deficiency, consequently increasing the content of polyphenol oxidase and quinine, which are harmful to plant growth (Hajiboland et al., 2013). Production of reactive oxygen species may be stimulated under B deficiency which significantly decreases the metabolism of ascorbic acid and glutathione (Marschner, 2011). B plays an important role in nitrogen metabolism. It has been reported that B increases nitrate concentrations and decreased nitrate reductase activity is reported under B-deficiency conditions (Shen et al., 1993). Other studies have reported the functions of B in rhizobial N fixation in leguminous crops (Bolaños et al., 2004). The direct role of B in photosynthesis has not been reported yet, but, through affecting ion transportation, it can indirectly affect photosynthesis (Wang et al., 2015). Mousavi and Motesharezadeh (2020) studied the direct and indirect effects of B on the growth, development, and yield of plants. They reported that B increases the uptake of essential plant nutrients by the plant though affecting the mobility and bioavailability of essential plant nutrients in the soil. B is also involved in the metabolism of phenolic, nucleic acid and indoleacetic acid, sugar transport and plant respiration. Table 12.1 shows a summary of different functions of B in different plant organs.

12.4 Importance of boron in agriculture and quality of production

Sufficient Boron is required for optimum growth and yield of plants. Based on the reports by Shorrocks (1997) studies in 80 countries have shown that B deficiency affects the yields for 132 different crops. Among the trace elements, boron is the only nonmetallic element available to

Table 12.1 Roles of boron in different organs of plants (Shireen et al., 2018).

Root	Stem	Leaf	Flower	Fruit
Formation of root hair	Devision and elongation of cell	Chlorophyll content	Production of flower	Plant productivity
Development of lateral root	Integration of cell wall and membrane	Photosynthesis	Production of anther	Seed setting
Biologic fixation of N	Translocation of ions across the membrane	Opening and closing of stomata	Germination of pollen	Fruit quality
Uptake of nutrients ions			Growth and development of pollen tube	Shelf life

Table 12.2 Necessity of boron in different plants (Keren et al., 1985).

Plants with a low need to B	Plants with a moderate need to B	Plants with a high need to B
Wheat, barley, oats, soybean, rye, rice, beans, peas, pecan, sugarcane, and sorghum	Corn, flax, cabbage, onion, sweet corn, lettuce, potatoes, tomatoes, tobacco, and spinach	Cotton, canola, sunflower, cauliflower beets, carrots, broccoli, apple, Peanuts alfalfa, sugar beet, turnip, and Lucerne

plants in the form of nonionized boric acid molecules (Singh et al., 2020). As shown in Table 12.2, sensitivity and tolerance to boron content in the soils varies between different plant species and varieties (Table 12.2). Usually, dicotyledons require higher boron content (often 4–7 times more) compared to monocotyledons. This means that the optimum range of boron content for dicotyledons and monocotyledons is equal to 20–70 and 5–10 mg/kg, respectively (Farooq et al., 2018).

While B affects both reproductive and vegetative growths in plants, it has a more considerable effect on reproductive growth and B deficiency can lead to cluster sterilization as a result of weak development of anthers and pollen grains as well as pollen grain germination, resulting in a decreased number of grains in clusters and lower yield (Rasheed, 2009).

B is also involved in cell wall development and strengthening as well as cell division and expansion, cell membrane formation and functions, lignin and xylem formation, plant–water relationships, development of fruits and seeds, sugar transport, cell membrane function, hormone production, RNA metabolism, respiration, and metabolism of indoleacetic acid, carbohydrates, proteins, and phenol (Power and Woods, 1997; Rasheed, 2009; Herrera-Rodríguez et al., 2010; Naqib and Jahan, 2017; Farooq et al., 2018; Singh et al., 2020).

More than 90% of boron present in plants can be found in cell walls. This means that B deficiency can alter the composition of cell walls (Rasheed, 2009). B deficiency changes the amount of cellulose, pectin, and lignin in the cell wall, increasing the concentrations of cellulose, hemicellulose, and uronic acid increase while decreasing the concentrations of water-soluble pectin. B, along with calcium, can form chemical bonds with some of the components of cell walls including pectin, polyhydroxy polymers, and polyol. The formation of these compounds shows the role of B in cell wall formation and stability through the formation of esters with the cis–diol groups (Rasheed, 2009). This means that young meristematic organs require more B than adult organs. In plants with B-deficiency release of potassium from the cell due to B deficiency means plasma membrane is overly permeable while lacking the necessary consistency (Naqib and Jahan, 2017).

B also facilitates phenol metabolism in plants, resulting in phenol accumulation under B deficiency. This accumulation of phenol is harmful to the plant due to enzymatic and nonenzymatic oxidation reactions and the formation of toxic quinones and free radicals. Under B-deficiency conditions, defense system of cells against free radicals is weakened because of the resulting decrease in ascorbic acid, free radical scavenging enzymes and SH-containing compounds (Ahmad et al., 2012). B is also involved in nitrogen fixation and development and normal functions of nitrogen-fixing nodes. Under B deficiency, nitrogen-fixing nodes show reduced weight and legume's nitrogen fixation capacity. B can also facilitate the stability of tubular membrane contamination and peribacterioids (Bolanos et al., 1994; Rasheed, 2009). Furthermore, B decreases diseases susceptibility of plants, while increasing phosphorus uptake, plant resistance to toxic effects of aluminum on the roots of dicotyledonous, and drought tolerance (Singh et al., 2020).

Boron uptake in plants occurs through a nonmetabolic process with mass flow being the main mechanism of plant B uptake which is then distributed in the plant through transpiration flow. In most plant species, boron is immobile but can move in the phloem of some species. In species with mobile B in the phloem (including stone fruits and vegetables such as celery and many ornamental species), B can accumulate developing tanks (fruits) instead of at the end of the plant evapotranspiration. Most species with B transportation in the phloem are susceptible to B toxicity (Nable et al., 1997). The mechanism of this mobility includes complex sugars such as sorbitol, which bond with B. These metabolites are transported to active reservoirs such as growing shoots and fruits after being formed in photosynthetic tissues. This mobility results in B accumulation in these tissues, leading to symptoms of B toxicity appearing in meristematic areas or fruits instead of mature leaves. In species with immobile boron, older leaves are suitable for determining the level of toxicity of B while young leaves or fruit tissues are suitable for species with mobile B (Brown and Hu, 1998).

Table 12.3 Some of the typic and main diseases in different plants (Singh et al., 2020).

Plant	Disease
Beat	Heart rot
Apple	Internal cork
Cabbage	Brown heart
Sweet potato	Internal brown spot
Tobacco	Terminal bud breakdown
Olive	Die-back
Alfalfa	Rosetting

The symptoms of boron deficiency in plants appear in different forms. Table 12.3 shows typic and main diseases resulting from B deficiency in different plants. Tissue expansion is the first process affected by B deficiency. B-deficiency symptoms in the shoot usually occur in meristematic tissue (such as terminal buds and young leaves) and include the death of growth points and cork spots below the petioles. These morphological disorders (which stop the growth of root tips and shoots, as well as flower development, while also reducing the formation and deformity of fruits and seeds) are related to the structural role of B in the cell wall and the low mobility of B in growing organs in most species (Singh et al., 2020).

12.5 Responses of different plants/varieties to the status of boron in the soil

Boron in plants is necessary for meristematic tissues and floral organs, for cell division, for lower male fertility, pollen tube germination and elongation, and formation of seed/fruit. Seeds with boron deficit generate abnormal seedlings (Marschner, 2011). Plants show different reactions depending on the B status of the soil, which will be discussed under the deficiency and toxicity conditions section.

12.5.1 Plant responses under boron deficiency condition

Under deficiency conditions, plants can ameliorate the stress using morphological, physiological, biochemical, and molecular strategies. All of these mechanisms could be classified into two categories: nutrient acquisition efficiency and nutrient utilization efficiency (Nourgholipour et al., 2019). Efficiency terms consist of B-utilization efficiency (BUTE), B-acquisition efficiency (BACE), and B-efficiency (BE).

BACE is affected by root characteristics. B affects root growth and mycorrhizal symbioses which is necessary for increasing root surface area and length for exploring large soil volumes (Tariq and Mott, 2006). Different crops show different levels of sensitivity to B deficiency. For example, wheat requires the highest amounts of B for the development of its root system followed by common beans, soybean, rice than maize (Fageria, 2000). González-Fontes et al. (2016) showed that under low B treatment, ethylene is involved in the induction of root hair formation and elongation.

BUTE is defined as the milligram of dry weight produced per milligram of nutrient element absorbed. BUTE depends on plants' uptake, utilization, or retranslocation of boron (Rose and Wissuwa, 2012). B transporters, which translocate B to the upper parts of plants under B-deficiency conditions, appear to be required for B homeostasis maintenance under low B conditions. Sugar alcohols are a vital part of B retranslocation and phloem B mobility (Hossain et al., 2017). Due to their capability to retranslocate B within the plant, polyols have been identified as transporter molecules and are useful for mitigating the short-term B deficiencies (Tanaka and Fujiwara, 2008).

A wide range of genotypic variation in relation to BE has been found in numerous crop species. Table 12.4 summarizes important strategies in the B-efficient genotypes for certain crops under B-deficiency conditions. In general, it is concluded that cell wall-bound pectin content is higher in B-efficient conditions, suggesting a relation between BUE and pectin content cell wall composition. These results indicate that the occurrence of potential genetic traits used in conventional breeding programs can improve the BE of crop plants.

12.5.2 Plant responses under boron toxicity condition

It is much more complicated to manage B toxicity than it is to avoid it. B toxicity sensitivity mainly depends on B release via roots, slow B uptake, and limited B translocation to the shoots and internal mechanisms (e.g., adsorption to cell walls and compartmentation of B in vacuoles). Two main mechanisms involved in B toxicity tolerance, include exclusion of B from the root, and redistribution of B within the leaf. Table 12.5 summarizes important strategies in the tolerant genotypes of some crops under B toxicity conditions.

Table 12.4 The important strategies in the boron-efficient genotypes of different crops under boron deficiency conditions.

Plant	Strategy	Reference
Wheat	Higher B transport to leaf and avoiding pollen sterility with supplying adequate B to the ear via xylem transport directly from the roots	Galindo et al. (2018), Nachiangmai et al. (2004)
Rapeseed plants	Retranslocation of B	Stangoulis et al. (2010)
Cotton	Accumulation of more B in different floral parts, especially in stamen and pistil from B-deficient medium	Jian et al. (2004)
Rice	Efficiency in biomass production	Rehman et al. (2018)
Pepper	Alterations in root growth pattern	Sarafi et al. (2018)
Citrus	Higher capacity to uptake and transport B from soil to the shoot under low B supply	Zhou et al. (2015)

Table 12.5 The important strategies in the tolerant genotypes of some plants under boron toxicity conditions.

Plant	Strategy	Reference
Wheat	Exclusion of B from root, redistribution of B within the leaf, and also the tolerant genotypes to B toxicity have a longer root system than the sensitive genotypes	Ghaffari Nejad et al. (2015); Kastori et al. (2008)
Barley	Restriction of B uptake	Nable et al. (1990);
Pomegranate	Accumulation of B in great quantities in the root, thus limiting its transport to the shoots	Olmo-Vega et al. (2021)

12.6 Effect of soil properties on bioavailability of boron

B chemistry in the soil is very simple compared to other nutrients since B is not subjected to oxidation, reduction or volatilization (Goldberg, 1997). Some soil and environmental factors can affect the bioavailability of boron, among which are parent materials, pH, texture, moisture, organic matter, clay mineralogy, temperature, calcium carbonates, irrigation sources, nutrient interactions, plant species, and soil salinity (Gupta et al., 1985; Goldberg et al., 1993; Hu and Brown, 1994).

Soil pH is the most important factor affecting B availability with the amount of B transferred to the soil inversely proportional to the soil pH. Increase in soil pH results in decreased B availability for the plant (Hu and Brown, 1994). B deficiency often occurs in light-textured soils (sandy loam and loamy sand) in wetlands. This is due to the fact that Borate anions are easily leached from the soil, exposing acidic sandy soils to B deficiency (Singh et al., 2020). B toxicity is observed in soils with high B content or high B content in irrigation water in arid and semiarid regions. Plants' B uptake is more affected by B concentration in the soil solution than by the total amount of B in the soil (Goldberg, 1997), making the amount of total B in the soil a poor indicator of the B availability in the soil. Boron availability decreases as the soil dries. In humid climates, the leaching of B from the soil and reducing the rate of transpiration reduces boron uptake and transfer to growing organs (Keren et al., 1985). The bioavailability of B is significantly lower in alkaline and calcareous soils and in the soils where salt or carbonate has been deposited, which can result in boron deficiency in plants (Ahmad et al., 2012; Singh et al., 2020). Organic matter acts as a reservoir of B in the soil and there is a direct relationship between the amount of available B and soil organic matter (Farooq et al., 2018).

12.7 Interactions of boron with other nutrients

The effects of B deficiency, sufficiency, and toxicity on the nutrients contents of plants are not well established. However, interactions of B (i.e., synergism and antagonism interactions), can alter the status of other nutrients in terms of phytoavailability under both deficiency and toxicity conditions. The reaction of plants to B differs depending on the plant species, genotype, environmental conditions, culture media, type of soil, type of plant tissue, and the growth stage (Tariq and Mott, 2006). Because both excess and deficiency of B may disturb metabolic processes of higher plants (Corey and Schulte, 1973), some researchers believe that B may act as a regulator or inhibitor in the uptake and utilization of other nutrients (Alvarez-Tinaut et al., 1979; Carpena Artes and Carpena Ruiz, 1987). In a study investigating the interactions of B and Zn with other nutrients in soybean root, leaf, and seed, Pawlowski et al. (2019) reported that the plants fertilized with low concentrations of B had lower content of Ca (leaf only for cv. Alpha), K (seed only for cv. Fairbault), and S (seeds and roots for cv. Fairbault) compared to the control plants. Furthermore, the plants which were fertilized with higher concentrations of B had lower content of Ca (leaves), Mg (leaves and roots), N (roots only for cv. Alpha), and S (seeds and roots for cv. Alpha) compared to the control treatments.

Hence, the relationship between B and other elements may be very complex. Singh reported that uptake of nutrients by wheat is decreased under high concentrations of B (i.e., antagonism relationship) (Singh et al., 1990). On the contrary, Carpena Artes and Carpena Ruiz (1987) reported a synergistic relationship between high content of B and uptake of nutrients by tomato. It is imagined that the behavior of B is related to the total content of cations and anions in the plant. An increase in uptake of B resulted in a reduction of the ratio of cation–anion equivalent in the leaf and an increase in the root of alfalfa (Wallace and Bear, 1949). Furthermore, the cation–anion contents were reduced in the sugar beet leaf by increasing the concentration of B in the nutrient solution (Vlamis and Ulrich, 1971). Some studies showed that B not only acts within the plant but also in the nutrient medium, accordingly affecting the uptake of nutrients (Santra et al., 1989). These reports show that B may also have a role in the control or delay of the phytoavailability of nutrients ions from the soil. Based on the report of Mousavi and Motesharezadeh (2020), a high deficient or toxic level of B may be responsible for secondary effects on account of the decline in plant growth and resulting in a change of nutrients uptake because of direct or indirect interactions of B with other nutrients.

There is little numerical information on the influence of B on the phytoavailability of other nutrients in the literature reviews. It is imagined that B, through affecting the permeability of plasmalemma at the root surface, increases the uptake of P (Morsy and Taha, 1986). The uptake of K is increased by B because of their mutual synergistic relationship but decreased uptake of Ca is due to their antagonistic relationship. The antagonism interaction of B and Ca was shown by Gupta et al. (1985) and Taban et al. (1995), who reported that the application of Ca could reduce the bioavailability of B. This interaction may explain that localization of Ca in the cell wall causes a reduction the cell wall B permeability. Turan et al. (2009) reported that the inhibitory effect of B toxicity on the shoot and root growth was decreased by the application of Ca. They concluded that B toxicity could be mitigated by the application of Ca in the plant growth media. The Ca translocation is directly affected by the content of B. Ganmore-Neumann and Davidov (1993) and Lopez-Lefebre et al. (2002) reported that excess of B increased the translocation of Ca in shoots of rose and tobacco, but the reduced concentration of Ca in the leaf of potato, whereas reduced B increased content of Ca (Abdulnour et al., 2000). Deficiency of B is shown to decrease the activity of nitrate transporter, which decreases concentrations of N in leaf of plants (Camacho-Cristóbal and González-Fontes, 2007; Kastori and Petrović, 1989; Ramón et al., 1989)

A greenhouse study reported that excess levels of Zn reduced B in the leaf of corn, but the results of a field experiment reported no relationship between B and Zn (Aref, 2012; Hosseini et al., 2007). Pawlowski et al. (2019) reported that the plants that were treated with high concentrations of Zn had lower contents of B in leaf and seed, but in the low concentrations of Zn, a reverse status happened. The antagonism interaction between B and K at excess levels of available K due maybe to cation–anion imbalance has been reported by Samet et al. (2015). Generally, it is concluded that the effect of B on phytoavailability of other nutrients to plants is related to the plant species, genotypes, type or valence of ions, type of the plant organ, stage of plant growth, and culture media. Table 12.6 are summarized the type of reaction between B and other nutrients in different plants.

12.8 Management of boron in the agricultural soils

12.8.1 Management of boron under deficiency condition

B fertilizers can be used for either soil or foliar applications. Soil applications are usually used on field crops, while foliar applications are often used to resolve deficiencies in fruit orchards. To stop localized toxicity, banding or seed placing of B fertilizers should be avoided unless banding is followed by cultivation. Broadcasting of B should be followed by integration to the full cultivation depth before seedbed preparation (Gupta et al., 1985).

In general, it is reported that foliar applications are more efficient than broadcast soil applications; however, due to the immobility of B within the plant a repeated application may be needed and judicious use is vital to avoid toxicity (Degryse, 2017).

B is a unique element among nutrients due to considerable variation among plant genotypes regarding B mobility. In species for which polyols

Table 12.6 Interactions of nutrients in relation to boron supply.

Plant	Plant organ	Nutrient	Correlation	Reference
Wheat		N, P, and K	Synergism	Yadav and Manchanda (1979)
Cabernet sauvignon vine	Stem and root	N, P, Ca, Mg, and B	Antagonism	Downton and Hawker (1980)
Cabernet sauvignon vine	Stem and root	K	Synergism	Downton and Hawker (1980)
Sunflower	Leaf	Mn	Antagonism	Rodriguez et al. (1981)
Bean		Fe and Mn	Synergism	Dave and Kannan (1981)
Bean		Zn	Antagonism	Dave and Kannan (1981)
Alfalfa, peanut, and soybean		K, Mn, Zn, Cu, Mo, and B	Synergism	Lombin and Bates (1982)
Barley		N, P, K, Na, and B	Synergism	Singh and Singh (1984)
Barley		Ca and Mg	Antagonism	Singh and Singh (1984)
Tomato	Leaf	B, P, K, and Mg	Synergism	Francois (1984)
Sugar plants	Roots and shoots	N, P, K, Mn, and B	Synergism	Morsy and Taha (1986)
Lettuce	Leaf	Mn, Fe	Synergism	Tariq and Mott (2006)
Lettuce	Leaf	Zn, Cu, and Mo	Antagonism	Tariq and Mott (2006)
Tobacco	Leaf	Mg and K	Antagonism	Camacho-Cristóbal and González-Fontes (2007)
Wheat	Root and shoot	Ca	Antagonism	Turan et al. (2009)
Cotton	different parts of the plant	Cu, Fe	Synergism	Ahmed et al. (2011)
Maize	Leaf	Mn	Antagonism	Aref (2011)
Rice		N	Antagonism	Koohkan and Maftoun (2015)
Pepper	Leaf	K	Antagonism	Samet et al. (2015)
Oilseed rape		S and Cu	Synergism	Sienkiewicz-Cholewa and Kieloch (2015)
Canola	Shoot	N	Antagonism	Koohkan and Maftoun (2016)
Pepper	Leaf	P, K, Ca	Antagonism	Sarafi et al. (2018)
Pepper	Leaf	Zn	Synergism	Sarafi et al. (2018)
Forages	Leaf	S	Antagonism	Mühlbachová et al. (2020)
Cherry tomato	Leaf	K	Synergism	Pogorzelski et al. (2020)

(e.g., sorbitol, mannitol, dulcitol) are the primary products of photosynthesis, B shows quick and significant phloem-mobility. This mobility is due to the formation of boron complexation with polyols (Brdar-Jokanović, 2020). A classification for different plants in respect to response to the content of B in the soil has been reported by Sutradhar et al. (2016)

Table 12.7 Classification of plants based on their response to the status of boron in the soil (Sutradhar et al., 2016).

Large response	Moderate response	Small response
Alfalfa, apple, broccoli, canola, cauliflower, celery, red beet, sugar beet, sunflower, turnip	Cabbage, carrot, clover, grape, lettuce, onion, radish, spinach, strawberry, tomato	Asparagus, barley, blueberries, field corn, cucumber, oats, pasture grasses, pea, pepper, potato, raspberry, rye, soybean, sweet corn, wheat

and presented in Table 12.7 while Table 12.8 shows the sufficiency ranges of B for some plants.

12.8.2 Management of boron under toxicity condition

Remediation of the soils containing excess concentrations of trace elements can occur through leaching, addition of various amendments, or growing tolerant plants (Mousavi et al., 2018a, 2018b, 2018c; Mousavi, 2021). For the soils contaminated with B, leaching with B-deficient water, for example, river water, is used following the successful remediation of saline soils (Brdar-Jokanović, 2020).

Gupta et al. (1985) recommended some strategies for avoiding B toxicities, including applying B early in a crop rotation; avoiding concentrating B in bands at levels recommended for broadcast applications; avoiding fertilizer overlaps in the field or at headlands when broadcasting B. Furthermore, banding should be followed by cultivation before drilling a sensitive crop and B must be applied at recommended rates.

12.9 Boron fertilizers

The fertilizers containing B are divided into two categories: refined and crushed ores (Table 12.9). Refined B fertilizers are soluble forms of B, delivered as solids or in solutions. The solubility of these chemicals facilitates B delivery to the plant and makes their use possible in foliar application. Their transportation cost is also low due to the high concentration of B in these materials. The main disadvantages of these fertilizers include losses through leaching, especially in coarse-textured soils and high prices. A high concentration of B can lead to B toxicity, especially in B toxicity susceptible plants like cereals. This means that B fertilizer management is very important due to the narrow range between B deficiency and toxicity. This has led to the production of slow-release and low B content B

Table 12.8 Sufficiency ranges of boron for some plants (Sutradhar et al., 2016).

Crop	Plant part	Time	Sufficiency range (ppm)
Alfalfa	Tops (6″ of new growth)	Prior to flowering	30–80
Apple	Leaf from the middle of current terminal shoot	July 15–August 15	30–50
Blueberry	Young mature leaf	The first week of harvest	25–70
Broccoli	Young mature leaf	Heading	30–100
Cabbage	The half-grown young wrapper leaf	Heads	27–75
Carrot	Young mature leaf	Mid-growth	30–100
Cauliflower	Young mature leaf	Buttoning	30–100
Edible bean	Most recently matured trifoliate	Bloom stage	5–24
Field corn	Whole tops	Less than 12″ tall	5–25
	Base of ear	Initial silk	5–25
Grape	Petiole from young mature leaf	Flowering	25–50
Pea	Recently mature leaflet	First bloom	25–60
Potato	Fourth leaf from the tip	40–50 days after emergence	20–40
Raspberry	Leaf 18″ from tip	The first week in August	25–300
Soybean	Trifoliate leaves	Early flowering	20–60
Spring wheat	Whole tops	As head emerges from the boot	6–10
Strawberry	Young mature leaf	Mid-August	25–60
Sweet corn	Ear leaf	Tasseling to silk	8–25
Sugar beet	Recently matured leaves	50–80 days after planting	31–200

fertilizers (crushed ores) (Table 12.9) while B-containing wastes are also used as B fertilizer (Degryse, 2017; Ghaffari Nejad and Etesami, 2020).

Surface spreading has the highest efficiency for B fertilizer. The B used in this method has a residual effect, with the residual B in the soil utilized by the next crop. The amount of soil application of B is determined by the type of the soil, type of the fertilizer, and the method of application and can vary between 0.25 and 3 kg/ha. On the other hand, the amount of B used, the type of soil, and the type of plant determine the frequency of B application. Borax or other soluble borates are usually applied to the

Table 12.9 Boron containing fertilizers (Degryse, 2017; Ghaffari Nejad and Etesami, 2020).

Refined B fertilizers	%B	Crushed B fertilizers	%B
Boric acid (H_3BO_3)	17	Ascharite ($2MgO \cdot B_2O_3 \cdot H_2O$)	Variable
Borax ($Na_2B_4O_7 \cdot 10H_2O$),	11	Colemanite ($Ca_2B6O_{11} \cdot 5H_2O$)	Variable
Sodium tetraborate ($Na_2B4O_7 \cdot 5H_2O$)	14	Dtolite ($2CaO \cdot B_2O_3 \cdot 2SiO_2 \cdot H_2O$)	Variable
Solubor ($Na_2B_8O_{13} \cdot 4H_2O$);	21	Hydroboracite ($CaO \cdot MgO \cdot 3B_2O_3 \cdot 6H_2O$)	Variable
Superphosphate containing B	0.18	Ulexite ($Na_2O \cdot 2CaO \cdot 5B_2O_3 \cdot 16H_2O$)	Variable
NPK fertilizers contain B	0.3	Kernite ($(Na_2B_4O_6)(OH)_2 \cdot 3H_2O$)	Variable

soil before planting (legumes and certain root crops require 2–4 kg B/ha). B fertilizer should not come in contact with seeds since it might result in B levels which are toxic to crops. The foliar application is used when root growth is limited and is unable to meet the total needs of the plant, and instead is used as a complementary method when deficiency symptoms appear. This method leads to uniform distribution of B in the plant in plants where B is mobile. However, since B is immobile in most plants, the foliar application should be used at different stages of plant growth (Farooq et al., 2018). In general, in case of a B-deficiency symptom, foliar application of 2–2.5 g/L of boric acid or solubor is recommended. Foliar application of 0.1%–0.25% solution of borax can also be used (Fageria and Nascente, 2014). B can be applied with some pesticide sprays in a combined foliar operation, which reduces application costs.

B use efficiency is defined as the portion of fertilizer taken up by the plant. B use efficiency depends on the age of crop, application rate, solubility of the fertilizer, the method of application, and the amount of leaching from soil (Ghaffari Nejad and Etesami, 2020).

12.10 Conclusion and future perspectives

As was stated, B is one of the essential micronutrients which has two main nutritional and physiological roles in different plants. B, along with calcium (Ca), is involved in cell wall synthesis, and its presence is important for cell division. Boron plays an important role in the growth and development of pollen tube and plant reproduction, and its deficiency can disturb the normal and optimum growth of plants. The bioavailability of B is significantly affected by the properties and status of the soil. For example, B is quickly leached in the coarse-textured soils, and also its mobility and phytoavailability are significantly decreased in the calcareous soils and low acidity conditions. Furthermore, despite the presence of B in the calcareous soils, the possibility of its absorption by the plant is very low due to its reaction with Ca. Therefore, it is concluded that B deficiency is an important challenge in most of the agricultural soils. Since the presence of B, especially during the fruit formation, is more necessary than other micronutrients, it is necessary to meet the need of plants for this element. However, it should be noted that excessive application of B can also cause plant toxicities. Therefore, its application should be based on the results of soil, plant, and water analysis. Due to the importance of B in agriculture, various management strategies have been recommended to manage its status under deficiency and toxicity conditions. However, there are still some unknown aspects about the nutrition and chemical behavior of B. For example, it is not yet clear whether Boron's role in improving plant health is due to its role in improving plant structural and physiological status or to its role in plant enzymatic activity. There is also a need to identify the important cultivars of agronomic and horticultural crops with vigorous root systems to utilize the available B, which can thrive best under B shortage.

References

Abdulnour, J.E., Donnelly, D.J., Barthakur, N.N., 2000. The effect of boron on calcium uptake and growth in micropropagated potato plantlets Potato Res. 43 (3), 287–295.

Ahmad, W., Zia, M.H., Malhi, S.S., Niaz, A., Ullah, S., 2012. Boron deficiency in soils and crops: a review. Crop Plant 2012, 65–97.

Ahmed, N., Abid, M., Ahmad, F., Ullah, M.A., Javaid, Q., Ali, M.A., 2011. Impact of boron fertilization on dry matter production and mineral constitution of irrigated cotton. Pak. J. Bot. 43 (6), 2903–2910.

Alvarez-Tinaut, M.C., Leal, A., Agui, I., Recalde-Martinez, L., 1979. Physiological effects of B-Mn interaction in tomato plants. II. The uptake and translocation of macro elements. Anal. Edafol. Agrobiol. 38 (5–6), 991–1012.

Aref, F., 2011. Influence of zinc and boron nutrition on copper, manganese and iron concentrations in maize leaf. Aust. J. Basic Appl. Sci. 5 (7), 52–62.

Aref, F., 2012. Evaluation of application methods and rates of zinc and boron on nitrogen, phosphorus and potassium contents of maize leaf. J. Plant Nutr. 35 (8), 1210–1224.

Bolanos, L., Esteban, E., de Lorenzo, C., Fernandez-Pascual, M., de Felipe, M.R., Garate, A., et al., 1994. Essentiality of boron for symbiotic dinitrogen fixation in pea (*Pisum sativum*) rhizobium nodules. Plant. Physiol. 104 (1), 85–90.

Bolaños, L., Lukaszewski, K., Bonilla, I., Blevins, D., 2004. Why boron? Plant. Physiol. Biochem. 42 (11), 907–912.
Brdar-Jokanović, M., 2020. Boron toxicity and deficiency in agricultural plants. Int. J. Mol. Sci. 21 (4), 1424.
Brown, P.H., Hu, H., 1998. Phloem boron mobility in diverse plant species. Botanica Acta 111 (4), 331–335.
Camacho-Cristóbal, J.J., González-Fontes, A., 2007. Boron deficiency decreases plasmalemma H + -ATPase expression and nitrate uptake, and promotes ammonium assimilation into asparagine in tobacco roots. Planta 226 (2), 443–451.
Carpena Artes, O., carpena Ruiz, R.O., 1987. Effects of boron in tomato plant. I: leaf evolutions. Agrochimica 31 (4–5), 391–400.
Corey, R.B., Schulte, E.E., 1973. Factors affecting the availability of nutrients to plants. Soil. Test. plant. Anal. 23–33.
Dave, I.C., Kannan, S., 1981. Influence of boron deficiency on micronutrients absorption by *Phaseolus vulgaris* and protein contents in cotyledons. Acta Physiol. Plant. 3, 27–32.
Degryse, F., 2017. Boron fertilizers: use, challenges and the benefit of slow-release sources—a review. J. Boron 2 (3), 111–122.
Dell, B., Huang, L., 1997. Physiological response of plants to low boron. Plant Soil 193 (1), 103–120.
Downton, W.J.S., Hawker, J.S., 1980. Interaction of boron and chloride on growth and mineral composition of cabernet sauvignon vines. Am. J. Enol. Viticult. 31 (3), 277–282.
Fageria, N.K., 2000. Níveis adequados e tóxicos de zinco na produção de arroz, feijão, milho, soja e trigo em solo de cerrado. Rev. Bras. Engenharia Agrícola Ambiental 4, 390–395.
Fageria, N.K., Nascente, A.S., 2014. Management of soil acidity of South American soils for sustainable crop production. Adv. Agron. 128, 221–275.
Farooq, M., Rashid, A., Nadeem, F., Stuerz, S., Asch, F., Bell, R.W., et al., 2018. Boron nutrition of rice in different production systems. A review. Agron. Sustain. Dev. 38 (3), 1–24.
Francois, L.E., 1984. Effect of excess boron on tomato yield, fruit size, and vegetative growth. J. Am. Soc. Horticult. Sci. 109 (3), 322–324.
Galindo, F.S., Filho, M.C.M.T., Buzetti, S., Rodrigues, W.L., Boleta, E.H.M., Santini, J. M.K., et al., 2018. Effects of boron (B) doses and forms on boron use efficiency of wheat. Austr. J. Crop. Sci. 12 (9), 1536–1542.
Ganmore-Neumann, R., Davidov, S., 1993. Uptake and distribution of calcium in rose plantlets as affected by calcium and boron concentration in culture solution. Plant Soil 155 (1), 151–154.
Ghaffari Nejad, Etesami, H., 2020. The Importance of B in Plant Nutrition. Metalloids in Plants: Advances and Future Prospects, 433–449.
Ghaffari Nejad, S.A., Savaghebi, G.R., Farahbakhsh, M., Maali Amiri, R., Rezaei, H., 2015. Tolerance of some wheat varieties to boron toxicity. Cereal Res. Commun. 43 (3), 384–393.
Goldbach, H.E., Wimmer, M.A., 2007. Boron in plants and animals: is there a role beyond cell-wall structure? J. Plant Nutr. Soil. Sci. 170 (1), 39–48.
Goldbach, H.E., Yu, Q., Wingender, R., Schulz, M., Wimmer, M., Findeklee, P., et al., 2001. Rapid response reactions of roots to boron deprivation. J. Plant Nutr. Soil. Sci. 164 (2), 173–181.
Goldberg, S., 1997. Reactions of boron with soils. Plant Soil. 193 (1), 35–48.
Goldberg, S., Forster, H.S., Heick, E.L., 1993. Boron adsorption mechanisms on oxides, clay minerals, and soils inferred from ionic strength effects. Soil Sci. Soc. Am. J. 57 (3), 704–708.

González-Fontes, A., Herrera-Rodríguez, M.B., Martín-Rejano, E.M., Navarro-Gochicoa, M.T., Rexach, J., Camacho-Cristóbal, J.J., 2016. Root responses to boron deficiency mediated by ethylene. Front. Plant. Sci. 6, 1103.

Graham, R.D., 2008. Micronutrient deficiencies in crops and their global significance. Micronutrient Deficiencies in Global Crop Production. Springer, Dordrecht, pp. 41–61.

Gupta, U.C., Jame, Y.W., Campbell, C.A., Leyshon, A.J., Nicholaichuk, W., 1985. Boron toxicity and deficiency: a review. Can. J. Soil. Sci. 65 (3), 381–409.

Gupta, U., Solanki, H., 2013. Impact of boron deficiency on plant growth. International Journal of Bioassays 2, 1048–1050.

Hajiboland, R., Bahrami-Rad, S., Bastani, S., 2013. Phenolics metabolism in boron-deficient tea [Camellia sinensis (L.) O. Kuntze] plants. Acta Biologica Hungarica 64 (2), 196–206.

Herrera-Rodríguez, M.B., González-Fontes, A., Rexach, J., Camacho-Cristobal, J.J., Maldonado, J.M., Navarro-Gochicoa, M.T., 2010. Role of boron in vascular plants and response mechanisms to boron stresses. Plant. Stress. 4 (2), 115–122.

Hossain, M.A., Kamiya, T., Burritt, D., Tran, L.S.P., Fujiwara, T. (Eds.), 2017. Plant Macronutrient Use Efficiency: Molecular and Genomic Perspectives in Crop Plants. Academic Press.

Hosseini, S.M., Maftoun, M., Karimian, N., Ronaghi, A., Emam, Y., 2007. Effect of zinc × boron interaction on plant growth and tissue nutrient concentration of corn. J. Plant Nutr. 30 (5), 773–781.

Hu, H., Brown, P.H., 1994. Localization of boron in cell walls of squash and tobacco and its association with pectin (evidence for a structural role of boron in the cell wall). Plant. Physiol. 105 (2), 681–689.

Ishii, T., Matsunaga, T., Hayashi, N., 2001. Formation of rhamnogalacturonan II-borate dimer in pectin determines cell wall thickness of pumpkin tissue. Plant. Physiol. 126 (4), 1698–1705.

Kastori, R., Petrović, N., 1989. Effect of boron on nitrate reductase activity in young sunflower plants. J. Plant Nutr. 12 (5), 621–632.

Kastori, R.R., Maksimović, I.V., Kraljević-Balalić, M.M., Kobiljski, B.Đ., 2008. Physiological and genetic basis of plant tolerance to excess boron. Zbornik Matice srpske za prirodne nauke 114, 41–51.

Keren, R., Bingham, F.T., Rhoades, J.D., 1985. Plant uptake of boron as affected by boron distribution between liquid and solid phases in soil. Soil. Sci. Soc. Am. J. 49 (2), 297–302.

Koohkan, H., Maftoun, M., 2015. Effect of nitrogen on the alleviation of boron toxicity in rice (*Oryza sativa* L.). J. Plant Nutr. 38 (9), 1323–1335.

Koohkan, H., Maftoun, M., 2016. Effect of nitrogen–boron interaction on plant growth and tissue nutrient concentration of canola (*Brassica napus* L.). J. Plant Nutr. 39 (7), 922–931.

Kumar, R., Shambhabhi, S., Kumari, S., Kumari, A., 2020. Effect of different sources of boron and it's different doses on boron fractions of the soil. J. Pharmac. Phytochem. 9 (2), 836–841.

Lawrence, K., Bhalla, P., Misra, P.C., 1995. Changes in NAD (P) H-dependent redox activities in plasmalemma-enriched vesicles isolated from boron-and zinc-deficient chick pea roots. J. Plant Physiol. 146 (5–6), 652–657.

Lombin, G.L., Bates, T.E., 1982. Comparative responses of peanuts, alfalfa, and soybeans to varying rates of boron and manganese on two calcareous Ontario soils. Can. J. Soil. Sci. 62 (1), 1–9.

Lopez-Lefebre, L.R., Rivero, R.M., Garcia, P.C., Sanchez, E., Ruiz, J.M., Romero, L., 2002. Boron effect on mineral nutrients of tobacco. J. Plant Nutr. 25 (3), 509–522.

Marschner, H., 2011. Marschner's Mineral Nutrition of Higher Plants. Academic Press.
Matoh, T., 1997. Boron in plant cell walls. Plant Soil 193 (1), 59–70.
Morsy, M.A., Taha, E.M., 1986. Effect of boron, manganese and their combination on sugar beet under El-Minia conditions. 2. Construction and uptake of N, P, K, B and Mn. Annals of Agricultural Science. Moshtohor, Egypt.
Mousavi, S.M., 2021. Silicon and nano-silicon mediated heavy metal stress tolerance in plants. In: Etesami, H., Anwar Hossain, M., Al-Saeedi, A.H., El-Ramady, H., Fujita, M., Pessarakli, M. (Eds.), Silicon and Nano-Silicon in Environmental Stress Management and Crop Quality Improvement: Recent Progress and Future Prospects. Elsevier publications.
Mousavi, S.M., Moshiri, F., Moradi, S., 2018a. Mobility of heavy metals in sandy soil after application of composts produced from maize straw, sewage sludge and biochar: discussion of Gondek et al. (2018). J. Environ. Manag. 222, 132–134.
Mousavi, S.M., Motesharezadeh, B., 2020. Boron deficiency in fruit crops. In: Srivastava, A.K., Hu, C. (Eds.), Fruit Crops. Elsevier, pp. 191–209.
Mousavi, S.M., Motesharezadeh, B., Hosseini, H.M., Alikhani, H., Zolfaghari, A.A., 2018b. Root-induced changes of Zn and Pb dynamics in the rhizosphere of sunflower with different plant growth promoting treatments in a heavily contaminated soil. Ecotoxicol. Environ. Saf. 147, 206–216.
Mousavi, S.M., Motesharezadeh, B., Hosseini, H.M., Alikhani, H., Zolfaghari, A.A., 2018c. Geochemical fractions and phytoavailability of zinc in a contaminated calcareous soil affected by biotic and abiotic amendments. Environ. Geochem. Health 40 (4), 1221–1235.
Mühlbachová, G., Čermák, P., Káš, M., Vavera, R., Pechová, M., Marková, K., 2020. Boron content in soils under increasing magnesium and sulphur doses in a field experiment. Plant, Soil. Environ. 66 (7), 366–373.
Nable, R.O., Bañuelos, G.S., Paull, J.G., 1997. Boron toxicity. Plant Soil 193 (1), 181–198.
Nable, R.O., Cartwright, B., Lance, R.C., 1990. Genotypic differences in boron accumulation in barley: relative susceptibilities to boron deficiency and toxicity. Genetic Aspects of Plant Mineral Nutrition. Springer, Dordrecht, pp. 243–251.
Nachiangmai, D., Dell, B., Bell, R., Huang, L., Rerkasem, B., 2004. Enhanced boron transport into the ear of wheat as a mechanism for boron efficiency. Plant Soil 264 (1), 141–147.
Naqib, S.A., Jahan, M.S., 2017. The function of molybdenum and boron on the plants. J. Agric. Res. 2 (3), 136–141.
Nourgholipour, F., Hosseini, H.M., Tehrani, M.M., Moteshare Zadeh, B., Moshiri, F., 2019. Phosphorus efficiency of winter canola cultivars and rhizosphere properties in rhizobox technique. Commun. Soil. Sci. Plant. Anal. 50 (1), 35–48.
Olmo-Vega, A., Simón-Grao, S., Simon, I., Alfosea-Simón, M., Martínez-Nicolás, J.J., García-Sánchez, F., 2021. The high tolerance of different pomegranate cultivars to the excess of Boron in irrigation water is due to their capacity to limit boron transport from the root to the leaves. J. Plant Nutr. Soil. Sci. 184 (1), 142–149.
O'Neill, M.A., Eberhard, S., Albersheim, P., Darvill, A.G., 2001. Requirement of borate cross-linking of cell wall rhamnogalacturonan II for Arabidopsis growth. Science 294 (5543), 846–849.
O'Neill, M.A., Ishii, T., Albersheim, P., Darvill, A.G., 2004. Rhamnogalacturonan II: structure and function of a borate cross-linked cell wall pectic polysaccharide. Annu. Rev. Plant. Biol. 55, 109–139.
Pawlowski, M.L., Helfenstein, J., Frossard, E., Hartman, G.L., 2019. Boron and zinc deficiencies and toxicities and their interactions with other nutrients in soybean roots, leaves, and seeds. J. Plant Nutr. 42 (6), 634–649.

Pogorzelski, D.Q., Pinheiro, D.T., de Queiroz, V., do Carmo Alexandre, J., de Faria, A. F., Martinez, H.E.P., 2020. The retranslocation of boron is influenced by the nutritional status of cherry tomato plants. Biosci. J. 36 (3).

Power, P.P., Woods, W.G., 1997. The chemistry of boron and its speciation in plants. Plant Soil 193 (1), 1–13.

Ramón, A.M., Ruiz, R.O.C., Gárate, A., 1989. In vitro stabilization and distribution of nitrate reductase in tomato plants. Incidence boron deficiency. J. Plant Physiol. 135 (1), 126–128.

Rasheed, M.K., 2009. Role of boron in plant growth: a review. J. Agric. Res. 47 (3), 329–338.

Rehman, A., Farooq, M., Rashid, A., Nadeem, F., Stuerz, S., Asch, F., Bell, R.W., Siddique, K.H., 2018. B nutrition of rice in different production systems. A review. Agronomy for Sustainable Development 38 (3), 1–24.

Rerkasem, B., Jamjod, S., 2004. Boron deficiency in wheat: a review. Field Crop. Res. 89 (2–3), 173–186.

Rodriguez, M.G., Gomez-Ortega, M., Alvarez-Tinaut, M.C., 1981. Boron, copper, iron, manganese and zinc contents in leaves of flowering sunflower plants (*Helianthus annuus* L.), grown with different boron supplies. Plant Soil 62 (3), 461–464.

Rose, T.J., Wissuwa, M., 2012. Rethinking internal phosphorus utilization efficiency: a new approach is needed to improve PUE in grain crops. Adv. Agron. 116, 185–217.

Ryden, P., Sugimoto-Shirasu, K., Smith, A.C., Findlay, K., Reiter, W.D., McCann, M. C., 2003. Tensile properties of Arabidopsis cell walls depend on both a xyloglucan cross-linked microfibrillar network and rhamnogalacturonan II-borate complexes. Plant. Physiol. 132 (2), 1033–1044.

Samet, H., Cikili, Y., Dursun, S., 2015. The role of potassium in alleviating boron toxicity and combined effects on nutrient contents in pepper (*Capsicum annuum* L.). Bulgarian J. Agric. Sci. 21 (1), 64–70.

Santra, G.H., Das, D.K., Mandal, B.K., 1989. Relationship of boron with iron, manganese, copper and zinc with respect to their availability in rice soil. Environment and ecology. Kalyani 7 (4), 874–877.

Sarafi, E., Siomos, A., Tsouvaltzis, P., Therios, I., Chatzissavvidis, C., 2018. The influence of boron on pepper plants nutritional status and nutrient efficiency. J. Soil. Sci. Plant. Nutr. 18 (3), 653–667.

Shen, Z., Liang, Y., Shen, K., 1993. Effect of boron on the nitrate reductase activity in oilseed rape plants. J. Plant Nutr. 16 (7), 1229–1239.

Shireen, F., Nawaz, M.A., Chen, C., Zhang, Q., Zheng, Z., Sohail, H., et al., 2018. Boron: functions and approaches to enhance its availability in plants for sustainable agriculture. Int. J. Mol. Sci. 19 (7), 1856.

Shorrocks, V.M., 1997. The occurrence and correction of boron deficiency. Plant Soil 193 (1), 121–148.

Sienkiewicz-Cholewa, U., Kieloch, R., 2015. Effect of sulphur and micronutrients fertilization on yield and fat content in winter rape seeds (*Brassica napus* L.). Plant Soil. Environ. 61 (4), 164–170.

Singh, A.K., Singh, A.K., Singh, J.P., 2020. Boron in crop production from soil to plant system: a review. Arch. Agricult. Environ. Sci. 5 (2), 218–222.

Singh, J.P., Dahiya, D.J., Narwal, R.P., 1990. Boron uptake and toxicity in wheat in relation to zinc supply. Fertiliz. Res. 24 (2), 105–110.

Singh, V., Singh, S.P., 1984. Effect of applied boron on nutrients and uptake by barley crop. Curr. Agricult. 8, 86–90.

Stangoulis, J., Tate, M., Graham, R., Bucknall, M., Palmer, L., Boughton, B., et al., 2010. The mechanism of boron mobility in wheat and canola phloem. Plant. Physiol. 153 (2), 876–881.
Sutradhar, A.K., Kaiser, D.E., Rosen, C.J., Lamb, J.A., 2016. Boron for Minnesota Soils. University of Minnesota Extension Publications AG FO-00723-C.
Taban, S., Alpaslan, M., Kutuk, C., Inal, A., Erdal, I., 1995. Relationship between boron and calcium on wheat (*Triticum aestivum* L.). In: 9th International Symposium of CIEC, "Soil Fertility and Fertilization Management-Bridge between Science, Industry and Practice", pp. 25–30.
Tanaka, M., Fujiwara, T., 2008. Physiological roles and transport mechanisms of boron: perspectives from plants. Pflügers Archiv-Eur. J. Physiol. 456 (4), 671–677.
Tariq, M., Mott, C.J.B., 2006. Effect of boron supply on the uptake of micronutrients by radish (*Raphanus sativus* L.). J. Agric. Biol. Sci. 1 (2), 1–8.
Turan, M.A., Taban, N., Taban, S., 2009. Effect of calcium on the alleviation of boron toxicity and localization of boron and calcium in cell wall of wheat. Not. Botanicae Horti Agrobotanici Cluj-Napoca 37 (2), 99–103.
Vlamis, J., Ulrich, A., 1971. Boron nutrition in the growth and sugar content of sugar-beets. Amer Soc. Sugar Beet Technol. J.
Voxeur, A., Fry, S.C., 2014. Glycosylinositol phosphorylceramides from Rosa cell cultures are boron-bridged in the plasma membrane and form complexes with rhamnogalacturonan II. Plant. J. 79 (1), 139–149.
Wallace, A., Bear, F.E., 1949. Influence of potassium and boron on nutrient-element balance in and growth of ranger alfalfa. Plant. Physiol. 24 (4), 664.
Wang, N., Yang, C., Pan, Z., Liu, Y., Peng, S.A., 2015. Boron deficiency in woody plants: various responses and tolerance mechanisms. Front. plant. Sci. 6, 916.
Warington, K., 1923. The effect of boric acid and borax on the broad bean and certain other plants. Ann. Botany 37 (148), 629–672.
Yadav, O.P., Manchanda, H.R., 1979. Boron tolerance studies in gram and wheat grown on a sierozem sandy soil. J. Indian. Soc. Soil. Sci. 27 (2), 174–180.
Zhou, G.F., Liu, Y.Z., Sheng, O., Wei, Q.J., Yang, C.Q., Peng, S.A., 2015. Transcription profiles of boron-deficiency-responsive genes in citrus rootstock root by suppression subtractive hybridization and cDNA microarray. Front. plant. Sci. 5, 795.
Jian, M., L. Wux, CaO and W. Liu. 2004. Study on physiological characteristics of floral organ in two cotton cultivars with different B efficiency. Cotton Sci., 16 (14): 206-210.

CHAPTER 13

Boron, hormones and secondary metabolites in plants: a molecular point of view

Faiçal Brini[1] and Marco Landi[2,3]
[1]Laboratory of Biotechnology and Plant Improvement, Center of Biotechnology of Sfax, Sfax, Tunisia
[2]Department of Agriculture, Food and Environment, University of Pisa, Pisa, Italy
[3]Centre for Climatic Change Impact, University of Pisa, Pisa, Italy

13.1 Introduction

Boron (B) is a chemical element belonging to group 13 of the periodic system of elements (III A according to Chemical Abstracts Service (CAS) and III B according to old International Union of Pure and Applied Chemistry (IUPAC) nomenclature), thus displaying the typical prerogatives of metalloids (i.e., metallic and nonmetallic properties), with the capability of producing both acids and bases. B is taken up by plant roots predominantly in the form of small uncharged boric acid molecules which easily enter the cell by passing through the wall's phospholipid bilayer (Brown et al., 2002). B can form complexes with a wide variety of organic compounds containing *cis*-diol groups (Gonzáles-Fontes and Fujiwara, 2020). In fact, there is clear evidence for a crucial role of B in the formation of diester bonds between the borate anion and the apiose residues of two rhamnogalacturonan II monomers, which provide mechanical strength to the cell wall (Kobayashi et al., 1996), but compelling evidence for the participation of B in other cellular activities is currently challenging (Matthes et al., 2020).

Vascular plants differ greatly in their ability to tolerate excess B, which causes different physiological and metabolic effects during their life cycle. Much of this variability can be explained by differences in the uptake and distribution of B in plants. B availability causes important alterations in root and shoot growth at both vegetative and reproductive stages, disturbs many metabolic and physiological processes such as membrane and cytoskeleton structure and function, oxidative stress and secondary metabolism, nitrogen assimilation, and gene expressions, among others (Hua et al., 2017; Liu et al., 2017; Camacho-Cristóbal et al., 2018). However,

Boron in Plants and Agriculture
DOI: https://doi.org/10.1016/B978-0-323-90857-3.00001-1

the mechanisms through which B is involved in these developmental processes are not well known.

An issue of increasing interest is how plants sense B availability. Ca^{2+} has been involved in signaling process associated with the sensing of B deficiency by plants. B starvation enhanced $[Ca^{2+}]$cyt as well as the expression of Ca^{2+}-related genes such as CNGC19 (Ca^{2+} channel), several ACAs (Ca^{2+}-ATPases), CAX3 (Ca^{2+}/H^+ antiporter), various CMLs (calmodulin-like proteins), and CDPKs (Ca^{2+}-dependent protein kinases) in *Arabidopsis* roots (Quiles-Pando et al., 2013).

This chapter aims to understand the mechanisms by which plants sense and convey signals of B deficiency or toxicity and modulate responses at various levels. A detailed knowledge of the role played by participants in these mechanisms, such as B transporters and channels, hormones, and their interactions with each other, or reactive oxygen species, among others, will be detailed.

13.2 Roles of boron in plant metabolism

A century ago, it was established that B is an essential micronutrient; however, its role in plant metabolism is still under debate (Lewis, 2019; Wimmer et al., 2019). The primary function of B is in cell-wall synthesis, and in maintaining its structure and integrity (Fig. 13.1). The amount of B stored in the cell wall depends on the plant species, organ, and B availability (Goldbach et al., 2000). Plants take up B in the form of small uncharged boric acid molecules (about 96%), as well as borate anions to a significantly lesser extent (Bolaños et al., 2004). Both these forms of B easily form complexes with cis-diols. Generally, monocotyledonous species do contain less pectin in their cell walls than dicotyledonous plants; therefore, they have comparatively lower B requirements and lower tolerance to excess B (Miwa and Fujiwara, 2010; Chormova and Fry, 2016). It was hypothesized that B primary role in plant metabolism may be the stabilization of molecules with cis-diol groups, no matter which function they have (Bolaños et al., 2004). It plays a role in membrane metabolism and function, thereby being involved in enzyme reactions, as well as the transport of ions, metabolites, and hormones (Fig. 13.1) (Goldbach and Wimmer, 2007). B stimulates, inhibits, or stabilize enzymes; it is involved in the transport of sugars across the membrane, lignin and flavonoid synthesis, and metabolism of auxins, nitrogen compounds, and phenols (Fig. 13.1) (Huang et al., 2014; Seth and Aery, 2017). Root growth

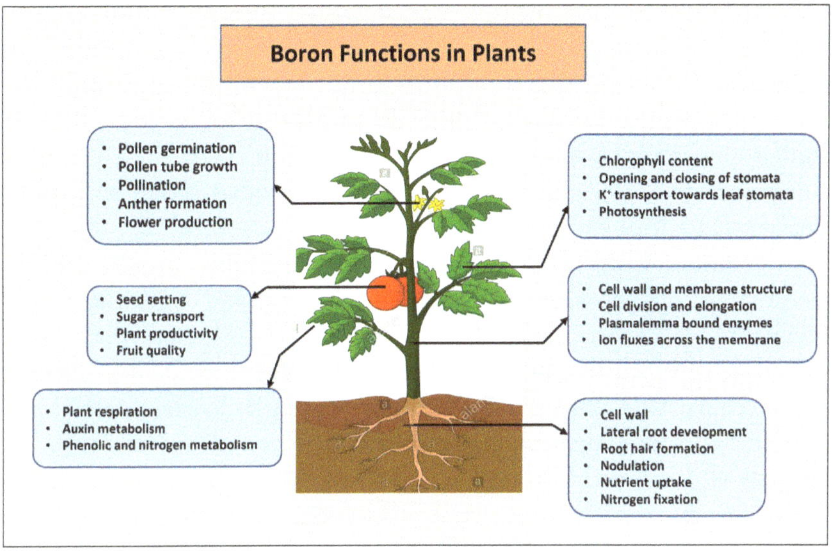

Figure 13.1 Functions of boron in different parts of the plant.

cessation observed in conditions of insufficient B supply results from the inhibition of DNA synthesis, as a result of pyrimidine base deficits (Blevins and Lukaszewski, 1998). The element is involved in cell division (Brown and Hu, 1997), tissue vegetative growth and differentiation (Cakmak and Romheld, 1997), translation and transcription (Dzondo-Gadet et al., 2002), and nitrogen fixation in legumes and mycorrhiza (Bolaños et al., 2004; Sonmez et al., 2009). The symptoms of B deficiency are multiple and include significant alterations in the roots and shoots during the vegetative and reproductive phases of development, as well as alteration of the xylem and phloem vessels and, consequently, in long-distance transport. Pommerrenig et al. (2019) have hypothesized that changes in the nutritional status of B in plants promote differential responses of the vasculature and the mesophyll. B deficiency in *Plantago major* affects reproductive but not vegetative growth. Vascular sucrose levels decrease, and the sucrose loading into the phloem is reduced in conditions of low B supply. Oikonomou et al. (2019) have been demonstrated that B excess altered leaf sugar proportion; increasing the accumulation of nontranslocating sugars such as glucose and fructose in Apple, which, in turn, can strongly affect fruit production of this worldwide-cropped species. El-Motaium et al. (1994) observed that in the condition of B excess, species that produce polyalcohols accumulated B principally in stems rather than in older leaves, where B typically accumulates in sucrose-translocating species. Papadakis et al. (2018) also observed that the phloem is the site of higher concentration of B in loquat plants, suggesting that it represents a mechanism of plant adaptation to B excess through binding and immobilizing B within the cell walls of the newly formed nonphotosynthesizing phloem tissues, preserving other tissues to harmful effects to B toxicity.

In addition, B deficiency affects the photosynthetic capacity and the transport of photosynthetic products (Goldbach and Wimmer, 2007; Wang et al., 2015; Shah et al., 2017; Li et al., 2017; Shireen et al., 2018). A decrease in stomatal conductance and CO_2 assimilation, and an increase in intercellular CO_2 concentrations were found in B-deficient *Citrus sinensis* leaves. Hexose and starch accumulation is increased and sucrose accumulation is decreased, implying feedback regulation in CO_2 assimilation. Antioxidant enzymes remain active; however, the system as a whole is not sufficiently effective in protecting from oxidative damage (Han et al., 2008). To maintain net carbon balance under deficiency, respiration, organic-acid metabolism, amino-acid biosynthesis, and the anaplerotic pathway are upregulated in leaves while downregulated in roots. The phenolic concentration is increased in both leaves and roots (Lu et al., 2014; Liu et al., 2015). Inadequate B supply causes numerous biochemical, physiological, and anatomic changes; therefore, it is extremely difficult to distinguish with certainty the primary and secondary effects of the deprivation. Interruption of cell division in the apical meristem is the most prominent effect that results in a reduction and even cessation of root growth. B deficiency decreases male fertility by reducing microsporogenesis, germination, and elongation of the pollen tube (Cheng and Rerkasem, 1993). After fertilization, the nutrition disorder affects embryogenesis, resulting in seed deterioration or the formation of incomplete or damaged embryos (Rerkasem et al., 2004; Rerkasem et al., 2019).

13.3 Boron transport mechanisms

B uptake was considered a controversial topic for a long time, inter alia due to various plant species, extremely high B concentrations, and different techniques used in the experiments (Hu and Brown, 1997). Plants take up B via the roots, predominantly in the form of boric acid. It is a small, soluble, undissociated, and uncharged molecule that easily migrates across the lipid bilayers. B is the only element that is not taken up from the soil as an ion. In conditions of sufficient supply, the element is transported by passive diffusion and without protein catalysis and energy consumption (Fig. 13.2). Because of the cell's high permeability to B, characteristic patterns of flux along the transpiration stream, and accumulation in the tips of the leaves, passive diffusion was long considered as the only mechanism of transport (Marschner, 1995). It was suggested that the element enters the cell partially via passive diffusion through the lipid bilayer, and partially via protein channels, aquaporins, or other channels that are Hg-sensitive (Fig. 13.2) (Dordas et al., 2000). The substantial differences among plant species regarding B mobility make the element unique among nutrients. B is rapidly and significantly phloem-mobile in species for which polyols (e.g., sorbitol, mannitol, dulcitol) are the primary products of photosynthesis. The mobility is due to B complexation with polyols. Such species are from *Prunus*, *Malus*, and *Pyrus* genera (prune, pear, apple, cherry, almond, plum, peach, apricot), as well as onion, celery, carrot, olive, bean, pea, cauliflower, cabbage, asparagus, and coffee. On the other hand, B has restricted phloem mobility in species with sucrose as a primary photosynthetic product. These species (e.g., wheat, barley) represent the majority.

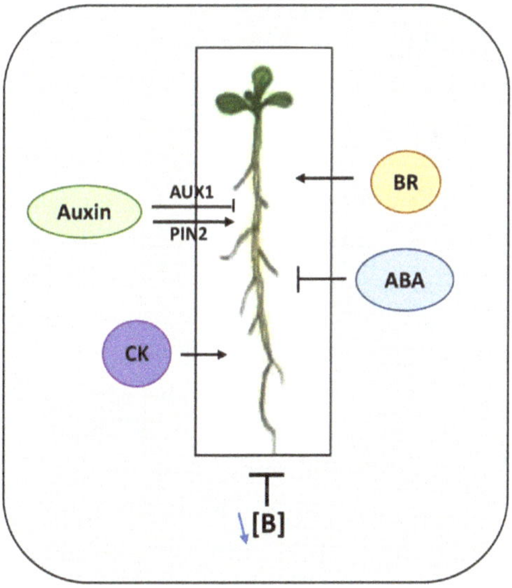

Figure 13.2 Boric acid (BA)/phenylboronic acid (PBA) uptake into plant cells. Passive and active uptake mechanisms of BA and PBA through either diffusion or the NIP5;1 channel protein. Due to the cytosolic pH, BA, and PBA are diverted into borate and phenylboronate anions. Export of borate and phenylboronate out of a plant cell occurs via diffusion or the BOR1 export transporter.

B moves along the transpiration stream and accumulates at its end. Therefore, the concentration of the element is dramatically higher in leaf tips and margins than in the rest of the leaf. On the contrary, B concentration is uniform across the leaves of phloem-mobile species. Consistently, comparatively higher B concentration in older leaves implies restricted phloem mobility. A higher concentration in young leaves and fruits indicates significant phloem mobility (Stangoulis et al., 2010). The dramatic inter- and intraspecies differences in plant tissue boron content, as well as in B tolerance and sensitivity, undermined the passive transport hypothesis. Plant tissue B content was generally considered to be lower in tolerant species and genotypes (Rehman et al., 2006), although exceptions were reported (Torun et al., 2006; Brdar-Jokanović et al., 2013). Nevertheless, differences in tolerance cannot be explained exclusively by differences in transpiration, because the above would mean that tolerant plants have to exhibit sevenfold higher water use efficiency than sensitive plants (Nable et al., 1997). Higher cell-wall pectin content (Hu et al., 1996), inactivation in the wall or cytoplasm (Reid and Fitzpatrick, 2009), separation into the vacuole (Wakuta et al., 2016) and cell wall (Papadakis et al., 2018), redistribution among plant organs (Nable, 1989), blocking them in the uptake by tolerant genotypes via decreased permeability of root cell plasma membranes (Huang and Graham, 1990), and efflux from the roots (Reid, 2007; Sutton et al., 2007) were hypothesized as possible mechanisms for plant B tolerance. Nevertheless, active transport has to be involved in tolerance (Miwa and Fujiwara, 2010). Because passive diffusion is not capable of satisfying plant B demands under deficiency, boric acid channels and borate exporters have to be involved in uptake and translocation to growing tissues (Fig. 13.2). Exporters are additionally involved in exclusion in conditions of excessive B. The channels transporting boric acid are of the major intrinsic protein (MIP) family, subfamily nodulin 26-like intrinsic proteins (NIPs). The first such channel was identified in *Arabidopsis* and named NIP5;1 (Takano et al., 2006). It functions in root cells exposed to B deprivation and it is downregulated if external B increases (Tanaka et al., 2011; Wang et al., 2017). NIP6;1 is involved in B transfer from the xylem to the phloem (Tanaka and Fujiwara, 2008). Homologous channels were found in barley, maize, and rice (Durbak et al., 2014; Shao et al., 2018). Borate transporters (BOR) functioning under B deficiency were firstly described in *Arabidopsis* (Tanaka and Fujiwara, 2008), and its homologs were reported for rice, wheat, maize, and other plant species (Yoshinari and Takano, 2017; Diehn et al., 2019). Seven BORs in *Arabidopsis* were identified so far.

BOR1 is involved in B transport from roots to the xylem and its translocation to leaves, collaborating with NIP5;1 and NIP6;1, respectively. BOR2 supports root growth by participating in cross-linking RGII molecules (Miwa et al., 2013; Chormova et al., 2014), whereas BOR4 functions primarily in tolerance to excessive B via exclusion from tissues. BOR4 homologs were found in barley, wheat, and other agricultural plants (Pallotta et al., 2014; Reids, 2014). In addition to the B availability in the medium, external factors such as CO_2 and irradiance affect the function of these proteins (Mishra et al., 2018). However, the elucidation of the mechanisms by which plants cope with B stresses remains a difficult task. For example, the proteins involved in the response to B deficiency in citrus roots belong to functional groups related to cell transport, biological regulation and signal transduction, stress responses, and other (protein, nucleic acid, carbohydrate and energy, cell wall and cytoskeleton, lipid) metabolic processes. The root adaptation to low B might include up- and downregulation of microRNA involved in decreased respiration, improved ability to scavenge reactive oxygen species, enhanced cell transport, increased lateral root number, improved osmoprotection, and other metabolic reactions (Yang et al., 2013; Lu et al., 2014). In addition, there are many differences between roots and leaves in adaptive mechanisms to low B at the transcriptional level; most of the differentially expressed fragments related to signal transduction and stress defense are down- and upregulated in roots and leaves, respectively (Lu et al., 2015). Several genes involved in cell-wall metabolism and transmembrane transport are highly regulated under B deficiency in citrus. Numerous metabolic pathways (lignin biosynthesis, nitrogen metabolism, and glycolytic pathway) are affected by stress (Zhou et al., 2015). The advanced knowledge on the physiological basis of B uptake and distribution should facilitate the breeding of plants tolerant to B deficiency and excess. Genes with transcript accumulations induced by B deficiency were identified in several studies (Takano et al., 2006; Kasajima and Fujiwara, 2007), as well as the first identification of a transcription factor gene induced by low boron (WRKY6), involved in normal *Arabidopsis* root growth under deficiency (Kasajima et al., 2010).

13.4 The boron nutritional status evokes contrasting changes in plant hormones metabolism

To sustain optimal shoot growth, B has to be continuously taken up, loaded into the xylem and delivered to aboveground plant organs

(Miwa and Fujiwara, 2010). Low B supply leads to a decrease in abscisic acid (ABA) and salicylic acid (SA) concentrations, and an increase in cytokinin (CK) and brassinosteroid (BR) levels in the vasculature and the mesophyll, respectively. Zhou et al. (2019) have carried out a genome-wide identification and characterization of long noncoding RNAs (IncRNAs) in the leaves of a widely used citrus rootstock (trifoliate orange or *Poncirus trifoliata*) subjected to B deficiency. Analysis by Gene Ontology and the Kyoto Encyclopedia of Genes and Genomes shows that pathways of secondary metabolite biosynthesis (including phenylpropanoid/lignin biosynthesis) and plant hormone signal transduction and the calcium signaling pathway are significantly enriched by this nutrition stress. These results provide valuable information on the role of lncRNAs in response to B-deficiency stress in the leaves of trifoliate orange. The involvement of B in signaling has also been proposed to explain some of the multitudes of changes in cellular activities induced by B deficiency. However, current knowledge on how plants perceive the signal of B deficiency is still very limited. Quiles-Pando et al. (2019) have revealed that B deficiency causes an increase in the cytosolic calcium concentration ($[Ca^{2+}]$cyt) in *Arabidopsis thaliana* roots after 6 and 24 h of this nutrient stress. This increase is mainly owed to Ca^{2+} movement across the plasma membrane from the apoplast, although it is possible that some Ca^{2+} comes from the vacuole through the tonoplast CNGC19 channel. The $[Ca^{2+}]$cyt is gradually restored when B-sufficient conditions are re-established. The results indicate that CAX3 would play an important role in the restoration of Ca^{2+} homeostasis after 24 h of B deficiency.

Potential interactions between B and in particular the hormones auxin, CK, ethylene (ET), and ABA have been hypothesized (Fig. 13.3) (Abreu et al., 2014; Li et al., 2015; Matthes and Torres-Ruiz, 2016; Macho-Rivero et al., 2017; Poza-Viejo et al., 2018; Gómez-Soto et al., 2019). While direct evidence for such interactions is missing, they would lead to the elucidation of additional roles of B beyond the cell wall and help determine if hormonal defects are a secondary effect of an altered cell wall and consequently an altered morphology, or if B interacts with hormones directly. Specifically, there is a long-standing interest in studying interactions of B with auxin, as B deprivation was shown to cause alterations of levels of the main active auxin indole 3-acetic acid (IAA) as reviewed in Blevins and Lukaszewski (1998). These alterations are hypothesized to be due to (1) enhancement of the activity of IAA oxidase, an enzyme involved in the catabolism of IAA (Jarvis et al., 1983); or (2) the

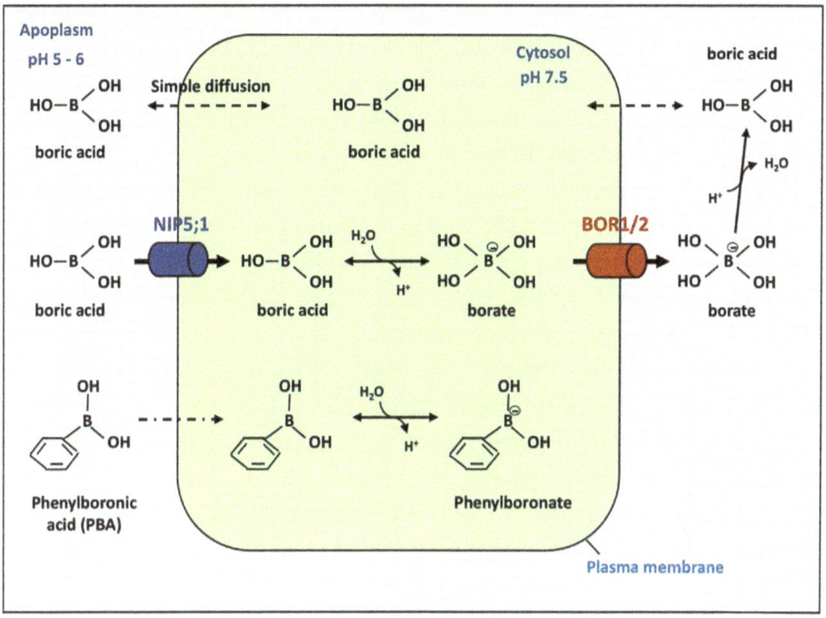

Figure 13.3 Interaction between plant hormones and boron over a primary root. Black arrows represent promotion whereas black blunt arrows show inhibition of primary root growth. In addition, blue arrows represent the regulation of B concentration. At the bottom of the figure is the overall effect of B on primary root growth. Primary root growth promotion or inhibition in auxin transport, signaling and concentration are marked. *ABA*, Abscisic acid; *BR*, brassinosteroids; *CK*, cytokinin.

involvement of B in auxin transport (Li et al., 2001, 2015; Wang et al., 2006; Matthes and Torres-Ruiz, 2016). Interestingly, numerous studies in various plant species have either directly or indirectly shown alterations of auxin levels upon B deficiency, but with different results (Fig. 13.3) (Blevins and Lukaszewski, 1998; Martín-Rejano et al., 2011; Li et al., 2015; Matthes and Torres-Ruiz, 2016; Zhou et al., 2016; Eggert and von Wiren, 2017), hindering general reasoning of the relationship between B deficiency and hormones, such as auxin. A recent study in Brassica seedlings (Eggert and von Wiren, 2017), which are highly sensitive to B deficiency (Zhang et al., 2014), highlights this importance. Eggert and von Wiren (2017) took a physiological approach, where they simultaneously measured active phytohormone species, including their precursors and catabolites, in association with the shoot B nutritional status to detect B-dependent growth responses. This study revealed that until the presence of B deficiency, symptoms are manifest, IAA levels slightly decrease, but afterward, the concentration of all tested auxin species increased with progressing B deficiency (Eggert and von Wiren, 2017). When crossing a certain threshold of B deficiency, however, IAA levels significantly dropped, although they were still higher than under B-sufficient conditions.

Although interactions between CK and B status are not as widely explored as those with B and auxin, there are a number of reports demonstrating involvement of CK in the regulation of the B stress response in plants. In *Pisum sativum* (pea) shoots, a reduction of CK was reported, when grown under B deficiency (hydroponics: 0 μM added B) (Li et al., 2001). In *Arabidopsis*, it was observed that a CK receptor gene, CYTOKININ RESPONSE 1 (CRE1)/WOODEN LEG (WOL)/ ARABIDOPSIS HISTIDINE KINASE4 (AHK4), is downregulated upon B deficiency and this downregulation is correlated with a phenotype of abnormal protoxylem differentiation, proposed to be due to failure in the transition from cell division to cell differentiation (Abreu et al., 2014). A dependence of cytokinin receptor expression on the B nutritional status was also shown in navel oranges (*Citrus sinensis* Osb.) (Yang et al., 2013). Interestingly, the cytokinin receptor gene WOL was upregulated in a B dose-dependent manner. It will be interesting to determine in the future whether the expression of the receptor genes depends on CK levels, which would allow the hypothesis that the transcriptional response of the CK receptors might be a secondary response of the downregulation of CK levels upon B deficiency. In support of this hypothesis, recent work by Poza-Viejo et al. (2018) showed increases in CK signaling components as early events after Arabidopsis seedlings had been transferred to B deficiency conditions (solid medium; 0.1 μM B).

Indications of B interactions with ABA initially came from B toxicity studies, where it was shown that upon B toxicity, ABA content in the shoot increased, which was correlated with an increased expression of the ABA biosynthesis gene AtNINE-CIS-EPOXYCAROTENOID DIOXYGENASE 3, in the root (Macho-Rivero et al., 2017). It was also recently shown that ABA and its glycosylester both increased upon severe B deficiency in Brassica (Eggert and von Wiren, 2017).

B deficiency can affect BR signaling, thereby altering root growth. RNA-sequencing analysis reveals a high coregulation between BR-regulated genes and B deficiency-responsive genes (Zhang et al., 2014). Low B negatively regulates BR signaling to control BR signaling-dependent root elongation, bes1-D exhibits insensitivity to low B stress, and bri1−301 mutants fails to respond to B depletion (Zhang et al., 2014).

Exogenous brassinolide (eBL) application can rescue the inhibition of root growth under B deficiency conditions, and application of BR biosynthesis inhibitor BRZ aggravates root growth inhibition of wild-type under B deficiency conditions. B deficiency reduces the nuclear signal of BRI1-EMS-SUPPRESSOR 1 (BES1). Moreover, B deficiency reduces the accumulation of brassinolide (BL) by reducing BR6ox1 and BR6ox2 mRNA levels to downregulate BR signaling and modulate root elongation (Zhang et al., 2014). Altogether, these results uncover a role of BR signaling in root elongation under B deficiency (Zhang et al., 2014).

Finally, recent studies indicate the participation of a signaling pathway involving ET in the response of root growth to B deficiency (Camacho-Cristóbal et al., 2015).

Altogether, approaches to separating out the complexity of primary and secondary effects of B deficiency include the analysis of hormone mutants under short-term B deprivation in solution culture or using a chemical approach by adding hormones to the solution culture and analyzing their effects on plant development under B deficiency (Camacho-Cristóbal et al., 2015; Li et al., 2015; Poza-Viejo et al., 2018; Gómez-Soto et al., 2019).

13.5 Effect of boron deficiency on plant development

Root growth is more sensitive to B deficiency than shoot growth. The cessation of cell division in the apical meristem resulting from B deficiency inhibits root elongation. Under severe deficiency, the root cap disappears, the growth stops, and root tips die out. The shoot/root ratio increases and plants become more susceptible to drought and nutritive imbalances (Fig. 13.4). Longitudinal splits close to the primary vein of young leaves and leaf margins deformed in shape were described in wheat by Snowball and Robson (Snowball and Robson, 1983); however, these symptoms are rarely seen in the field (Rerkasem et al., 2004). In the case of severe and prolonged deficiency, the internodes are shortened, and necrotic spots appear on leaves. B requirements are generally higher during the reproductive phase of a plant's life cycle. Consequently, in field conditions, the yield can be significantly reduced without the appearance of visual symptoms of the deficiency (Dell and Huang, 1997). The critical level for deficiency in the vegetative phase is 1 ppm in flag leaves of wheat; however, similar to toxicity, leaf B concentration is not necessarily a reflection of B utilization efficiency (Wongmo et al., 2004). On the other hand,

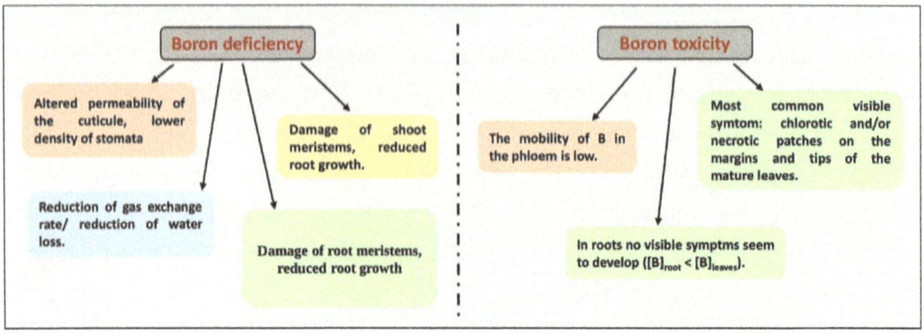

Figure 13.4 Effects of boron deficiency and toxicity in different parts of plants.

7–8 ppm in anthers and 5–6 ppm in carpels result in wheat sterility (Rerkasem and Jamjod, 1997). The most common effect of deficiency during the reproductive phase is male sterility. If fertilization does occur, seeds may be aborted. Accordingly, the most pronounced effect of B deficiency in, for example, wheat is a reduced number of seeds per spike (Rerkasem et al., 1993) and, in extreme cases, yield can be 100% reduced due to nutrition disorder (Pant et al., 1998). Genotypes with stable yield in the conditions that reduce yield in other genotypes are regarded as having more efficient B utilization (Rerkasem et al., 2004). Phloem-mobile species rarely experience a deficiency, thanks to their ability to retranslocate internal B. Therefore, the increased production of polyols may be an effective strategy for improving B utilization efficiency. This may be achieved through transformation (e.g., tobacco, rice), or via the selection of genotypes with high polyol production (Bellaloui et al., 2003). There is a wide variation among wheat genotypes in terms of B utilization efficiency, and it is probably the widest variation of all agricultural plants and nutrients. Therefore, breeding for utilization efficiency was proposed as the most feasible way to overcome the deficiency problem (Miwa et al., 2006; Rerkasem et al., 2019). It was found that B efficiency in wheat is controlled by two independent major loci. Dominant loci (Bod1, Bod2) imply high efficiency; thus, it is believed that the standards for utilization efficiency feature the following combinations of loci: Bod1 Bod1, Bod2Bod2 (Fang60—efficient cultivar), bod1bod1bod2bod2 (Bonza—inefficient), and bod1bod1Bod2 Bod2 (SW 41—medium efficient), according to Jamjod et al. (2004). The utilization efficiency is conditioned by a high capacity for B transport from roots to aboveground plant organs (Nachiangmai et al., 2004). Takano et al. (2002) carried out the first identification of the *Arabidopsis thaliana* membrane protein BOR1 responsible for loading B in the xylem and its translocation in conditions of deficiency. A generation of *Arabidopsis* that is tolerant to B deficiency was produced (Miwa et al., 2006). This was achieved via increased BOR1 production causing an increased boron translocation from roots to aboveground organs. The transgenic plants successfully yielded in deficiency conditions that reduced wild-type yield by 100%, and they did not exhibit increased sensitivity to excess B. The transporters that were later identified in major agricultural plants represent the basis for further research aimed at breeding or designing varieties with improved B utilization efficiency (Yoshinari and Takano, 2017; Yuan et al., 2017). For example, transgenic tomato lines (cultivar Micro-Tom) with improved growth in B deficiency conditions were designed (Uraguchi et al., 2014). Tomato is often used as a model of fruit-bearing crops. The lines are boron-efficient due to a strong expression of AtBOR1 (*Arabidopsis thaliana* BOR1). When compared to nontransgenic plants, transgenic plants grown in deficiency conditions had higher shoot dry weight, as well as shoot and fruit B concentration. The results of this study imply that transgenic crops may be a sustainable solution for B-deficient areas. Another approach may be marker-assisted selection breeding for B utilization efficiency. The trait is polygenic in most agricultural plants. In *Brassica napus*, for example, efficiency includes B uptake, transport, and utilization. Thanks to the previously constructed high-density genetic map, several quantitative trait loci (QTL) involved in B utilization efficiency were identified. The development of molecular markers associated with the QTLs should help breeders to identify efficient genotypes. However, substantial research remains to be done to transfer the knowledge on the physiology and genetics of B utilization efficiency into B-efficient plant varieties with good growth, quality, and yield (Hua et al., 2016; Yuan et al., 2017; Pommerrenig et al., 2018).

13.6 Alleviation of the effects of boron toxicity

Symptoms of B toxicity differ between species with restricted and significant phloem mobility. In phloem-immobile species, B moves via the xylem and accumulates at the end of the transpiration stream. Accordingly, foliar symptoms in barley and wheat include chlorosis and necrosis spreading from the leaf tips, with brown lesions that are at first formed on the margins, and then cover much of the leaf surface. The oldest leaves are first affected by the disorder, which further spreads to the top of the plant (Fig. 13.4). In severe cases, brown lesions are present at leaf sheaths, stems, spikes, and awns. Delayed emergence and delayed foliation, as well as a reduction in stem height, dry matter weight, 1000-kernel weight, number of spikes per plant, and yield, were reported (Cartwright et al., 1984; Nable, 1988; Bañuelos et al., 1999; Rehman et al., 2006). Generally, symptoms are more pronounced and earlier exhibited in barley than in wheat (Yau and Saxena, 1997). If B concentration increases with soil depth, symptoms are manifested later, and yield is less reduced (Brennan and Adcock, 2004). Excessive B does not affect leaf area, width, or length (Yau and Saxena, 1997). General root weakness and decreased growth of lateral roots were described in hydroponically grown wheat and barley (Nable, 1988). The severity of all B toxicity symptoms varies among genotypes, for example, there are those with lower, the same, or higher dry matter weight in B treatments (Torun et al., 2006). Reduced yield, as well as significant variability among genotypes in terms of both yield reduction and visual symptoms of B toxicity, was observed in wheat and barley (Paull et al., 1988; Yau and Saxena, 1997; Yau et al., 1997; Yau and Ryan, 2008). Furthermore, there are genotypes with stable yields in conditions of elevated soil B, despite pronounced toxicity symptoms and high shoot B concentration (Avci and Akar, 2005). In maize, tomato, carrot, and alfalfa, excess B decreases emergence (Bañuelos et al., 1999). In celery, irregular stem shapes and deformed young leaves occur as symptoms of B toxicity (Brown and Shelp, 1997). Dramatical differences in literature data concerning plant B concentrations may be explained by plant species (B mobility), organ [e.g., concentration in wheat flag leaf is about 10-fold higher than in grains (Campbell et al., 1993)], stage of life (older plants contain more B (Moody et al., 1988)), treatment B concentration, genotype, and experimental technique (Furlani et al., 2003; Turan et al., 2018). Generally, in wheat, 68% of plant B is located in the leaves, 16% in roots, 10% in spikes, and 6% in stems (Subedi et al., 1999). Field trials are under the influence of numerous environmental factors; however, the experimental design of the trials performed in controlled conditions also affects B uptake. Pot size and type, temperature, watering regime, and the intensity of the applied treatments are among the factors influencing nutrient uptake in controlled conditions (De Vries, MPC, 1980). Considering the narrow range between B toxicity and deficiency, the costs and constraints related to B fertilization and amelioration of B-excessive soils, the effects of environmental conditions on its uptake and utilization, the identified vast intraspecies variability in terms of the utilization efficiency and tolerance, and the recent findings related to its physiology and genetics, the best solution which may be possible to accomplish in the future would be to breed varieties of agricultural plants that are adjusted to the wide range of soil B concentrations.

13.7 Conclusions and future prospects

Boron is in many ways unique among plant nutrients; it differs among species in terms of mobility, the narrow range between deficiency and toxicity, and differential inter- and intraspecies response to an inadequate supply. Both B deficiency and toxicity are major management issues in agriculture and there is clearly a need for greater research to understand how B supply can best be manipulated to optimize crop production. As Brdar-Jokanović (2020) highlighted in her review, inadequate B supply has detrimental effects on agricultural crop yields. Plant species and genotypes drastically differ in terms of B requirements. Thus, a B concentration in soil that causes deficiency for one crop may be toxic for another less tolerant species. It has been proposed to exploit varieties that are efficient in using B, along with B fertilization, to solve the problems associated with B deficiency. Both transgenic and marker-assisted selection methods can be effective strategies to improve the efficiency of B utilization in crops. Unlike B deficiency, soil B toxicity is much more difficult to ameliorate. Physiological and genetic advances in B toxicity tolerance may facilitate the development of tolerant varieties to cope with the problem of excess B in the soil.

References

Abreu, I., Poza, L., Bonilla, I., Bolaños, L., 2014. Boron deficiency results in early repression of a cytokinin receptor gene and abnormal cell differentiation in the apical root meristem of *Arabidopsis thaliana*. Plant. Physiol. Biochem. 77, 117–121.

Avci, M., Akar, T., 2005. Severity and spatial distribution of boron toxicity in barley cultivated areas of Central Anatolia and Transitional zones. Turkish J. Agric. Forestry 29, 377–382.

Bañuelos, G.S., Ajwa, H.A., Caceres, L., Dyer, D., 1999. Germination responses and boron accumulation in germplasm from Chile and the United States grown with boron-enriched water. Ecotoxicol. Environ. Saf. 43, 62–67.

Bellaloui, N., Yadavc, R.C., Chern, M.S., Hu, H., Gillen, A.M., Greve, C., et al., 2003. Transgenically enhanced sorbitol synthesis facilitates phloem-boron mobility in rice. Physiol. Plant. 117, 79–84.

Blevins, D.G., Lukaszewski, K.M., 1998. Boron in plant structure and function. Annu. Rev. Plant Physiol. Plant Mol. Biol. 49, 481–500.
Bolaños, L., Lukaszewski, K., Bonilla, I., Blevins, D., 2004. Why boron? Plant Physiol. Biochem. 42, 907–912.
Brdar-Jokanović, M., 2020. Boron toxicity and deficiency in agricultural plants. Int. J. Mol. Sci. 21, 1424.
Brdar-Jokanović, M.D., Maksimovic, I., Marija, K.B., et al., 2013. Boron concentration vs. content as criterion for estimating boron tolerance in wheat. J. Plant Nutri. 36 (3), 470–480.
Brennan, R.F., Adcock, K.G., 2004. Incidence of boron toxicity in spring barley in Southwestern Australia. J. Plant Nutr. 27, 411–425.
Brown, P.H., Hu, H., 1997. Does boron play only a structural role in the growing tissues of higher plants? Plant Soil 196, 211–215.
Brown, P.H., Shelp, B.J., 1997. Boron mobility in plants. Plant Soil 193, 85–101.
Brown, P.H., Bellaloui, N., Wimmer, M.A., Bassil, E.S., Ruiz, J., Hu, H., et al., 2002. Boron in plant biology. Plant Biol. 4, 205–223.
Cakmak, I., Romheld, V., 1997. Boron deficiency-induced impairments of cellular functions in plants. Plant Soil 193, 71–83.
Camacho-Cristóbal, J.J., Martín-Rejano, E.M., Begoña Herrera-Rodríguez, M., Navarro-Gochicoa, M.T., Rexach, J., González-Fontes, A., 2015. Boron deficiency inhibits root cell elongation via an ethylene/auxin/ROS-dependent pathway in Arabidopsis seedlings. J. Exp. Bot. 66 (13), 3831.
Camacho-Cristóbal, J.J., Navarro-Gochicoa, M.T., Rexach, J., González-Fontes, A., Herrera-Rodríguez, M.B., 2018. Plant response to boron deficiency and boron use efficiency in crop plants. In: Hossain, M.A., Kamiya, T., Burritt, D.J., Phan Tran, L. S., Fujiwara, T. (Eds.), Plant Micronutrient Use Efficiency: Molecular and Genomic Perspectives in Crop Plants. Academic Press, San Diego, CA, pp. 109–121.
Campbell, T.A., Moody, D.B., Jefferies, S.P., Cartwright, B., Rathjen, A.J., 1993. Grain yield evaluation of near isogenic lines for boron tolerance. In: Proceedings of the 8th International Wheat Genetics Symposium, vol. 2. Beijing, China, pp. 1021–1027.
Cartwright, B., Zarcinas, B.A., Mayfield, A.H., 1984. Toxic concentrations of boron in a red-brown earth at Gladstone, South Australia. Austr J. Soil Res. 22, 261–272.
Cheng, C., Rerkasem, B., 1993. Effects of boron on pollen viability in wheat. Plant and Soil 155/156, 313-315.
Chormova, D., Messenger, D.J., Fry, S.C., 2014. Boron bridging of rhamnogalacturonan-II, monitored by gel electrophoresis, occurs during polysaccharide synthesis and secretion but not post-secretion. Plant J. 77, 534–546.
Chormova, D., Fry, S.C., 2016. Boron bridging of rhamnogalacturonan-II is promoted in vitro by cationic chaperones, including polyhistidine and wall glycoproteins. N. Phytol. 209, 241–251.
Dell, B., Huang, L., 1997. Physiological response of plants to low boron. Plant Soil 193, 103–120.
De Vries, MPC, 1980. How reliable are results from pot experiments? Commun. Soil Sci. Plant Anal. 11, 895–902.
Diehn, T.A., Bienert, M.D., Pommerrening, B., Liu, Z., Bienert, M.D., Pommerrening, B., et al., 2019. Boron demanding tissues of *Brassica napus* express specific sets of functional nodulin26-like intrinsic proteins and BOR1 transporters. Plant J. 100, 68–82.
Dordas, C., Chrispeels, M.J., Brown, P.H., 2000. Permeability and channel mediated transport of boric acid across membrane vesicles isolated from squash roots. Plant Physiol. 124, 1349–1362.
Durbak, A.R., Phillips, K.A., Pike, S., O'Neill, M.A., Mares, J., Gallavotti, A., et al., 2014. Transport of boron by the tassel-less1 aquaporin is critical for vegetative and reproductive development in maize. Plant Cell 26, 2978–2995.

Dzondo-Gadet, M., Mayap-Nzietchueng, R., Hess, K., Nabet, P., Belleville, F., Dousset, B., 2002. Action of boron at the molecular level: Effects on transcription and translation in an acellular system. Biol. Trace Element Res. 85, 23–33.
Eggert, K., von Wiren, N., 2017. Response of the plant hormone network to boron deficiency. N. Phytol. 216, 868–881.
El-Motaium, R., Hu, H., Brown, P.H., 1994. The relative tolerance of six Prunus rootstocks to boron and salinity. J. Am. Soc. Horticult. Sci. 119, 1169–1175.
Furlani, Â.M.C., Carvalho, C.P., De Freitas, J.G., Verdial, M.F., 2003. Wheat cultivar tolerance to boron deficiency and toxicity in nutrient solution. Sci. Agric. 60, 359–370.
Goldbach, H.E., Wimmer, M.A., Findeklee, P., 2000. Discussion paper: boron—how can the critical level be defined? J. Plant Nutr. Soil Sci. 163, 115–121.
Goldbach, H.E., Wimmer, M., 2007. Boron in plants and animals: is there a role beyond cell-wall structure? J. Plant Nutr. Soil Sci. 170, 39–48.
Gómez-Soto, D., Galván, S., Rosales, E., Bienert, P., Abreu, I., Bonilla, I., et al., 2019. Insights into the role of phytohormones regulating pAtNIP5;1 activity and boron transport in Arabidopsis thaliana. Plant Sci. 287, 110198.
Gonzáles-Fontes, A., Fujiwara, T., 2020. Advances in plant boron. Int. J. Mol. Sci. 21, 1–3.
Han, S., Chen, L.S., Jiang, H.X., Smith, B.R., Yang, L.T., Xie, C.Y., 2008. Boron deficiency decreases growth and photosynthesis, and increases starch and hexoses in leaves of citrus seedlings. J. Plant Physiol. 165, 1331–1341.
Hu, H., Brown, P.H., Labavitch, J.M., 1996. Species variability in boron requirement is correlated with cell wall pectin. J. Exp. Bot. 47, 227–232.
Hu, H., Brown, P.H., 1997. Absorption of boron by plant roots. Plant Soil 193, 49–58.
Hua, Y., Zhou, T., Ding, G., Yang, Q., Shi, L., Xu, F., 2016. Physiological, genomic and transcriptional diversity in responses to boron deficiency in rapeseed genotypes. J. Exp. Bot. 67 (19), 5769–5784.
Hua, Y., Feng, Y., Zhou, T., Xu, F., 2017. Genome-scale mRNA transcriptomic insights into the responses of oilseed rape (*Brassica napus* L.) to varying boron availabilities. Plant Soil 416, 205–225.
Huang, C., Graham, R.D., 1990. Resistance of wheat genotypes to boron toxicity is expressed at cellular level. Plant Soil 126, 295–300.
Huang, J.H., Cai, Z.J., Wen, S.X., Guo, P., Ye, X., Lin, G.Z., et al., 2014. Effects of boron toxicity on root and leaf anatomy in two citrus species differing in boron tolerance. Trees 28, 1653–1666.
Jamjod, S., Niruntrayagul, S., Rerkasem, B., 2004. Genetic control of boron effciency in wheat (*Triticum aestivum* L.). Euphytica 135, 21–27.
Jarvis, B., Ali, A., Shaheed, A., 1983. Auxin and boron in relation to the rooting response and ageing of mung bean cuttings. N. Phytol. 95, 509–518.
Kasajima, I., Fujiwara, T., 2007. Identification of novel Arabidopsis thaliana genes which are induced by high levels of boron. Plant Biotechnol. 24, 355–360.
Kasajima, I., Ide, Y., Yokota Hirai, M., Fujiwara, T., 2010. WRKY6 is involved in the response to boron deficiency in *Arabidopsis thaliana*. Physiol. Plant 139, 80–92.
Kobayashi, M., Matoh, T., Azuma, J., 1996. Two chains of rhamnogalacturonan II are cross-linked by borate-diol ester bonds in higher plant cell walls. Plant Physiol. 110, 1017–1020.
Lewis, D.H., 2019. Boron: the essential element for vascular plants that never was. N. Phytol. 221, 1685–1690.
Li, C., Pfeffer, H., Dannel, F., Römheld, V., Bangerth, F., 2001. Effects of boron starvation on boron compartmentation, and possibly hormone-mediated elongation growth and apical dominance of pea (*Pisum sativum*) plants. Physiol. Plant 111, 212–219.
Li, K., Kamiya, T., Fujiwara, T., 2015. Differential roles of PIN1 and PIN2 in root meristem maintenance under low-B conditions in *Arabidopsis thaliana*. Plant Cell Physiol. 56, 1205–1214.

Li, M., Zhao, Z., Zhang, Z., Zhang, W., Zhou, J., Xu, F., et al., 2017. Effect of boron deficiency on anatomical structure and chemical composition of petioles and photosynthesis of leaves in cotton (*Gossypium hirsutum* L.). Sci. Rep. 7, 4420.

Liu, G.D., Dong, X.C., Liu, L., Wu, L., Peng, S.A., Jiang, C.C., 2015. Metabolic profiling reveals altered pattern of central metabolism in navel orange plants as a result of boron deficiency. Physiol. Plant 153, 513–524.

Liu, X., Zhang, J.W., Guo, L.X., Liu, Y.Z., Jin, L.F., Hussain, S.B., et al., 2017. Transcriptome changes associated with boron deficiency in leaves of two citrus scion-rootstock combinations. Front. Plant Sci. 8, 317.

Lu, Y.B., Yang, L.T., Qi, Y.P., Li, Y., Li, Z., Chen, Y.B., et al., 2014. Identification of boron-deficiency-responsive microRNAs in *Citrus sinensis* roots by Illumina sequencing. BMC Plant Biol. 14, 123.

Lu, Y.B., Qi, Y.P., Yang, L.T., Lee, J., Guo, P., Ye, X., et al., 2015. Long-term boron-deficiency-responsive genes revealed by cDNA-AFLP differ between *Citrus sinensis* roots and leaves. Front. Plant Sci. 6, 585.

Macho-Rivero, M.Á., Camacho-Cristóbal, J.J., Herrera-Rodríguez, M.B., Müller, M., Munné-Bosch, S., González-Fontes, A., 2017. Abscisic acid and transpiration rate are involved in the response to boron toxicity in Arabidopsis plants. Physiol. Plant 160, 21–32.

Marschner, H., 1995. Mineral Nutrition in Higher Plants, second ed. Academic Press Inc, London, UK, p. 889.

Martín-Rejano, E.M., Camacho-Cristóbal, J.J., Herrera-Rodríguez, M.B., Rexach, J., Navarro-Gochicoa, M.T., González-Fontes, A., 2011. Auxin and ethylene are involved in the responses of root system architecture to low boron supply in Arabidopsis seedlings. Physiol. Plant 142, 170–178.

Matthes, M., Torres-Ruiz, R.A., 2016. Boronic acid treatment phenocopies monopteros by affecting PIN1 membrane stability and polar auxin transport in *Arabidopsis thaliana* embryos. Development 143, 4053–4062.

Matthes, M.S., Robil, J.M., McSteen, P., 2020. From element to development: the power of the essential micronutrient boron to shape morphological processes in plants. J. Exp. Bot. 71, 1681–1693.

Mishra, S., Heckathorn, S.A., Frantz, J.M., Krause, C., 2018. The effect of boron availability, CO2, and irradiance on relative accumulation of the major boron transport proteins, BOR1 and NIP5;1. Biol. Plant 62, 121–128.

Miwa, K., Fujiwara, T., 2010. Boron transport in plants: co-ordinated regulation of transporters. Ann. Bot. 105 (7), 1103–1105.

Miwa, K., Takano, J., Fujiwara, T., 2006. Improvement of seed yields under boron-limiting conditions through overexpression of BOR1, a boron transporter for xylem loading, in *Arabidopsis thaliana*. Plant J. 46 (6), 1084–1091.

Miwa, K., Fujiwara, T., 2010. Boron transport in plants: co-ordinated regulation of transporters. Ann. Bot. 105 (7), 1103–1108.

Miwa, K., Wakuta, S., Takada, S., Ide, K., Takano, J., Naito, S., et al., 2013. Roles of BOR2, a boron exporter, in cross linking of rhamnogalacturonan II and root elongation under boron limitation in Arabidopsis. Plant Physiol. 163, 1699–1709.

Moody, D.B., Rathjen, A.J., Cartwright, B., Paull, J.G., Lewis, J., 1988. Genetic diversity and geographical distribution of tolerance to high levels of soil boron. In: Proceedings of the 7th International Wheat Genetics Symposium, vl. 2. Cambridge, UK, pp. 859–865.

Nable, R.O., 1988. Effects of B toxicity among several barley wheat cultivars: a preliminary examination of the resistance mechanism. Plant Soil. 112, 45–52.

Nable, R.O., 1989. Effects of Boron Toxicity upon the Mineral Nutrient Composition of Barley and Wheat Cultivars. Divisional Report No. 104, CSIRO Division of Soils: Glen Osmond, Australia, pp. 1–10.

Nable, R.O., Banuelos, G.S., Paull, J.G., 1997. Boron toxicity. Plant Soil 193, 181–198.

Nachiangmai, D., Dell, B., Bell, R., Huang, L., Rerkasem, B., 2004. Enhanced boron transport into ear of wheat as a mechanism for boron efficiency. Plant Soil 264, 141–147.

Oikonomou, A., Ladikou, E.V., Chatziperou, G., et al., 2019. Boron excess imbalances root/shoot allometry, photosynthetic and chlorophyll fluorescence parameters and sugar metabolism in apple plants. Agronomy 9 (11), 731.

Pallotta, M., Schnurbusch, T., Hayes, J., Hay, A., Baumann, U., Paull, J., et al., 2014. Molecular basis of adaptation to high soil boron in wheat landraces and elite cultivars. Nature 514, 88–91.

Pant, J., Rerkasem, B., Noppakoonwong, R., 1998. Effect of water stress on the boron response of wheat genotypes under low boron field conditions. Plant Soil 202, 193–200.

Papadakis, I.E., Tsiantas, P.I., Tsaniklidis, G., Landi, M., Psychoyou, M., Fasseas, C., 2018. Changes in sugar metabolism associated to stem bark thickening partially assist young tissues of Eriobotrya japonica seedlings under boron stress. J. Plant Physiol. 231, 337–345.

Paull, J.G., Rathjen, A.J., Cartwright, B., 1988. Genetic control of tolerance to high concentrations of soil boron in wheat. In: Koebner, R.M.D., Miller, T.E. (Eds.), Proceedings of the 7th International Wheat Genetics Symposium, vol. 2. England Institute of Plant Science Research, Cambridge, UK, pp. 871–877.

Pommerrenig, B., Junker, A., Abreu, I., Bieber, A., Fuge, J., Willner, E., et al., 2018. Identification of rapeseed (Brassica napus) cultivars with a high tolerance to boron-deficient conditions. Front. Plant Sci. 9, 1142.

Pommerrenig, B., Eggert, K., Bienert, G.P., 2019. Boron deficiency ects on sugar, ionome, and phytohormone profiles of vascular and non-vascular leaf tissues of common plantain (*Plantago major* L.). Int. J. Mol. Sci. 20, 3882.

Poza-Viejo, L., Abreu, I., González-García, M.P., Allauca, P., Bonilla, I., Bolanos, L., et al., 2018. Boron deficiency inhibits root growth by controlling meristem activity under cytokinin regulation. Plant Sci. 270, 176–189.

Quiles-Pando, C., Rexach, J., Navarro-Gochicoa, M.T., Camacho-Cristóbal, J.J., Herrera-Rodríguez, M.B., González-Fontes, A., 2013. Boron deficiency increases the levels of cytosolic Ca2 + and expression of Ca2 + -related genes in *Arabidopsis thaliana* roots. Plant Physiol. Biochem. 65, 55–60.

Quiles-Pando, C., Navarro-Gochicoa, M.T., Herrera-Rodríguez, M.B., Camacho-Cristóbal, J.J., González-Fontes, A., Rexach, J., 2019. Boron deficiency increases cytosolic Ca2 + levels mainly via Ca2 + influx from the apoplast in *Arabidopsis thaliana* roots. Int. J. Mol. Sci. 20, 2297.

Rehman, S., Park, T.I., Kim, Y.J., Seo, Y.W., Yun, S.J., 2006. Inverse relationship between boron toxicity tolerance and boron contents of barley seed and root. J. Plant Nutrit. 29, 1779–1789.

Reid, R., 2007. Identification of boron transporter genes likely to be responsible for tolerance to boron toxicity in wheat and barley. Plant Cell Physiol. 48, 1673–1678.

Reid, R., 2014. Understanding the boron transport network in plants. Plant Soil 385, 1–13.

Reid, R., Fitzpatrick, K., 2009. Influence of leaf tolerance mechanisms and rain on boron toxicity in barley and wheat. Plant Physiol 151, 413–420.

Rerkasem, B., Lordkaew, S., Yimyam, M., Jamjod, S., 2019. Evaluating boron efficiency in heat tolerant wheat germplasm. Int. J Agri. Biol. 21 (2), 385–390.

Rerkasem, B., Netsangtip, R., Lordkaew, S., Cheng, C., 1993. Grain set failure in boron deficient wheat. Plant Soil 155, 309–312.

Rerkasem, B., Jamjod, S., 1997. Boron deficiency induced male sterility in wheat (*Triticum aestivum* L.) and implications for plant breeding. Euphytica 96, 257–262.

Rerkasem, B., Nirantrayagul, S., Jamjod, S., 2004. Increasing boron effciency in international bread wheat, durum wheat, triticale and barley germplasm will boost production on soils low in boron. Field Crop. Res. 86, 175–184.

Seth, K., Aery, N.C., 2017. Boron induced changes in biochemical constituents, enzymatic activities, and growth performance of wheat. Acta Physiol. Plant 39, 244.

Sonmez, O., Aydemir, S., Kaya, C., 2009. Mitigation effects of mycorrhiza on boron toxicity in wheat (Triticum durum) plants. N. ZealJ. Crop. Horticult. Sci. 37, 99–104.

Shah, A., Wu, X., Ullah, A., Fahad, S., Muhammad, R., Yan, L., et al., 2017. Deficiency and toxicity of boron: alterations in growth, oxidative damage and uptake by citrange orange plants. Ecotoxicol. Environ. Saf. 145, 575–582.

Shao, J.F., Yamaji, N., Liu, X.W., Yokosho, K., Shen, R.F., Ma, J.F., 2018. Preferential distribution of boron to developing tissues is mediated by the intrinsic protein OsNIP3. Plant Physiol. 176, 1739–1750.

Shireen, F., Nawaz, M.A., Chen, C., Zhang, Q., Zheng, Z., Sohail, H., et al., 2018. Boron: functions and approaches to enhance its availability in plants for sustainable agriculture. Int. J. Mol. Sci. 19, 1856.

Snowball, K., Robson, A.D., 1983. Symptoms of Nutrient Deficiencies: Subterranean Clover and Wheat; Department of Soil Science and Plant Nutrition. University of Western Australia, Nedlands, Australia, p. 73.

Stangoulis, J., Tate, M., Graham, R., Bucknall, M., Palmer, L., Boughton, B., et al., 2010. The mechanism of boron mobility in wheat and canola phloem. Plant Physiol 153, 876–881.

Subedi, K.D., Gregory, P.J., Gooding, M.J., 1999. Boron accumulation and partitioning in wheat cultivars with contrasting tolerance to boron deficiency. Plant Soil 214, 141–152.

Sutton, T., Baumann, U., Hayes, J., Collins, N.C., Shi, B.J., Schnurbusch, T., et al., 2007. Boron-toxicity tolerance in barley arising from eux transporter amplification. Science 318, 1446–1449.

Takano, J., Noguchi, K., Yasumori, M., Kobayashi, M., Gajdos, Z., Miwa, K., et al., 2002. Arabidopsis boron transporter for xylem loading. Nature 420, 337–340.

Takano, J., Wada, M., Ludewig, U., Schaaf, G., Von Wirén, N., Fujiwara, T., 2006. The Arabidopsis major intrinsic protein NIP5; 1 is essential for efficient boron uptake and plant development under boron limitation. Plant Cell 18 (6), 1498–1509.

Tanaka, M., Fujiwara, T., 2008. Physiological roles and transport mechanisms of boron: perspectives from plants. Pflug. Arch. 456, 671–677.

Tanaka, M., Takano, J., Chiba, Y., Lombardo, F., Ogasawara, Y., Onouchi, H., et al., 2011. Boron-dependent degradation of NIP5;1 mRNA for acclimation to excess boron conditions in Arabidopsis. Plant Cell 23 (9), 3547–3559.

Torun, A., Yazici, A., Erdem, H., Çakmak, I., 2006. Genotypic variation in tolerance to boron toxicity in 70 durum wheat genotypes. Turkish J. Agric. For. 30, 49–58.

Turan, M.A., Taban, S., Kayin, G.B., Taban, N., 2018. Effect of boron application on calcium and boron concentrations in cell wall of durum (*Triticum durum*) and bread (*Triticum aestivum*) wheat. J. Plant Nutr. 41, 1351–1357.

Uraguchi, S., Kato, Y., Hanaoka, H., Miwa, K., Fujiwara, T., 2014. Generation of boron-deficiency tolerant tomato by overexpressing an *Arabidopsis thaliana* borate transporter AtBOR1. Front. Plant Sci. 5, 125.

Wakuta, S., Fujikawa, T., Naito, S., Takano, J., 2016. Tolerance to excess-boron conditions acquired by stabilization of a BOR1 variant with weak polarity in Arabidopsis. Front. Cell Dev. Biol. 4, 4.

Wang, G., Römheld, V., Li, C., Bangerth, F., 2006. Involvement of auxin and CKs in boron deficiency induced changes in apical dominance of pea plants (*Pisum sativum* L.). J. Plant Physiol. 163, 591–600.

Wang, N., Yang, C., Pan, Z., Liu, Y., Peng, S., 2015. Boron deficiency in woody plants: various responses and tolerance mechanism. Front. Plant Sci. 6, Article 916.

Wang, S., Yoshinari, A., Shimada, T., Hara-Nishimura, I., Mitani-Ueno, N., Feng Ma, J., et al., 2017. Polar localization of the NIP5;1 boric acid channel is maintained by endocytosis and facilitates boron transport in Arabidopsis roots. Plant Cell 29, 824–842.

Wimmer, M.A., Abreu, I., Bell, R.W., Bienert, M.D., Brown, P.H., Dell, B., et al., 2019. Boron: An essential element for vascular plants. A comment on Lewis 'boron: the essential element for vascular plants that never was'. N. Phytol.

Wongmo, J., Jamjod, S., Rerkasem, B., 2004. Contrasting responses to boron deficiency in barley and wheat. Plant Soil 259, 103–110.

Yang, L.T., Qi, Y.P., Lu, Y.B., Guo, P., Sang, W., Feng, H., et al., 2013. iTRAQ protein profile analysis of *Citrus sinensis* roots in response to long-term boron-deficiency. J. Proteomic 93, 179–206.

Yau, S.K., Nachit, M.M., Ryan, J., 1997. Variation in boron-toxicity tolerance in a durum wheat core collection. In: Bell, R.W., Rerkasem, B. (Eds.), Boron in soils and plants. Kluwer Academic Publ, Dordrecht, the Netherlands, pp. 117–120.

Yau, S.K., Ryan, J., 2008. Boron toxicity tolerance in crops: A viable alternative to soil amelioration. Crop Sci 48, 854–865.

Yau, S.K., Saxena, M.C., 1997. Variation in growth, development, and yield of durum wheat in response to high soil boron. I. Average effects. Austr. J. Agric. Res. 48, 945–949.

Yoshinari, A., Takano, J., 2017. Insights into the mechanisms underlying boron homeostasis in plants. Front. Plant Sci. 8, 1951.

Yuan, D., Li, W., Hua, Y., King, G.J., Xu, F., Shi, L., 2017. Genome-wide Identification and characterization of the aquaporin gene family and transcriptional responses to boron deficiency in *Brassica napus*. Front. Plant Sci. 8, 1336.

Zhang, D., Hua, Y., Wang, X., Zhao, H., Shi, L., Xu, F., 2014. A high-density genetic map identifies a novel major QTL for boron effciency in oilseed rape (*Brassica napus* L.). PLoS One 9, e112089.

Zhou, G.F., Liu, Y.Z., Sheng, O., Wei, Q.J., Yang, C.Q., Peng, S.A., 2015. Transcription profiles of boron-deficiency-responsive genes in citrus rootstock root by suppression subtractive hybridization and cDNA microarray. Front. Plant Sci. 5, 795.

Zhou, T., Hua, Y., Huang, Y., Ding, G., Shi, L., Xu, F., 2016. Physiological and transcriptional analyses reveal differential phytohormone responses to boron deficiency in *Brassica napus* genotypes. Front. Plant Sci. 7, 221.

Zhou, G.F., Zhang, L.P., Li, B.X., Sheng, O., Wei, Q.J., Yao, F.X., et al., 2019. Genome-wide identification of long non coding RNA in trifoliate orange (*Poncirus trifoliata* (L.) Raf) leaves in response to boron deficiency. Int. J. Mol. Sci. 20, 5419.

CHAPTER 14

An overview on boron and pollen germination, tube growth and development under in vitro and in vivo conditions

Muhammad Aasim, Özlem Akgür and Büşra Yıldırım
Department of Plant Protection, Faculty of Agricultural Sciences and Technologies, Sivas University of Science and Technology, Sivas, Turkey

14.1 Introduction

Pollen grains are essential for fruit and seed set, fertilization, pollination, and play a key role in plant growth by transmitting male genetic material. Pollen grains are male gametophytes with two or three cell numbers at the time of release from the anthers (Borg et al., 2009). In nature, pollen grains germinate and form pollen tubes on the stigma. The first major morphogenetic process is pollen germination, which completes their final task of expelling male gametes into the fetal sac (Pal and Mondal, 2016). However, the whole process of pollen grain development, release and pollen tube development is affected and regulated by a number of internal and external factors. The pollen grains' fertility and viability are essential for successful fertilization and fruit development in plants. Germination of pollen occurs over the receptive stigmas to discharge the male gametes into the embryo sac, however in vivo investigations are difficult due to difficulties in the pistillate tissues (Pal and Mondal, 2016). Pollen tubes are regarded as the fastest-growing cells in the plant world because they may develop to a significant length in a short period of time under ideal conditions (Malik, 1977). Pollen germination and pollen tube development are therefore valuable study resources for morphological, physiological, biotechnological, ecological, evolutionary, biochemical, and molecular biology studies (Ottavio et al., 1992). Pollen germination is a crucial step after pollination, and this process is necessary for healthy seed germination and development. The most accurate method of determining pollen viability is to conduct an in vitro pollen germination test (Chatterjee et al., 2014).

Boron in Plants and Agriculture
DOI: https://doi.org/10.1016/B978-0-323-90857-3.00014-X

B is a mineral ingredient that plants require to grow and the significance of B for enhanced plant growth and development has already been proven (Shelp, 1993; Marschner, 1995; Day and Aasim, 2020). The major biochemical, physiological morphological processes controlled or regulated by B include cell wall growth and strength, cell division, sugar transfer and hormone development and fruit and seed development (Day et al., 2017; Day and Aasim, 2020). On the other hand, the presence or absence of B is also linked with the availability of other nutrients like nitrogen, phosphorus, potassium, and calcium in plants. Other functions of B in plant growth and development like cell wall development, stimulation or inhibition of specific metabolic pathways, sugar transport, membrane integrity and RNA, cell wall production and maintenance, indoleasetik acid (IAA), and phenol metabolism (Loomis and Durst, 1992; Bonilla et al., 1997; Dordas and Brown, 2000; Brown et al., 2002). Boron has been discovered to play a role in flowering plant fertilization as well as pollen germination and tube development. In most situations, boron promotes pollen germination and pollen tube growth in vitro (Wang et al., 2003).

Nutrition is essential for an organism's cell, tissue, organ, and system growth and development and some of them are required as respiratory substrates, while others are required for ion transport and the control of several metabolic processes. Similarly, pollen germination and pollen tube development are also regulated by a number of external factors like balanced nutrition and climatic conditions. Plants require generally more B at the reproductive stage rather than the vegetative growth stage (Loomis and Durst, 1992). Plants exhibit B sensitivity for pollen development and it may vary with species. A study reported by Dell and Huang (1997) revealed more B requirement for compact inflorescences rather than axillary inflorescences. Furthermore, anther and pollen tube development also needs B (Rawson, 1996) and deficiency of B led to burst of pollen tubes (Brown et al., 2002). However, plants do not show a similar level of B sensitivity at the reproductive stage (Huang et al., 2000). Deficiency of B may lead to empty, deformed and shriveled pollens (Dell and Huang, 1997; Herrera-Rodríguez et al., 2010).

The present study highlights the impact of B on pollen germination, growth and pollen tube development and growth of different plants ranging from field crops to fruit plants (Fig. 14.1). The study also presents the efficacy of B either alone or in combination with other minerals or salts under in vitro conditions or field conditions (in vivo).

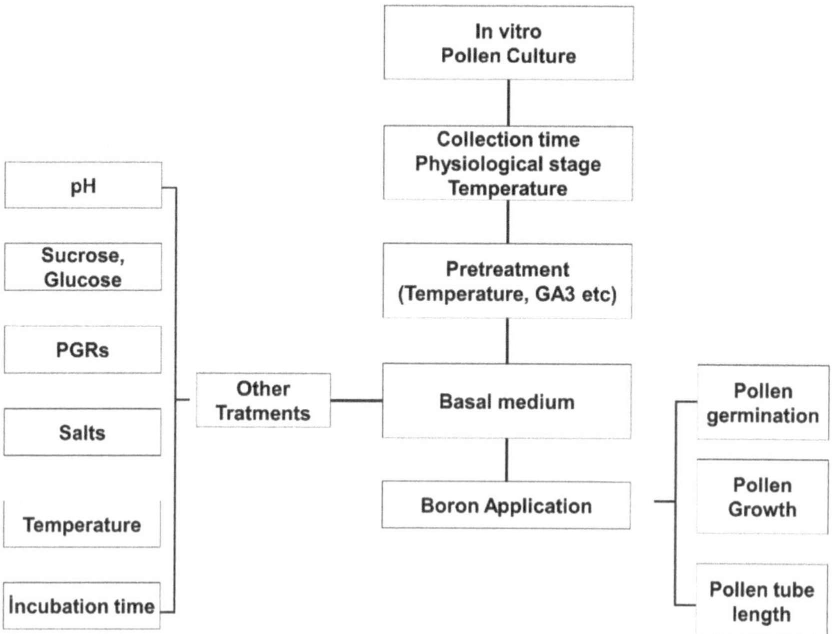

Figure 14.1 An overview of boron application for pollen germination and pollen tube development.

14.2 In vitro studies on boron and pollen

Pollination is the most significant factor for yield improvement of fruit trees due to regulating fruit and seed set (Acar and Eti, 2008; Nogueira et al., 2016). Therefore, viable pollens are a prerequisite for pollination and it can be measured by using the novel approach of germinating pollens under in vitro conditions (Shivanna et al., 1991; Soares et al., 2008). Pollen grains are generally cultured on basal medium-enriched with different nutrients and salts like (B, Ca, K, sucrose) provided with variable culture conditions (temperature, exposure time etc.). To date, a number of studies revealed the use of B alone or in combination with other salts, nutrients and culture conditions for estimating the pollen viability, germination and pollen tube growth and development of plant species. Table 14.1 presents an overview of B concentration applied for pollen germination under in vitro conditions. Furthermore, the collection of pollen samples from plants has also been summarized in the given table.

The results summarized in Table 14.1 revealed the different B doses and pollen collection time of different plant species used in these studies. The

Table 14.1 An overview of in vitro pollen germination of different crops.

Plant	B concentration	Collection time	References
Chinese fir	0.01%, 0.1%, and 0.2%	End February to mid-March	Fragallah et al. (2019)
Mitragyna parvifolia (Roxb.) Korth	25, 50, 100, 200, 300, 400, and 500 ppm	Evening time (19.00–20.30)	Dey et al. (2016)
Almond hybrids	0 and 100.0 mg/L	—	Moheb et al. (2016)
Jacquinia ruscifolia Jacq.	25, 50, 100, 200, 300, 400, and 500 ppm	Anthesis stage; Evening (08:00–09:30)	Pal and Mondal (2016)
Olive	0 and 100 ppm	Before flower opening (phenological stage)	Mehri et al. (2015)
Calabash tree	25, 50, 100, 200, and 300 ppm	Newly opened flowers; evening (18.30–19.30)	Ghanta and Mondal (2013)
Areca palm	0, 0.1, 0.2, 0.4, 0.6, 0.8, 1.0, 2.0, 3.0, 4.0, and 5.0 g/L	July to August	Liu et al. (2013)
Lychee	0, 10, 100, and 500 ppm	May to June, Morning (9.00–10.00)	Matsuda and Higuchi (2013)
Henna tree	25, 50, 100, 200, 300, and 400 ppm	Newly opened flowers; morning (6.00–8.00)	Mondal and Ghanta (2012)
Almond and peach	0 and 100.0 mg/L	—	Imani et al. (2011)
Maize hybrid (NS72)	0 and 20 μM	—	Lordkaew et al. (2011)
Pistachio (Male)	10, 25, 50, 75, and 100 ppm	Beginning of blooming from inflorescences	Acar et al. (2010)
Kiwifruit	0, 25, 50, 100, and 500 mg/L		Sotomayor et al. (2010)

comprehensive review of studies also revealed the significance of other factors like the use of sucrose, elements and salts in combination with B or used singly (Fig. 14.1). Therefore, the work on in vitro pollen germination, growth and development has been summarized according to the plant species.

14.2.1 Almond (*Prunus amygdalus*; Rosaceae)

Almond is an important nut fruit and different studies revealed the significant impact of B in combination with other input variables. The impact of cold stress and boron on pollen germination of almond hybrids under in vitro culture was studied by Moheb et al. (2016). They conducted their research by using boron (0 and 100.0 mg/L), agar (1.0%) and sucrose (10.0%) in the culture medium under a dark environment with a temperature of 24°C. Their findings revealed the variable germination percentages in response to inputs used. In H24 and almond hybrids, the greatest percentage of pollen germination was 98.35% and 97.0% respectively from the medium enriched with 100.0 ppm H_3BO_3. Whereas, germination (%) were 87.56% (H24) and 89.46% (S8) without using boric acid. As a result, almond pollen germination in vitro is inhibited in boron-free and cold-stressed settings, but they grow in boron-containing circumstances. In another study, Samiee Rad et al. (2016), used "Saba," "Rabie," and "Padre" almond cultivars with various plant growth regulators (PGRs). They used gibberellic acid (0, 50.0 and 100.0 mg/L GA3), naphthaleneacetic acid (0, 50.0 and 100.0 mg/L NAA) along with variable boron concentrations (0, 50.0 and 100.0 mg/L H_3BO_3) and 10.0% sucrose to examine pollen germination medium at 24°C temperature. The greatest pollen germination rate (90.37%) for all three almond cultivars tested were recorded from a medium containing 100.0 mg/L boric acid, 1.0% agar and 10.0% sucrose, On the contrary, supplementation of 50.0 mg/L of NAA resulted in lower germination rate (3.59%). This ratio was discovered in all kinds in an environment containing 50.0 mg/L of NAA, 1.0% agar and 10.0% sucrose. Pollen germination rates in almond types declined dramatically as the number of growth regulators rose in the culture medium.

14.2.2 Almond (*Prunus amygdalus*; Rosaceae) and peach (*Prunus persica*; Rosaceae)

Investigation on pollen germination and pollen's variability of two different plants (almond and peach) was done by Imani et al. (2011). They used 48 different media in this context. KNO_3 (0 and 100.0 mg/L), $MgSO_4$ (0 and

200.0 mg/L), $Ca(NO_3)_2$ (0, 150.0 and 300.0 mg/L) and H_3BO_3 (0 and 100.0 mg/L) were all employed in various doses. Furthermore, the media contained 1.0% agar, 10.0% and 15.0% sucrose. The pollen viability was determined by keeping it at 4°C, −20°C, and −80°C for 3 months, and then optimizing and evaluating the germination %ages. B2M2K1C2S medium (100 mg/L B, magnesium sulfate 100.0 mg/L, sucrose 15.0%, agar 1.0%, potassium nitrate 0.0 mg/L and calcium nitrate 0.0 mg/L) had the highest pollen germination rate (99.80%). On the contrarr, the lowest germination rate (30.57%) was attributed to B1K1M2C3S1 medium (10.0% sucrose, 1.0% agar, 150.0 mg/L $Ca(NO_3)_2$, 100.0 mg/L $MgSO_4$, 0.0 mg/L H_3BO_3 and 0.0 mg/L KNO_3). According to the results obtained, the Rabie Pollen stored at −80°C had the highest germination yield (90.66%) and the Anjiri pollen stored at +4°C had the lowest germination percentage (64.00%).

14.2.3 Apocynaceae family (Allamanda, Alstonia, Catharanthus, Nerium, Plumeria, Thevetia, and Tabernaemontana)

In vitro germination frequency (%) of seven taxa (Thevetia, Plumeria, Nerium, Catharanthus, Alstonia, Allamanda, and other species) belonging to the Apocynaceae family was investigated by Chatterjee et al. (2014). The anthers were picked from newly opened flower buds and surface sterilized by using a 10.0% NaOCl solution. The hanging drop method was used to assess germination in vitro. The grains were then transferred to microslides with solution culture medium (Brewbaker medium; 3.0% H_3BO_3, 4.0% Ca $(NO_3)_2$ and 6.0% Glucose). The pollen density was kept at an optimal level to avoid the population effect and to avoid taking nutrients as a limiting factor. The average pollen germination was measured after the specified time. Application of 4.0% $Ca(NO_3)_2$ was shown to be particularly efficient for inducing the germination of all taxa studies. After 1 and 2 h, Thevetia grains exhibited a proportion of germination. The germination rate after 3 h is 37.5%. Tabernaemontana and Nerium grains had the lowest germination rates. The germination of Nymphaea sp. has been discovered to be stimulated by H_3BO_3 solution. Because H_3BO_3 has a direct impact on metabolic processes, it is now routinely employed in many cultural contexts to stimulate pollen germination rates. However, it appears that lower quantities of H_3BO_3 favored toward the start of germination in Apocynaceae pollen grains. Alstonia grains exhibited a 100.0% germination rate in 1 and 2 h. Pollen grain germination occurred at a high rate when the nutritional medium was high in glucose and sucrose. Whereas, the presence of a 6.0%

glucose solution demonstrated the high degree of germination of all cereals. The results of an in vitro study of Aposinas pollen grains in the presence of a Brewbaker medium are positive.

14.2.4 Areca palm (*Areca catechu* L.; Arecaceae)

In a research published in 2013, Liu et al. examined the effects of sucrose and boric acid on the pollen germination and growth of *Areca catechu* L (areca palm). The germination and expansion of pollen tube growth were determined to be greatest at 0.4–0.6 g/L H_3BO_3. The medium enriched with 40.0 g/L sucrose content had the highest germination rate in Areca pollen. It was concluded that a medium containing 0.4–0.6 g/L H_3BO_3, 5.0 g/L agar, 40 g/L sucrose, and water is highly acceptable for the growth of Areca pollen.

14.2.5 Calabash tree (*Crescentia Cujete* L.; Bignoniaceae)

The effects of several nutrients (sucrose, Ca, Mg, K salts, and H_3BO_3) on pollen germination of *Crescentia cujete* were investigated in vitro (Ghanta and Mondal, 2013). In the evening, the newly opened flowers were picked (18.30–19.30) and immediately transferred to polyethylene bags for in vitro pollen germination testing. Fresh pollen samples were sown on grooved slides containing an H_3BO_3 and sucrose solution at various proportions individually and in combinations, as well as Ca, Mg, K salts. After that, the slides were stored in Petri Dishes lined with wet filter paper. In a 20.0% sucrose solution supplemented with 100 ppm H_3BO_3 and among the salts, maximum pollen germination (97.0%) and pollen tube development (4173 μm) was observed. Whereas, maximum of 70.0% of pollen germination with pollen tube size of 1456 μm was observed on 100 ppm $Ca(NO_3)_2$ solution

14.2.6 Chinese fir (*Cunninghamia lanceolata* L.; Cupressaceae)

Pollen germination and tube development of neglected Chinese fir plant was investigated by In Fragallah et al. (2019). The researchers investigated the impact of H_3BO_3 (0%, 0.1%, and 0.2%), sucrose (0%, 10.0%, and 15.0%), pH (4.5, 5, and 7) and incubation time (1, 12, 24, and 48 h) effect on pollen germination and tube development of nine different Chinese fir clones. Sucrose has been found to have a considerable influence on pollen germination and tube expansion with the largest impact was found at 15.0% sucrose contents. Whereas the application of boric acid dramatically improved the pollen germination and tube growth and 0.01% H_3BO_3 exerted the exhibited significant impact. The optimum pH for pollen germination and tube expansion was 7.0. The incubation time was optimized as 24–48 h with 48 h being ideal for pollen germination and tube development.

14.2.7 *Jacquinia ruscifolia* Jacq. (Theophrastaceae)

Application of boric acid, sucrose and various nutrients such as magnesium sulfate, potassium nitrate, and calcium nitrate salts on *Jaquinia roseifolia* Jacq pollen germination in vitro were studied (Pal and Mondal, 2016). The fresh flowers were collected in the evening and then stored in polyethylene bags. Solutions of H_3BO_3 (25–500 ppm), sucrose (1.0%–50.0%), $MgSO_4$, KNO_3, and $Ca(NO_3)_2$ (50–500 ppm) were used at various concentrations. Thereafter, fresh pollen samples were mounted on multiple slotted slides containing solutions at different concentrations and combinations. The slides were then covered with wet filter paper at room temperature (25°C) and then stored in Petri dishes. Finally, the seeds were inspected under a microscope at various intervals to determine the pollen germination (%) and length. Pollen grains begin to germinate in 2.0% sucrose solution, with a 2.0% germination rate and tube length of 52 μm. whereas, maximum pollen germination with longer pollen tube development (312 μm) occurred at 20.0% sucrose solution. The application of 100 ppm H_3BO_3 exhibited higher germination (60.0%) and long pollen tubes (234 μm). The combination of sucrose (20.0%) solution and H_3BO_3 (100 ppm) generated the highest pollen germination (92.0%) with a longer pollen tube (455 μm long). Application of different salts also proved to be effective for pollen germination and development. Pollens exposed to 100 ppm $Ca(NO_3)_2$ solution produced a maximum of 12.0% pollen germination and 182 μm long pollen tubes Whereas, exposing pollens to 200 ppm MgSO solution resulted in 10.0% pollen germination and 130 μm long pollen tube. On the other hand, the use of 100 ppm KNO_3 solution resulted in 8.0% pollen germination and 130 μm long pollen tubes.

14.2.8 Kiwifruit (*Actinidia deliciosa* Cultivar Matua; Actinidiaceae)

In vitro pollen germination and tubular growth of the "Matua" kiwifruit in response to different B concentrations was investigated (Sotomayor et al., 2010). For this study, pollen samples were gathered from 50 different flowers. The germination media prepared with different dosages of boric acid (0, 25.0, 50.0, 100.0, and 500.0 mg/L H_3BO_3), 2.0% agar and 15.0% sucrose. The prepared germination media were sprayed onto the pollen samples with a sterile brush. The samples were incubated for 48 h at 20°C and pollen tubes length was measured after 24 h. According to the findings, pollen germination was observed as 67.5% and 73.5%, respectively, from pollens incubated at 25.0 and 50.0 mg/L H_3BO_3. At B concentrations of 100.0 and 500.0 mg L^{-1}, a significant decrease in pollen germination was recorded. Whereas, medium prepared without B yielded only 16.7% of pollen germination. Based on the findings, it was determined that high H_3BO_3 concentrations of 100.0 mg/L or more in the nutrient medium inhibit pollen tube germination and growth.

14.2.9 Henna tree (*Lawsonia inermis* Linn.; Lythraceae)

Lawsonia inermis Linn. is a medicinal and dye-producing significant plant belonging to the Lythraceae family. In vitro pollen germination in response to boric acid and sucrose either single or in combinations was studied (Mondal and Ghanta, 2012). The newly opened flowers were collected in the morning time 6–8 a.m. and immediately transferred to a polythene bag for in vitro pollen germination studies. Fresh pollen samples were sown on grooved slides supplemented with boric acid and sucrose solutions. The slides were stored in Petri dishes lined with damp filter paper followed by examining samples using an Olympus microscope. Samples were checked after different time intervals at low magnification (10x × 15x) to determine the germination (%) and pollen tube length. For germination, double pollen tube length compared to the pollen diameter was considered as germinated pollen. In vitro pollen germination studies revealed 72.0% pollen germination with an average of 858 μm long pollen tube development at 15.0% sucrose solution. Whereas, 84.0% germination and 897 μm pollen tube length was attributed to the 100 ppm boric acid. Though either sucrose or boric acid had a positive effect on pollen germination and tube development, sucrose in combination with boric acid improved pollen germination and tube development. Therefore, the application of 15.0% sucrose solution along with 100 ppm boric acid yielded 92.0% pollen germination after 3 h. The possible reason is the development of the sugar–borate complex which exhibits better translocation than nonborate or nonionized sugar molecules. On the other hand, the pollen grains samples collected between 6:00 and 7:00 a.m. exhibited better performance.

14.2.10 Lychee (*Litchi chinensis* Sonn.; Sapindaceae)

Incorporation of agar, sucrose and H_3BO_3 concentrations in the "Chakrapat" germination media and optimal temperatures for in vitro germination of "Bengal" and "Chakrapat" lychee pollen were studied by Matsuda and Higuchi (2013). Maximum pollen germination was optimized as 25°C–35°C for "Bengal" and 25°C–35°C for "Chakrapat," pollen germination was maximum. Pollen tube elongation was good in both cultivars at 25°C–30°C. When incubated at 25°C, the pollen germination percentage and pollen tube elongation in "Chakrapat" were maximum on 1.0% agar medium containing 10 ppm H_3BO_3 and 15.0% sucrose. Pollen germination lasted for another 12 h after being placed on the medium, and intense pollen-tube elongation lasted for another 9 h.

14.2.11 Maize (*Zea mays* L.; Poaceae)

Maize is one of the most commercialized and economic plants all over due to its use directly or indirectly by providing raw material to different industries. Therefore, new varieties with desired elite traits are the major target of the researchers and pollen become the major target for breeding activities. Lordkaew et al. (2011) conducted four experiments in a maize hybrid grown in sand culture. The study used a randomized complete block (RCB) strategy, with two boron levels (B0: 0 and B2: 20 μM B) added to the nutrient solution and replicated thrice. The parameters studied were vegetative, anthesis, and maturity. Their results revealed that B deficiency reduced maize grain output while having no impact on the vegetative dry weight. B0 plants generated 0.4 grain/ear when pollen was manually applied to the silk of each plant, compared to 410 grains/ear in B20 plants. Only B0 plants displayed indications of B deficiency, including narrow white to translucent longitudinal stripes on leaves, many but small and deformed ears with extremely short silk, little tassels with some branches, and small, shriveled anthers empty of pollen. B0 plants exhibited 3–4 mg B/kg DW and B20 exhibited almost double B in their tassels, silk, and pollen tissues. A cross-fertilization experiment revealed the sensitivity of the silk to B deficiency than the tassels and pollen, despite the fact that the tassels and pollen were more damaged.

14.2.12 *Mitragyna parvifolia* (Roxb.) Korth.; (Rubiaceae)

In vitro pollen germination test on Mitragyna parvifolia (Roxb.) Korth. plant by different nutrients was conducted by Dey et al. (2016). The nutrients used were boric acid and sucrose at various concentrations either singly or in combination with salts like KNO_3, $Ca(NO_3)_2$ and $MgSO_4$. Flowers were collected during evening time (19.00–20.30 h) and placed in polythene bags for in vitro pollen germination studies. Maximum pollen germination (96.0%) with pollen tube development (1105 μm) was observed on medium supplemented with 10.0% sucrose and 100 ppm H_3BO_3. Application of 50 ppm $Ca(NO_3)_2$ produced a maximum of 90.0% pollen germination with a pollen tube of 910 μm long. The incorporation of 500 ppm $MgSO_4$ generated 728 μm pollen tube length with 85.0% pollen germination.

14.2.13 Olive (*Olea europaea* L.; Oleaceae)

The work on four cultivars named Nabali, Manzanille, Frantoio and Koroneiki were used for improving pollen germination with a purpose to be used in cross-pollination experiments. Pollen samples collected were subjected to two different treatments of H_3BO_3 application and storage treatments to check the germination fertility (germination, viability, tube growth). Pollen from each cultivar was treated with 0 and 100 ppm H_3BO_3 in agar gelled media. The other set of pollen grains were stored at different temperatures like room temperature (25°C; control), in the refrigerator (+10°C) and freezer (−20°C) for 0, 1, 6, and 12 months. The addition of 100 ppm H_3BO_3 to the culture media enhanced the pollen germination by 12.0%–24.9% but it was cultivar dependent. Low-temperature treatment storage was more beneficial than high temperature. Increasing the storage period from one to twelve months was detrimental for pollen germination and development for all cultivars. Storage of pollen grains at room temperature (25°C) induced rapid cell death after 1 month, and pollen survived up to 6 months when stored at −20°C and +10°C.

14.2.14 Pistachio (*Pistacia vera* L.; Anacardiaceae)

The impact of boron and gibberellin on in vitro pollen germination was investigated by Acar et al. (2010), with using male pistachio species such as Uyghur, Horse, Kaska, Cengel, and Poplar, Pollen was seeded individually in a germination medium containing sucrose (20.0%), five different dosages of boric acid (10, 25, 50, 75, and 100 ppm H_3BO_3) as well as gibberellic acid (GA_3). The increasing concentration of GA_3 in the germination medium has been observed to significantly lower pollen germination for all types, with the lowest value at 100 ppm. In H_3BO_3, GA3 germination diminishes by up to 25 ppm before progressively increasing to 100 ppm. The degree and pattern of reactivity in pollen germinability differed substantially amongst cultivars with boric acid and gibberellic acid environments ranging from 0 to 100 ppm. GA_3 hindered pollen germination substantially, whereas boron aided it marginally. The findings suggest that gibberellic acid has a deleterious impact on pistachio pollen germination. In another study, six different pistachio varieties were used to determine the in vitro pollen viability and germination (%) age of germination by Aldahadha et al. (2019). In vitro germination medium was consisted of sucrose (15.0%), agar (1.0%), and 100 ppm H_3BO_3. Sucrose (60.0%) and triphenyl tetrazolium chloride (1.0, 2.0, 3.0, 5.0% TTC) were used to control pollen viability. From the results obtained, it was found that there is an important correlation between the viability of the pollen and the shelf life of pistachio varieties. The viability and in vitro pollen germination of the Batouri and Lazuardi varieties was recorded as 87.0% and 69.7%, respectively. Based on this data, it has been determined that the Batouri and Lazuardi varieties can be used as good pollinators. Whereas, the lowest pollen viability and in vitro pollen germination was observed as 43.7% and 40.3%, respectively. It has been observed that pollen varieties that are stirred at 40 degrees for a month significantly lose their pollen viability totally.

14.2.15 Pomegranate (*Punica granatum*; Lythraceae)

Korkmaz and Güneri (2019), aimed to see how different boron concentrations affected pollen germination rates in the Hicaz Nar pomegranate cultivar at a full yield under in vitro conditions. The pollens from the A and B flower types of pomegranate trees were harvested separately and germinated using the "agar Petri" method. As a germination medium, four distinct boron doses (0, 10.0, 20.0, and 30.0 mg/L H_3BO_3), 1.0% agar and 15.0% sucrose were utilized. At 10.0 mg/L boron treatment, both types of flower's-based pollens exhibited the maximum germination rate of 62.17% (Type A flowers) and 41.87% (Type B flowers). Pollen germination rates of both flowers increased significantly with boron treatments compared to control treatments.

14.3 In vivo studies on boron and pollen

In vitro studies are more commonly employed for investigating the impact of B and other salts and stress factors. However, there are studies that revealed the relationship between B and pollen (viability, germination and pollen tube development of different fruit trees like mango, peach, apple and blueberry, etc.). Table 14.2 presents the impact of B concentration on pollen growth, germination and pollen tube development of different plant species.

14.3.1 Apple (*Malus domestica* L.; Rosaceae)

The effects of different B concentrations using four distinct apple cultivars (Fuji, Gala, Red Delicious, and Golden Delicious) were studied by Sharafi and Raina (2021) for in vivo pollen germination and tube penetration. H_3BO_3 (1000.0 and 2000.0 mg/L H_3BO_3) was sprayed on the leaves prior to flowering. Fluorescence microscopy was used to assess pollen–pistil interaction in terms of pollen germination and tube penetration 72 and 120 h after hand pollination. Among the different crosses, Golden Delicious x Red Delicious treated with 1000.0 mg/L H_3BO_3 demonstrated maximum pollen germination (27.39%). Only 6.49% of seeds germinated in the boron-free control group. The results obtained in this study reveals that foliar boron application of more than 1000.0 mg/L on several apple cultivars inhibits pollen germination and pollen tube development down the style.

14.3.2 Lowbush Blueberry (*Vaccinium angustifolium* Ait.; Ericaceae)

Around 12 different *Vaccinium angustifolium* clones were used as plant material to investigate the impact of B and Ca on pollen germination (Chen et al., 1998). The clones were subjected to foliar spray of

Table 14.2 An overview of pollen germination of different crops under field conditions.

Material	B concentration	References
Apple	1000 and 2000 mg/L	Sharafi and Raina (2021)
Peach	400, 800, and 1.200 mg/L	Souza et al. (2017)
Mango	0.3%	Muengkaew et al. (2017)
Picea meyeri	0%, 0.001%, 0.01%, and 1%	Wang et al. (2003)
Vaccinium angustifolium	400.0 mg/L	Chen et al. (1998)

(1) 400.0 mg/L B, (2) 4000.0 mg/L Ca, (3) 400.0 mg/L B + 4000.0 mg/L Ca, and (4) control (water). The parameters studied were pollen germination, flower number, fruit set and a number of harvestable berries. As a result, from untreated clones, pollen germination was on average 17.4% and the remaining three treatments (B, Ca, and B + Ca) resulted in increased pollen germination up to 8.0%. With the B alone treatment, more seeds per berry mean more ovules fertilized when B deficiency is remedied. In vitro and in vivo pollen germination were found to be unrelated. Increased berry size alone did not compensate for lower yields due to fewer blooms and berries per stem.

14.3.3 Mango (*Mangifera indica* L. cv.; Anacardiaceae)

Mango is one of the most popular fruits, especially in Asia and tropical and subtropical regions of the world. Muengkaew et al. (2017), studied the effects of various concentrations of B and Ca on pollen germination and fruit setting during off-season production of mango. For pollen germination analysis, randomly selected 20 hermaphrodite flowers from "Mahachanok" mango tree for pollen collection at the period of flowering. At least 300 pollen grains were cultured per Petri dishes containing 10.0% sucrose, 0.3% H_3BO_3 and 40.0% calcium-enriched media. The solutions were prepared as 1.0, 2.0, 3.0, and 4.0 mL/L and deionized water as the control. Light microscopy was used to assess the pollen germination. The experiment was performed as RCB design with five treatments and six replicates per treatment (1.0, 2.0, 3.0, and 4.0 mL/L 40% Ca and 0.3% B with water as a control) and added 0.1% Tween 20. The solution was sprayed (5 L/plant) on the trees at bud swelling, floral development, and before flower opening stages. The addition of B and Ca to the media increased the pollen germination significantly with maximum pollen germination was recorded at a concentration of 3.0 mL/L of Ca—B. On the contrary, the lowest pollen germination (%) was recorded for control treatment with the slow expansion of germinated pollens compared to Ca-B solutions. Application of Ca—B spray at three flowering stages followed by culturing of pollens on the media for 24 h resulted in the highest pollen germination. Application of 3.0 mL/L Ca—B spray resulted in the longest inflorescences and the most fruit set per mango tree. Whereas, spraying Ca—B at 3.0 or 4.0 mL/L concentrations resulted in the greatest number of hermaphroditic blooms, as well as the maximum yield.

14.3.4 Peach (*Prunus persica*; Rosaceae)

Application of H_3BO_3 on the buds of 18 peach cultivars were investigated for germination of pollen grains and fruit set in a subtropical environment (Souza et al., 2017). Three concentrations of boric acid (400, 800, and 1.200 mg/L H_3BO_3) were used in the initial spray treatments, along with a water control treatment. On early maturing cultivars Aurora 2 and Doçura 2, treatments were sprayed on the shoots as the flower buds were swelling. Results revealed the sensitivity of cultivars to H_3BO_3 and negative effect on the proportion of germinated pollen grains and fruit set in "Aurora 2" was obvious Whereas, positive outcomes from "Doçura 2" cv was recorded at 400 mg/L H_3BO_3 application. Boron promotes pollen grain germination and fruit set for some cultivars in a follow-up experiment for 18 cultivars using only the water control and 400 mg/L H_3BO_3 treatment. The rate of growth in germination and fruit set varies greatly amongst cultivars. H_3BO_3 concentrations of 400 mg/L promote pollen germination and fruit set in some cultivars, but not in all cultivars.

14.3.5 Petunia and Agapanthus (*Petunia* Juss.; Solanaceae and *Agapanthus* L Herit.; Amaryllidaceae)

Application of styles and the semivivo techniques were used for investigating the impact of B on pollen tube development of Petunia Juss. (Robbertse et al., 1990). Whereas, Agapanthus L'Herit. pollens were used for in vitro germination trials. Petunia pollen tubes were protruded from the cut ends of styles and cultivated on boron-enriched media. When styles were cut between the cut end and the pollen tube front, and the cut end of the incision was exposed to a boron-containing media while the other wounded area was exposed to a boron-free media, the direction of pollen tube growth was altered. Results revealed that lesions exposed to boron-containing media protruded pollen tubes only and growth was attributed toward the higher boron concentrations gradient, established on agar strips. Agapanthus pollens were germinated on either the low or high boron end in vitro. In this study, the pollen tube's putative chemotropic reaction to boron was proven for the first time.

14.3.6 *Picea meyeri* (Pinaceae)

To investigate the efficacy of boron on pollen germination and pollen tube growth of *Picea meyeri* Rehd. et Wils., Wang et al. (2003), cultivated pollen grains in B-deficient media or standard media. The detection of pectins and callose position pollen tubes were detected by using staining agents (aniline blue or immunolabeling with antibodies JIM5 and JIM) and laser scanning confocal microscopy. Fourier transform infrared microspectroscopy was used for investigating the possible changes in the architectures of pectins and phenolics. Pollen germination varied from 18.0% to 24.0% in B-deficient media and 61.0% on standard media. Whereas, accumulation of callose in pollen tubes was observed only in boron-deficient media.

14.4 Conclusion

B is highly significant for plant growth and development, especially in soils with deficiency or excessive B. Therefore, optimum provision of B to plants is important for optimum plant growth but may vary with plant species. In recent years, exposing pollens to B singly or in combination with other nutrients, salts and growth conditions under in vitro or field conditions directly on plants revealed a significant role of B on pollen germination and pollen tube growth. Application of excessive B is also detrimental for pollens and therefore, it is direly needed to a screening of plant species to B for optimum plant growth in respect to pollens.

References

Acar, I., Eti, S., 2008. Effect of pistil receptivity, pollen mixtures, and pollen application distances on fruit set of pistachios (*Pistacia vera*). N. ZealJ. Crop. Horticul. Sci. 36 (4), 295–300.

Acar, I., Ak, B.E., Sarpkaya, K., 2010. Effects of boron and gibberellic acid on in vitro pollen germination of pistachio (*Pistacia vera* L.). Afr. J. Biotechnol. 9 (32), 5126–5130.

Aldahadha, A.M., Al Sane, K., Bataineh, A., Alloush, A.A., Hammouri, Z., 2019. Pollen viability and in vitro germination of six pistachio (*Pistacia vera* L.) cultivars grown in northern Jordan. Adv. Horticult. Sci. 33 (3), 441–446.

Bonilla, I., Mergold-Villasenor, C., Campos, M.E., Sanchez, N., Perez, H., Lopez, L., et al., 1997. The aberrant cell walls of boron-deficient bean root nodules have no covalently bound hydroxyproline-/proline-rich proteins. Plant Physiol. 115 (4), 1329–1340.

Borg, M., Brownfield, L., Twell, D., 2009. Male gametophyte development: a molecular perspective. J. Exp. Bot. 60 (5), 1465–1478.

Brown, P.H., Bellaloui, N., Wimmer, M.A., Bassil, E.S., Ruiz, J., Hu, H., et al., 2002. Boron in plant biology. Plant Biol. 4 (02), 205–223.

Chatterjee, R., Sarkar, S., Rao, G.N., 2014. Improvised Media for in vitro pollen germination of some species of Apocynaceae. Int. J. Environ. 3 (3), 146–153.

Chen, Y., Smagula, J.M., Litten, W., Dunham, S., 1998. Effect of boron and calcium foliar sprays on pollen germination and development, fruit set, seed development, and

berry yield and quality in lowbush blueberry (*Vaccinium angustifolium* Ait.). J. Am. Soc. Horticult. Sci. 123 (4), 524–531.
Day, S., Aasim, M., 2020. Role of boron in growth and development of plant: deficiency and toxicity perspective. Plant Micronutrients. Springer, Cham, pp. 435–453.
Day, S., Çıkılı, Y., Aasim, M., 2017. Screening of three safflower (*Carthamus tinctorius* L.) cultivars under boron stress. Acta Scientiarum Polonorum-hortorum Cultus 16 (5), 109–116.
Dell, B., Huang, L., 1997. Physiological response of plants to low boron. Plant Soil 193 (1), 103–120.
Dey, K., Mondal, S., Mandal, S., 2016. Studies on in vitro pollen germination *of Mitragyna parvifolia* (Roxb.) Korth. Int. J. Curr. Microbiol. Appl. Sci. 5 (1), 768–777.
Dordas, C., Brown, P.H., 2000. Permeability of boric acid across lipid bilayers and factors affecting it. J. Membr. Biol. 175 (2), 95–105.
Fragallah, S.A.D.A., Lin, S., Li, N., Ligate, E.J., Chen, Y., 2019. Effects of sucrose, boric acid, pH, and incubation time on in vitro germination of pollen and tube growth of chinese fir (*Cunnighamial lanceolata* L.). Forests 10 (2), 102.
Ghanta, R., Mondal, S., 2013. Effect of some nutrients on in vitro pollen germination of *Withania somnifera* (L.) Dunal. Chronic Illn. 6, 7.
Herrera-Rodríguez, M.B., González-Fontes, A., Rexach, J., Camacho-Cristobal, J.J., Maldonado, J.M., Navarro-Gochicoa, M.T., 2010. Role of boron in vascular plants and response mechanisms to boron stresses. Plant Stress. 4 (2), 115–122.
Huang, L., Pant, J., Dell, B., Bell, R.W., 2000. Effects of boron deficiency on anther development and floret fertility in wheat (*Triticum aestivum* L.'Wilgoyne'). Ann. Bot. 85 (4), 493–500.
Imani, A., Kargar, M.H., PIRI, P.S., Asgari, F., Masomi, S.H., 2011. Evaluation of germination capacity of stored pollen of almond and peach.
Korkmaz, N., Güneri, M., 2019. Effect of different boron levels on pollen germination of hicaz nar pomegranate (*Punica granatum* L.) cultivar. Int. J. Agricult. For. Life Sci. 3 (1), 151–156.
Liu, L., Huang, L., Li, Y., 2013. Influence of boric acid and sucrose on the germination and growth of areca pollen. Am. J. Plant Sci. 2013.
Loomis, W.D., Durst, R.W., 1992. Chemistry and biology of boron. BioFactors (Oxford, Engl.) 3 (4), 229–239.
Lordkaew, S., Dell, B., Jamjod, S., Rerkasem, B., 2011. Boron deficiency in maize. Plant Soil 342 (1), 207–220.
Malik, C.P., 1977. Enzymes in pollen development and pollen tube growth. Adv. Pollen-Spore Res.
Marschner, H., 1995. Mineral Nutrition of Higher Plants. Academic Press, San Diego, CA, pp. 379–396.
Matsuda, H., Higuchi, H., 2013. Effects of temperature and medium composition on pollen germination of 'Bengal'and 'Chakrapat'lychee (*Litchi chinensis* Sonn.) in vitro. Trop. Agric. Dev. 57 (4), 120–125.
Mehri, H., Mhanna, K., Feleh, E., 2015. In vitro pollen germination of four olive cultivars (*Olea europaea* L.): effect of boric acid and storage. Am. J. Plant Physiol. 10 (2), 55–67.
Moheb, M.B., Imani, A., Shamili, M., 2016. Effects of boron and cold stress on germination of almond pollen in vitro culture. J. Nuts 7 (2), 149–159.
Mondal, S., Ghanta, R., 2012. Effect of sucrose and boric acid on in vitro pollen germination of Solanum macranthum Dunal. Indian J. Fundam. Appl. Life Sci. 2 (2), 202–206.
Muengkaew, R., Chaiprasart, P., Wongsawad, P., 2017. Calcium-boron addition promotes pollen germination and fruit set of mango. Int. J. Fruit. Sci. 17 (2), 147–158.

Nogueira, P.V., Coutinho, G., Pio, R., Silva, D.F.D., Zambon, C.R., 2016. Establishment of growth medium and quantification of pollen grains and germination of pear tree cultivars1. Rev. Ciência Agronômica 47, 380–386.

Ottavio, E., Mulcahy, D., Sari Goria, M., Mulcahy, G.B., 1992. Angiosperm Pollen and Ovules. Springer-Verlag.

Pal, S., Mondal, S., 2016. An investigation on in vitro pollen germination and tube development of *Jacquinia ruscifolia* Jacq. Int. J. Adv. Res. Biol. Sci. 3 (9), 71–77.

Rawson, H.Á., 1996. The developmental stage during which boron limitation causes sterility in wheat genotypes and the recovery of fertility. Funct. Plant Biol. 23 (6), 709–717.

Robbertse, P.J., Lock, J.J., Stoffberg, E., Coetzer, L.A., 1990. Effect of boron on directionality of pollen tube growth in Petunia and Agapanthus. South Afr. J. Bot. 56 (4), 487–492.

Samiee Rad, Z., Imani, A., Salmani, M., 2016. The interaction effects of boron and plant growth regulators on pollen germination of almond. J. Nuts 7 (01), 45–50.

Sharafi, Y., Raina, M., 2021. Effect of boron on pollen attributes in different cultivars of *Malus domestica* L. Natl. Acad. Sci. Lett. 44 (3), 189–194.

Shelp, B.J., 1993. Physiology and biochemistry of boron in plants. Boron its role. Crop. Prod. 53–85.

Shivanna, K.R., Linskens, H.F., Cresti, M., 1991. Pollen viability and pollen vigor. Theor. Appl. Genet. 81 (1), 38–42.

Soares, T.L., Silva, S.O., Costa, M.A.P.C., Santos-Serejo, J.A., Souza, A.D.S., Lino, L.S. M., et al., 2008. In vitro germination and viability of pollen grains of banana diploids. Crop. Breed. Appl. Biotechnol. 8 (2), 111–118.

Sotomayor, C., Ruiz, R., García-Huidobro, J., 2010. Boron concentration affects in vitro pollen germination and tube growth in kiwifruit (Actinidia deliciosa) cultivar 'Matua'. In XXVIII International Horticultural Congress on Science and Horticulture for People (IHC2010): International Symposium on Plant 932, pp. 517–521.

Souza, F.B.M.D., Pio, R., Tadeu, M.H., Zambon, C.R., Gregory, L.R., 2017. Boric acid in germination of pollen grains and fruit set of peach cultivars in subtropical region1. Rev. Ciência Agronômica 48, 496–500.

Wang, Q., Lu, L., Wu, X., Li, Y., Lin, J., 2003. Boron influences pollen germination and pollen tube growth in Picea meyeri. Tree Physiol. 23 (5), 345–351.

CHAPTER 15

Impact of boron nutrition on pollen stigma interaction and seed quality

Archana[1], Preetam Verma[2] and Nalini Pandey[1]
[1]Plant Nutrition and Stress Physiology Laboratory, Department of Botany, University of Lucknow, Lucknow, India
[2]Government Polytechnic, Department of Biotechnology, Ayodhya, India

15.1 Introduction

Boron, a vital microelement, shows an important functioning in the cell wall structure, membrane integrity (Brown et al., 2002), sugar transportation, and in metabolic pathways of phenol, carbohydrate, nucleic acid and indole acetic acid (Marschner, 1995). In addition to its participation in numerous metabolic processes, boron has a great impact on reproductive growth with special reference to microsporogenesis (Huang et al., 2000), pollen germination (Wang et al., 2003) and seed growth (Brown et al., 2002). Various studies divulged that low, as well as excess boron status, affects the reproductive yield more than biomass yield, even in the absence of any visible sign of deficiency and toxicity symptoms (Asad et al., 2003).

For agricultural output, boron is found to be important in both from the view point of its effects in deficiency and excess. The deficiencies, as well as toxicities of boron, exist in a great variety of crops and climatic conditions than any other micronutrient element (Reisenauer et al., 1973; Nable et al., 1997). Boron deficiency is the most widespread of micronutrient deficiencies in plants (Blevins and Lukaszewski, 1998). In many countries of the world, boron deficiency is commonly found in light sandy coarse-textured soils where leaching and heavy cropping have depleted soil boron reserves and in soils that have been limed to correct soil acidity (Reisenauer et al., 1973). Boron toxicity is also a significant problem in agricultural regions across the world (Nable et al., 1997), where high levels of soil boron can severely affect the growth and yield of much agricultural produce. Boron toxicity is found in arid and semiarid areas both

Boron in Plants and Agriculture
DOI: https://doi.org/10.1016/B978-0-323-90857-3.00002-3

because of high levels of boron in soils and from additions of boron in irrigation waters. The major reason for boron pollution in the soil is the irrigation with boron laden waters (Nable et al., 1997). Boron is also probably more important than any other micronutrient in obtaining high-quality agricultural yields.

The most censorious event in a plant's life is a pollination phenomenon in which there is a fusion of cell to cell interactions that accounts for the pollen–pistil interaction (Heslop-Harrison, 1978). At the beginning of the fertilization, first of all, pollen has to exhibit molecular compatibility with the stigma and after that pollen tube germination is started that invades the stigma and grows via transmitting tissue of the style to pinpoint an ovule within the ovary. Commencement and fruitful completion of this sequence of events depend upon the stigma and style providing the favorable conditions for pollen germination and stimulated growth and guidance of the pollen tube through the pistil and ovary (Heslop-Harrison, 2000; Swanson et al., 2004). The pollen must therefore be programmed to respond appropriately at every step of this interaction.

There is an interchange of chemical signals between the male and female reproductive structures in flowering plants during the pollen–stigma interaction which shows a physical association among them. For the preliminary adhesion before the pollen hydration and pollen tube germination a molecular interlinkage between the components of the pollen coat and stigma surface is pivotal (Shakya and Bhatla, 2010). Proteinaceous, lipidic, carbohydrate, and ionic components secreted by the pollen coat and stigma surface during the onset of the early events in pollen–stigma interaction are the governing components for the acceptance of well-matched pollen grains on the stigmatic surface and their adhesion at stigma maturity. For the determination of the specificity of pollen, the receptive stigma surface is signalized by the expression of biomolecules, such as peroxidases, esterases, and reactive oxygen species (ROS) (predominantly H_2O_2) (McInnis et al., 2006).

Seed yield is considered to be a hard trade as it is the product of several individual yield components such as number of inflorescences per plant, number of peroxidases (pods) per inflorescence, number of pods per plant, number of seeds per inflorescence, seed weight per pods, seed weight per inflorescence and mean seed weight (Yassin, 1973). In oilseed crops, oil is an essential component obtained from seeds. There are many factors that influence seed yield and seed quality such as genotype, agronomic techniques nutritional disorder and the environment. Seed quality is reflected in seedling density, seedling vigor, the competitiveness and uniformity of crop growth (Hampton, 1991). The quality of grains is basically depending on the storage proteins present in their seeds. These proteins are mainly stored in seeds and came into existence at the time of seed development and act as a nitrogen source during germination.

To the best of our knowledge, limited information is available on the effect of boron nutrition on reproductive biology with particular reference to pollen–stigma interaction as well as on seed quality in terms of carbohydrate, proteins and oil content of oilseed crops. So, the work presented in this thesis is an attempt to generate information regarding the role of boron in pollen–stigma interaction and seed quality of the plants.

15.2 Experimental design

The design of the experiment was completely randomized. Plants were raised in sand culture under glasshouse conditions till maturity (Sharma, 1996). The composition of nutrient solution used for growing the plants excluding B was: macronutrients: 4 mM KNO_3, 4 mM Ca $(NO_3)_2$, 2 mM $MgSO_4$, 1.33 mM NaH_2PO_4, 0.1 mM Fe EDTA, 0.1 mM NaCl, micronutrients: 10 μM $MnSO_4$, 1 μM $CuSO_4$, 1 μM $ZnSO_4$, 0.1 μM Na_2MoO_4, 0.1 μM $CoSO_4$, and 0.1 μM $NiSO_4$. Boron was supplied as H_3BO_3 at three concentrations- 0.033 mg B/L (deficient), 0. 33 mg B/L (sufficient) and 3.3 mg B/L (toxic). For the localization of enzymes (acid phosphatase, peroxidase and esterase) on the stigmatic surface, stigma from 10 to 20 flowers were gently excised without injuring the stigma and style. Stigmas were placed in a cavity slide so that the styles did not dip into the solution and incubated in the reaction solution as mentioned below for 10–20 min at 25°C–350°C in a humid chamber. After the appropriate reaction, each pistil was removed separately, rinsed thoroughly in phosphate buffer and mounted in glycerin jelly. The stained samples were observed in a Nikon E-400 light microscope and were photographed by Nikon F-60 camera using a 35 mm (100 ASA) color film. Staining for peroxidase was done by the method of Raa (1973). The stigma was stained for 30 min in 0.5% paraphenylenediamine and 0.5% H_2O_2 and washed thoroughly in phosphate buffer before mounting. For localization of acid phosphatase the stigma was placed for 15 min in a reaction mixture containing α- naphthyl phosphate (sigma) as the substrate, fast garnet GBC (sigma) as the coupling agent in 0.1 M acetate buffer (pH 4.0) and 10% MgCl2 (Shivana and Rangaswamy, 1992). Staining for localization of esterase was done in a freshly prepared solution of α-naphthyl acetate (Sigma) in 0.15 M phosphate buffer, 10% sucrose and 25 mg Fast Blue B salt (Sigma). For control α-naphthyl acetate was excluded from the reaction mixture (Shivana and Rangaswamy, 1992). The sugars and starch in seeds were estimated by the method of Nelson (1944) and Montgomery (1957), respectively. The proteins in the seeds were estimated by the method of Lowry et al. (1951). Oil extraction was done in mature seeds by the solvent extraction method.

15.3 Pollen–stigma interaction

The principal step for the advantageous pollination and fertilization in plants is the pollen–pistil interaction. It has a significant role in the reproductive biology of flowering plants. This phenomenon is initiated with the landing of pollen on the stigmatic epidermis and arrives finally in the embryonic sac where the fusion of sperm with egg occurs. This phenomenon is known as siphonogamy.

In the pollen–pistil interaction mechanism, first of all, the male pollen comes into the contact of the female stigma, which might be particular for coming pollen. Later on, various processes were going on like rapid pollen recognition, pollen hydration and pollen tube formation (Gregory and Daphne, 2000). If the nearby states are favorable the compatible pollen starts germinating. For the inhalation of the compatible pollen grains, the stigmatic secretion on the surface of the stigma performs an essential function.

15.4 Steps in pollen–stigma interaction

Capture and adhesion of pollen on stigmatic surface

↓

Pollen hydration

↓

Pollen germination

↓

Pollen tube formation

↓

Stigma invasion by pollen tube

↓

Pollen tube movement in transmitting tissue of the style

↓

Pollen tube entry into ovule

↓

Discharge of male gamete

The first response shown by stigma in self and cross-pollinated angiospermic plants is to catch the pollen grains on its surface. To make this mechanism successful, receptive stigma all-important has an adhesive surface. Pollen–stigma interaction is initiated after the adhesion of pollen grains on the stigmatic head and numerous events taking place. A very first step in the pollen stigma interface is the hydration of the pollen grain and the release of wall proteins that bind to receptors on the stigma surface. Several workers (Nothnagel, 1997; Hiscock and Allen, 2008) have installed the association of proteins in the recognition phenomenon. According to Allen et al. (2011), a complex array of pollen-wall material and stigmatic extracellular secretion (proteins, glycoproteins and glycolipids) at the stigma surface and from the pollen wall are involved in the pollen–stigma interaction mechanism.

In angiospermic plants, stigma may be dry or wet and its receptivity is defined as the ability to pick up pollen by adhesion followed by its hydration and therefore germinate a pollen tube. Edlund et al. (2004) have explained that wet stigmatic surface cells often release viscous surface secretions of proteins, lipids, polysaccharides, and pigments; and that of dry stigmas, having intact surface cells which protrude as papillae and have a primary cell wall covering a waxy cuticle and a proteinaceous pellicle.

Once the pollen hydration and germination occurs then the formation of pollen tube within the stigma is administered. Pollen tube growth arises via papilla cell wall in dry stigmas and pollen tube penetration is associated with the cell wall expansion or loosening in the stigma (Elleman et al., 1992). Wall-degrading enzymes produced either by the pollen itself or by the stigma or by both seem to be facilitated the tube growth in the stylar tissues. Heslop-Harrison (2000) reported that the receptivity of stigma to pollens can remain from one hr to several days in different species and is altered by several factors.

15.5 Enzymes responsible for pollen–stigma interaction

The compatible stigmatic receptivity is reflected by the appropriate activities of enzymes involved in this phenomenon. The previous studies in this direction suggested that various enzymes like esterase, acid phosphatase, peroxidases etc. are found to be vital for the pollen–stigma interaction.

15.5.1 Esterase

Numerous enzymes are released by a mature pollen grain after getting exposure to the stigmatic surface (Brewbaker, 1971; Heslop-Harrison, 1987). The hydrolysis of carboxylic acid esters is catalyzed by carboxylic-ester hydrolases (EC 3.1.1; Webb, 1992) are esterases (EC 3.1) which might be the cause of the establishment of an alcohol and a carboxylic acid anion (Mäkinen and Brewbaker, 1967; Mäkinen and MacDonald, 1968; Lavithis and Bhalla, 1995). Various workers in their research were observed the presence of three active cutinases (EC 3.1.1.74) in the pollen of *Tropaeolum majus* (Shaykh et al., 1977; Maiti et al., 1979), Brassica napus (Hiscock et al., 1994) and Arabidopsis thaliana (Takahashi et al., 2010). These specialized esterases break down the waxy polymers of cutin present in the stigma cuticle, allowing the pollen tube to breach this barrier (Hiscock et al., 1994; Takahashi et al., 2010). Once the cuticle has been breached, the pollen tube penetrates the stigma tissues and grows through the intercellularspaces of the stylar transmitting tissue. Earlier it was also explained by Hiscock et al. (2002a,b) that the higher activity of esterase probably could be a response to the germinating pollen and must be facilitating the penetration of the pollen tube into the stigma by hydrolyzing the cutin layer. Mature olive pollen grains contain 14 different esterase enzymes with a capacity to hydrolyze nonspecific ester substrates such as a-naphthyl acetate. These enzymes were functionally classified as carboxylesterases, acetylesterases and AChEs. Olive pollen did not show arylesterase activity (Rejón et al., 2012). Hiscock et al. (1994, 2002a) reported up to six carboxylesterases in pollen grains of Brassica. Shakya and Bhatla (2010) were detected five isoenzymes in pollen grains of sunflower.

The subcellular localization pattern of nonspecific esterase activity in the olive pollen grain and pollen tube was similar to that previously reported in Vicia faba (Bednarska, 1992). The vital function of the esterase enzyme is the hydrolysis of cutin and dissolution of the stigmatic cuticle prior to entry of the pollen tube into the papillar cell wall (Hiscock et al., 2002a,b). Pollen and/or stigma cutinases are serine esterases and mediate pollen tube penetration. It is suggested that the activities of cutinases (Edlund et al., 2004) and esterases (Hiscock et al., 2002a,b; Pandey et al., 2009) are possibly involved in the rupture of the stigmatic cuticle. Diminished pollen development within the anther, and hampered pollen germination and pollen tube elongation was due to the overexpression of this enzyme which might be the cause of severe male sterility.

15.5.2 Acid phosphatase

In angiospermic plants, the enhanced activity of acid phosphatase during pollen-stigma interaction provides evidence regarding a role for this enzyme in pollen tube growth in the style. The motive of this enzyme is to provide inorganic phosphates via hydrolysis of phosphate esters in response to the penetrating pollen tube (Renjumol and Radhamany, 2017). Ibrahim et al. (2002) reported the release of acid phosphatase activity during pollen germination and pollen tube growth in *Lilium longiflorum Thunb* and also explained that the export of the acid phosphatase activity was affected and tube growth stopped by inhibiting the components of the secretory pathway. Through their experiments, they claimed that the Ca^{+2} channel blocker gadolinium retarded the acid phosphatase secretion and tube growth, whereas 3 mM caffeine, another Ca^{+2} uptake inhibitor, stimulated the acid phosphatase release, while tube growth was inhibited. This might suggest that the boron stress conditions in plants can cause Ca^{+2} uptake inhibitions which in turn elevate the acid phosphatase activity and retard the pollen tube growth.

15.5.3 Peroxidase (POD)

Peroxidase may have a regulatory function in pollen–stigma interaction and also facilitate the fertilization phenomenon. POD generated by the pollens may function in the formation of the pollen tube wall and its regulation and that of stigmatic POD may facilitate the pollen tube growth in the transmitting tissue of the style (Bredemeijer, 1984). McInnis et al. (2006) reported that the peroxidases may perform a shielding mechanism against pathogen attack when the stigma can receive pollens (McInnis et al., 2006). They have also observed that the breakdown of H_2O_2 is catalyzed by peroxidases and produced highly oxidizing intermediates which in turn oxidizes a variety of organic and inorganic reducing substrates. Various workers (Cheong et al., 2002; Delannoy et al., 2003; Do et al., 2003; McInnis et al., 2006) have proven that peroxidase enzyme performs a defence mechanism at the time of pollen stigma interaction. When pollen grains arrive on the stigmatic surface, they act as a signal that might stimulate the enzyme peroxidase as defence mechanisms in plants.

15.6 Impact of boron stress on floral development, pollen growth and enzymatic activities

Apart from affecting vegetative growth boron deficiency as well as toxicity was found to retard reproductive development in almost all crops. Flowering was delayed in boron deficient and toxic plants compared to plants receiving sufficient (0.33 mg B/L) boron supply. Flower formation was not only delayed but also reduced in number per plant. A decrease in flowering was more prominent under deficiency of boron. Boron stress caused reduced development of both male and female reproductive parts of the flower. Compared to sufficient plants the size of anther was found to be decreased under boron stress conditions.

In soybean, under boron deficiency, there was delayed flowering and the number and size of the flower formed were also decreased. Development of androecium and gynoecium was also retarded in flowers of plants grown in the deficient supply of boron (Archana and Pandey, 2015a,b). Decreased diameter of capitulum with a reduction in the size of ray and disc floret was observed in sunflower plants grown under boron deficiency. The size of anther and its pollen-producing capacity was also reduced along with a decrease in pollen viability. In sunflower, the toxicity of boron also retarded the reproductive growth and development of pollen and stigma, but the effect was not as severe as in deficient plants (Archana and Pandey, 2021). Decreased size of the anther, pollen-producing capacity, number and size of pollen grains, and poor pollen viability were observed in boron deficient plants of mustard (data not published).

In vitro germination studies clearly indicated the poor viability of pollen grains in boron deficient and toxic plants (Cheng and Rerkasem, 1993; Wang et al., 2003). Boron stress inhibits the pollen germination and tube growth by affecting H + -ATPase activity (Feijo et al., 1995). Boron is also known to bind with hydroxyl-rich organic molecules like sugar more rapidly than sugar (sucrose) alone and is thus essential for proper pollen tube formation and elongation.

Hydrolytic enzymes peroxidase and acid phosphatase, essential for pollen–stigma interaction was also altered by boron deficiency and toxicity. Observations made based on histochemical staining of stigmatic surface for the enzymes peroxidase, acid phosphatase and esterase in boron deficient, sufficient and toxic plants showed that, under boron stress the activity of peroxidase and acid phosphatase was increased and activity of esterase was decreased (Fig. 15.1). The stigmatic region of boron deficient and toxic plants stained darker for peroxidase and acid phosphatase and lesser for esterase as compared to the stigmatic region of boron sufficient plants. Decreased esterase activity on stigma under boron deficiency and toxicity diminished the pollen germination and the growth of pollen tube within

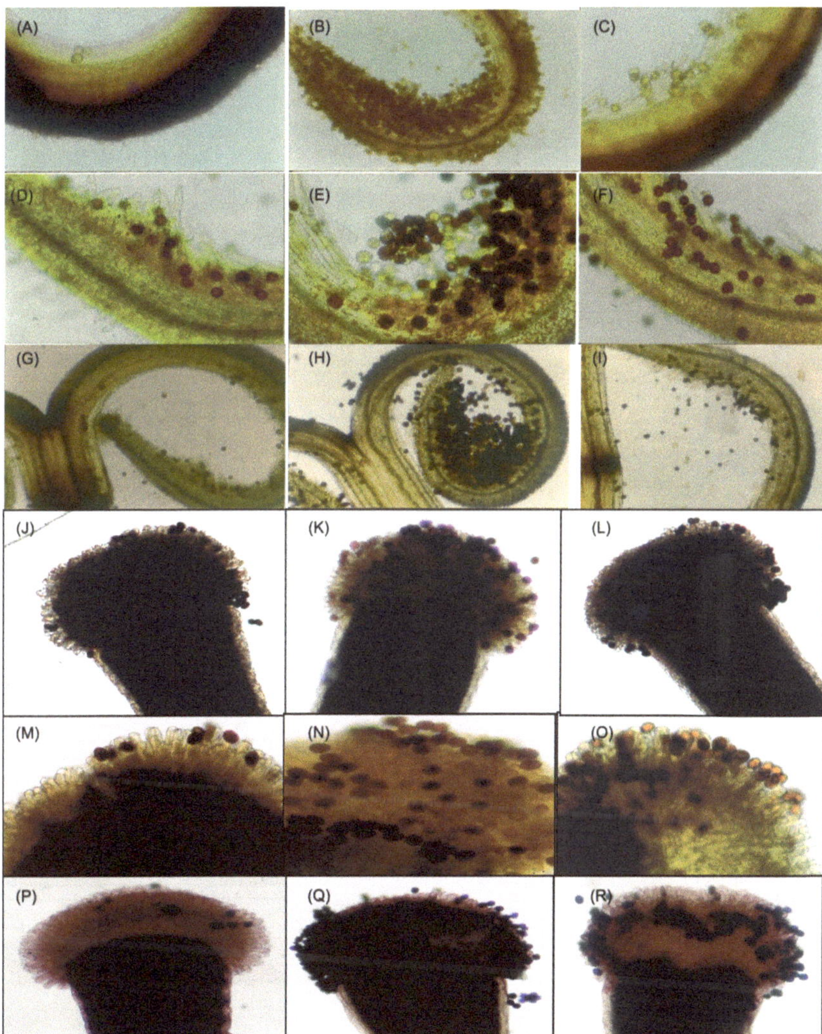

Figure 15.1 Effect of deficient (A, D, G, J, M, P), sufficient (B, E, H, K, N, Q) and toxic (C, F, I, L, O, R) boron supply on the peroxidase (A, B, C, J, K, L), acid phosphatase (D, E, F, M, N, O) and esterase (G, H, I, P, Q, R) in the stigma of sunflower (A–I) and mustard (J–R) plants.

the transmitting tissue of style in soybean, sunflower and mustard leading to poor fertilization (Archana and Pandey, 2015a,b, 2021).

Enhanced activity of acid phosphatase impaired the pollen tube germination at the time of pollen stigma interaction (Roggen and Stanley, 1969). The actual functioning of acid phosphatase in pollen stigma interaction and fertilization phenomenon is still a matter of debate. The sunflower plants experiencing the boron stress condition showed increased peroxidase activity on the stigmatic surface (Archana and Pandey, 2021). As earlier reported that the boron stress creates oxidative stress conditions in plants (Archana and Pandey, 2015a,b) and accelerated ROS/H_2O_2 generation in the stigma of boron deficient plants (Archana and Pandey, 2015a,b) creates oxidative stress conditions and increased POD activity might be to overcome this by detoxification of H_2O_2. It has been investigated by several workers (Rentel and Knight, 2004; McInnis et al., 2006) that H_2O_2 and other ROS are involved in cell signaling in plants and synchronized manifold features of plant metabolism and cell growth.

The mechanism of sporopollenin synthesis, pollen hydration, pollen–stigma recognition, pollen tube entry into the pectin-cellulose cell wall of stigma and to facilitate the growth of pollen tube within the transmitting tissue of style for successful fertilization, all are regulated by the esterase (Hiscock et al., 2002a,b). Thus esterase is found to be a very important enzyme for plants and decreased esterase activity under boron deficiency and toxicity might be the cause of poor pollen germination on the stigmatic surface and poor fertilization. Our studies in soybean and sunflower are suggested that there was diminished esterase activity under boron stress conditions which result in poor pollen germination and pollen tube formation (Archana and Pandey, 2015a,b, 2021).

15.7 Impact of boron stress on seed yield and its quality

Boron malnutrition had a more severe effect on reproductive yield than vegetative yield. The number, length and weight of both pods and seed were reduced due to boron stress. Both boron deficiency, as well as toxicity, caused a reduction in seed setting. There was a reduction in a total number of seeds formed (developed + undeveloped) per inflorescence/pod. Postharvested seeds of boron deficient as well as excess supplied plants were reduced in size and showed the poor rate of germination and vigor index as compared to seeds of boron sufficient plants. Length of embryonic axes of germinating seedlings was also decreased up to 50% in seeds of plants grown with insufficient boron supply.

Fruit and seed setting per plant under boron deficiency and toxicity was reduced in plants grown in boron stress conditions due to poor pollen viability and fertilization. The quality of seeds of plants grown in boron stress conditions was also diminished. Both sugar and starch was found to be decreased in postharvested seeds suggesting the poor translocation of carbohydrates from source (leaves) to sink (seeds). There was a marked decrease in total protein soluble content in postharvested seeds, which indicates the role of boron in the protein metabolism of seeds. Oil content was found to be low in seeds of plants grown under boron stress conditions. This suggested the role of boron in lipid biosynthesis and a study on the fatty acid composition of oils would provide valuable information (Fig. 15.2).

Deformed seed structure and poor storage capacity for reserves such as carbohydrates, proteins, and oil contents might be the cause of retarded

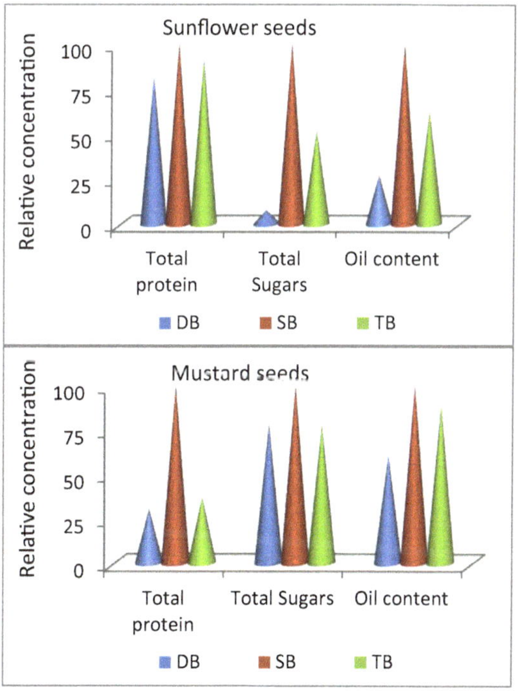

Figure 15.2 Effect of deficient, sufficient, and toxic boron supply on the seed quality of sunflower and mustard seeds.

seed viability and vigor in postharvested seeds of plants subjected to boron stress. The poor seed formation might be due to impaired embryogenesis resulting in seed abortion, the production of malformed, damaged embryo sac which ultimately reduced the seed yield. Furthermore, the reduction in seed weight under boron deficiency might be due to the formation of lesser number of endosperm cells/seeds and lowered concentration of stored carbohydrate, protein, and seed size. These results are in accordance with the report of Dell and Huang (1997).

Thus seeds developed in plants with adequate nutrient supply show higher germination percentage and are successful under all conditions exhibiting high yield and quality compared to the seeds of plants experiencing inadequate nutrient supply (Welch, 1999).

15.8 Conclusion

The present study indicates that there is an optimum quantity of boron nutrient supply is needed for appropriate reproductive development and seed quality must be met at the commencement of the reproductive phase. Boron deficiency, as well as toxicity, might be a crucial component responsible for poor seed yield and quality of agricultural crops widely cultivated on boron deficient and toxic soils the world over. The remarked changes in the functioning of stigmatic and pollen peroxidase, acid phosphatase and esterase influenced by the boron deficiency and toxicity create an unfavorable condition for the pollen–pistil interaction and limits fertilization providing a major cause for poor seed setting and seed quality.

References

Allen, A.M., Thorogood, C.J., Hegarty, M.J., Lexer, C., Hiscock, S.J., 2011. Pollen-pistil interactions and self incompatibility in the asteraceae: new insights from studies of Senecio squalidus (Oxford ragwort). Ann. Bot. 108 (4), 687–698.

Archana, Pandey, N., 2015a. Boron deficiency induced retardation in pollen-stigma interaction in soybean plants. Int. J. Appl. Pure Sci. Agric. 1 (7), 14–21.

Archana, Pandey, N., 2015b. Oxidative damage and osmotic stress in periwinkle (*Catharanthus roseus* L.) plants subjected to B nutrition. J. Global Biosci. 4 (7), 2751–2762.

Archana, Pandey, N., 2021. Reproductive development and pollen stigma interaction in sunflower plants receiving boron deficient and toxic supply. J. Plant Nutr. 2157–2166.

Asad, A., Blamey, F.P.C., Edwards, D.G., 2003. Effects of boron foliar applications on vegetative and reproductive growth of sunflower. Ann. Bot. 92, 565–570.

Bednarska, E., 1992. The localization of nonspecific esterase and cholinesterase activity in germinating pollen and in pollen tube of *Vicia faba* L. The effect of actinomycin D and cycloheximide. Biol. Plant 34, 229–240.

Blevins, D.G., Lukaszewski, K.M., 1998. Boron in plant structure and function. Ann. Rev. Plant Physiol. Plant Mol. Biol. 49, 481–501.

Bredemeijer, G.M.M., 1984. The role of peroxidase in pistil and pollen interaction. Sex. Plant Repro. 68, 193–206.

Brewbaker, J.L., 1971. Pollen enzymes and isoenzymes. In: Heslop- Harrison, J. (Ed.), Pollen: Development and Physiology. Butterworths, London, pp. 156–170.

Brown, P.H., Bellaloui, N., Wimmer, M.A., Bassil, E.S., Ruiz, J., Hu, H., et al., 2002. Boron in plant biology. Plant Biol. 4, 205–223.

Cheng, C., Rerkasem, B., 1993. Effect of boron on pollen viability in wheat. Plant Soil 155/156, 313–315.

Cheong, Y.H., Chang, H.S., Gupta, R., Wang, X., Zhu, T., Luan, S., 2002. Transcriptional profiling reveals novel interactions between wounding, pathogen, abiotic stress, and hormonal responses. Arabidopsis. PlantPhysiol 129, 661–677.

Delannoy, E., Jalloul, A., Assigbetse, K., Marmey, P., Geiger, J.P., Lherminier, J., et al., 2003. Activity of class III peroxidases in the defense of cotton to bacterial blight. Molec. Plant Microbe Interact. 16, 1030–1038.

Dell, B., Huang, L., 1997. Physiological responses of plants to low boron. Plant Soil 193, 103–120.

Do, H.M., Hong, J.K., Jung, H.W., Kim, S.H., Ham, J.H., Hwang, B.K., 2003. Expression of peroxidase-like genes, H2O2 production, and peroxidase activity during the hypersensitive response to *Xanthomonas campestris* pv. vesicatoria in *Capsicum annuum*. Mol. Plant Microbe Interact. 16(3), 196–205.

Edlund, A.F., Swanson, R., Preuss, D., 2004. Pollen and stigma structure and function: the role of diversity in pollination. Plant Cell 16, S84–S97.

Elleman, C.J., Franklin-Tong, V., Dickinson, H.G., 1992. Pollination in species with dry stigmas: the nature of the early stigmatic response and the pathway taken by pollen tubes. N. Phytol. 121, 413–424.

Feijo, J.A., Malho, R., Obermeyer, G., 1995. Ion dynamics and its possible role during in-vitro pollen germination and tube growth. Protoplasma 187, 155–167.

Gregory, M.Z., Daphne, P., 2000. Dissecting Arabidopsis pollen–stigma interactions reveals novel mechanisms that confer mating specificity. Ann. Bot. 85, 15–21.

Hampton, J.G., 1991. Herbage seed lot vigour. Do problems start with seed production? J. Appl. Seed Prod. 9, 87–93.

Heslop-Harrison, J., 1987. Pollen germination and pollen tube growth. Int. Rev. Cytol. 107, 1–78.

Heslop-Harrison, J., 1978. Recognition and response in the pollen-stigma interaction. In. Curtis, A. (Ed.), Cell-cell Recognition. SEB Symposium Series No. 32. Cambridge University Press, pp. 121–138.

Heslop-Harrison, Y., 2000. Control gates and micro-ecology: the pollen–stigma interaction in perspective. Ann. Bot. 85, 5–13.

Hiscock, S.J., Allen, A.M., 2008. Diverse cell signalling pathways regulate pollen-stigma interactions: the search for consensus. N. Phytol. 179 (2), 286–317.

Hiscock, S.J., Dewey, F.M., Coleman, J.O.D., Dickinson, H.G., 1994. Identification and localization of an active cutinase in the pollen of *Brassica napus* L. Planta 193, 377–384.

Hiscock, S.J., Bown, D., Sarah, J., Dickinson, H.G., 2002a. Serine esterases are required for pollen tube penetration of the stigma in Brassica. Sex. Plant Reprod. 15, 65–74.

Hiscock, S.J., Hoedemaekers, K., Friedman, W.E., Dickinson, H.G., 2002b. The stigma surface and pollen–stigma interactions in *Senecio squalidus* L. (Asteraceae) following cross (compatible) and self (incompatible) pollinations. Int. J. Plant Sci. 163, 1–16.

Huang, L., Pant, J., Dell, B., Bell, R.W., 2000. Effect of boron deficiency on anther development and floral fertility in wheat (*Triticum aestivum* L. 'Wilgoyne'). Ann. Bot. 85, 493–500.

Ibrahim, H., Pertl, H., Pittertschatscher, K., Fadl-Allah, E., El-Shahed, A., Bentrup, F.W., et al., 2002. Release of an acid phosphatase activity during lily pollen tube growth involves components of the secretory pathway. Protoplasma .

Lavithis, M., Bhalla, P.L., 1995. Esterases in pollen and stigma of Brassica. Sex. Plant Reprod. 8, 289–298.

Lowry, O.H., Roserbrough, N.J., Farr, A.L., Randall, R.J., 1951. Protein measurement with folin-phenol reagent. J. Biol. Chem. 193, 256–275.

Mäkinen, Y., Brewbaker, J.L., 1967. Isozyme polymorphism in flowering plants I. Diffusion of enzymes out of intact pollen grains. Physiol. Plant 20, 477–482.

Mäkinen, Y., MacDonald, T., 1968. Isoenzyme polymorphism in flowering plants II. Pollen enzymes and isoenzymes. Physiol. Plant 21, 477–486.

Maiti, I.B., Kolattukudy, P.E., Shaykh, M., 1979. Purification and characterization of a novel cutinase from nasturtium (*Tropaeolum majus*) pollen. Arch. Biochem. Biophys. 196, 412–423.

Marschner, H., 1995. Mineral Nutrition of Higher Plants, second ed. Academic Press, San Diego, CA, pp. 379–396.

McInnis, M.S., David, C.E., Robert, P., Radhika, D., John, T.H., Simon, J.H., 2006. The role of stigma peroxidases in flowering plants: insights from further characterization of a stigma-specific peroxidase (SSP) from *Senecio squalidus* (Asteraceae). J. Exp. Bot. 57, 1835–1846.

Montgomery, R., 1957. Determination of glycogen. Arch. Biochem. Biophys. 67, 378–386.

Nable, R.O., Banuelos, G.S., Paull, J.G., 1997. Boron toxicity. Plant Soil 193, 181–198.

Nelson, N., 1944. Photometric adaptation of Somogyi's method for determination of glucose. J. Biol. Chem. 153, 375–378.

Nothnagel, E.A., 1997. Proteoglycans and related components in plant cells. Int. Rev. Cytol. 174, 195–291.

Pandey, N., Pathak, G.C., Sharma, C.P., 2009. Impairment in reproductive development is a major factor limiting reproductive yield of black gram under zinc deficiency. Biol Plant 53, 723–727.

Raa, J., 1973. Cytochemical localization of peroxidase in plant cells. Physiol. Plant 28, 132–133.

Reisenauer, H.M., Walsh, L.M., Hoeft, R.G., 1973. Testing soil for sulphur, boron, molybdenum and chlorine. In: Walsh, L.M., Beaton, J.D. (Eds.), Soil Testing and Plant Analysis. Soil Sci. Soc. Am. Inc., Madison, WI, pp. 173–200.

Rejón, J.D., Zienkiewicz, A., María Isabel Rodríguez-García, Castro, A.J., 2012. Profiling and functional classification of esterases in olive (*Olea europaea*) pollen during germination. Ann. Bot. 110, 1035–1045.

Renjumol, R., Radhamany, P.M., 2017. Differential activity of four selected enzymes in the pistils and pollen grains of two varieties of *Hamelia patens* jacq. (rubiaceae) following compatible and incompatible pollination. Int. J. Adv. Res. 5 (3), 636–644.

Rentel, M.C., Knight, M.R., 2004. Oxidative stress-induced calcium signaling in Arabidopsis. Plant Physiol. 135, 1471–1479.

Roggen, H.P.J.R., Stanley, R.G., 1969. Cell-wall-hydrolyzing enzymes in wall formation as measured by pollen tube extension. Planta 84 (4), 295–303.

Shakya, R., Bhatla, S.C., 2010. A comparative analysis of the distribution and composition of lipidic constituents and associated enzymes in pollen and stigma of sunflower. Sex. Plant Repro. 23 (2), 163–172.
Sharma, C.P., 1996. Deficiency Symptoms and Critical Concentration of Micronutrients in Crop Plants. Lucknow University Center, Lucknow.
Shaykh, M., Kolattukudy, P.E., Davis, R., 1977. Production of a novel extracellular cutinase by the pollen and the chemical composition and ultrastructure of the stigma cuticle of nasturtium (*Tropaeolum majus*). Plant Physiol. 60, 907–915.
Shivana, K.R., Rangaswamy, N.S., 1992. Pollen Biology: A Laboratory Manual. Narosa Publishing House, New Delhi, pp. 45–50.
Swanson, R., Edlund, A.F., Preuss, D., 2004. Species specificity in pollen-pistil interactions. Ann. Rev. Gen. 38, 793–818.
Takahashi, K., Shimada, T., Kondo, M., Tamai, A., Mori, M., Nishimura, M., 2010. Ectopic expression of an esterase, which is candidate for the unidentified plant cutinase, causes cuticular defects in *Arabidopsis thaliana*. Plant Cell Physiol. 51, 123–131.
Wang, Q., Lu, L., Wu, X., Li, Y., Lin, J., 2003. Boron influences pollen germination and pollen tube growth in *Picea meyeri*. Tree Physiol. 23, 345–351.
Webb, E.C., 1992. Enzyme Nomenclature 1992: Recommendations of the Nomenclature Committee of the International Union of Biochemistry and Molecular Biology on the Nomenclature and Classification of Enzymes. Academic Press, San Diego, CA.
Welch, R.M., 1999. Importance of seed mineral nutrient reserves in crop growth and development. In: Rengel, Z. (Ed.), Mineral Nutrition of Crops Fundamental Mechanisms and Implications. Food Product Press, NewYork, pp. 205–226.
Yassin, T.E., 1973. Genotypic and phenotypic variances and correlations in field beans (*Vicia faba* L). J. Agric. Sci. 81, 445–448.

Index

Note: Page numbers followed by "*f*" and "*t*" referred to figures and tables, respectively.

A

ABC transporters, 200
Abiotic stress relief, 59–61, 62*t*
Abscisic acid (ABA), 101–102
ACA10, 193
Actinomycete symbiosis, 7–8
AGP deposition, 8
Agricultural soils
  boron management under deficiency condition, 260–262, 262*t*, 263*t*
  boron management under toxicity condition, 262
Almond *(Prunus amygdalus; Rosaceae)*, 297–298
Aluminum (Al), 237–238
Aminolevulinic acid, 178–179
Antioxidants, 31, 138–139
  enzymes, 7, 156–157, 274
Apocynaceae family, 298–299
Apple *(Malus domestica)*, 31–32, 305
Aquaporins, 100–101
Arabidopsis, 71–74, 97–98, 101, 110
Arabidopsis, 56, 190
  A. thaliana, 152–153, 209–210
Arabinogalactan proteins (AGPs), 56–57, 136–137
Areca palm, 299
Ascorbate (AsA), 138–139
Ascorbate/glutathione cycle, 7
Asparagine synthetase (AS) transcript-level, 134
Autophagy, 197–198
Auxin, 95
  synthesis, distribution, and transport of, 95
Auxin response factors (ARFs), 194
Avocados *(Persea americana)*, 169–170

B

B-acquisition efficiency (BACE), 256
Barley, 210–212
Barley *(Hordeum vulgare)*, 110
B-deficient Citrus sinensis leaves, 6
B-efflux BOR4 transporter, 99–100
B-inefficient (W10) cultivars, 191–192
Biochemical alteration, 18*t*
bis-N-acetyl-borate, 10
bis-Suc borate, 10
Borate crosslinking, 155
Borate exporters, 152–154, 188
Borate transporters (BOR) functioning, 274–277
BOR4-GFP aggregation, 61–63
BOR1 homologs, in plants, 188–192
Boric acid, 38
Boric acid channels, 151–152
Boron (B)
  aberrant seedlings, 52
  in agriculture, 2–4, 252–255, 253*t*, 255*t*
  biochemical and physiological changes, 104–108
  biochemical, physiological, anatomical, and molecular changes, 92
  boron nutritional status, 277–281
  cis-form configuration, 52
  deficiency
    cessation of plant growth, 75–76
    membranes, cytoskeleton, and cell wall, 55–57
    nitrate uptake and fixation, 57–58
    oxidative stress and secondary metabolism, 58
    on plant development, 281–283
    symptoms of, 53–55
    and transcript level changes, 82–83
  distribution and redistribution in plant adaptation, 100–104
  fertilizers, 262–264, 264*t*
  in fruit crops
    mineral sources and fertilization, 37–45, 38*t*, 39*t*, 45*t*

Boron (B) (*Continued*)
and nutrient boron, 30*t*
physiology and transport, 31–32
plants, fertilization, and boron sources, 34–37
sandy soils, 31
toxicity and deficiency symptoms, 32–34, 33*t*, 35*f*, 36*f*
functional role of, 29, 71–73
gene expression and, 64
in higher plants
B deficiency, 12–16
dynamics and transport, 9–10
reproductive phase, 8–9
toxicity, 16–19
translocation and distribution, 10
vegetative phase, 4–8
hydrosphere and lithosphere, 51
lignin and phenolic compounds, 29
metalloid subgroup III, 51
micronutrients, 169
mur1–1 and mur1–2, 52
non-essential elements, 169
nutrients interaction, 258–260, 261*t*
nutritional status, 277–281, 279*f*
in photosynthesis, 52
with phytohormones interactions, 73, 77*f*
boron–abscisic acid (ABA) interactions, 80–81
boron–auxin interactions, 76–79
boron–brassinosteroid interactions, 81
boron–cytokinin interactions, 79–80
boron–ethylene interactions, 80
boron–jasmonic acid interactions, 81–82
plant metabolism, 155–157
in plants
importance and functions of, 251–252
metabolism, 272–274
role, 154–155
plant tolerance to, 157–158
plasma membrane integrity and function, 1–2
pptake and transport mechanisms, 150
public attributes and chemistry of, 148
reduced B transport, 237–239
regulation of uptake, 97–100
rhamnogalacturonan II (RG-II), 52
root system, 94–97
sensitive and tolerant plants, 11–12
in soil, 128
susceptibility and resilience, 161–162
toxicity
for abiotic stress relief, 59–61, 60*f*, 62*t*
tolerance to, 61–63
transport mechanisms, 274–277, 275*f*
vegetative and reproductive phases, 1–2
Boron scarcity, 156
Bot1, 63
B-ramnogalacturonan-II (B-RG-II), 91
Brassica napus, 134–135
Brefeldin A-induced compartments, 64
Brown mustard *(Brassica juncea)*, 106–107
B-stress-mediated carbohydrate accumulation, 107–108
B supplementation, 61
B uptake efficiency, 3
B-utilization efficiency (BUTE), 256

C

Calabash tree, 299
Calvin cycle, 176
Carbohydrate biosynthesis and degradation, 107–108
Carbohydrate metabolism, 139
Carbon skeleton, 175
Carrots *(Daucus carota)*, 105–106
Cell division, 75–76
Cell elongation, 75–76
Cellulose, 136
Cell wall, 55–57
formation and functionality, 91
Chinese fir, 299–300
Chlorophyll fluorescence, 108
Chloroplasts, 130–132, 174–175
Chlorosis, 16–17
Cis-polyols, 10
Citrus macrophylla, 106
Climatic zone, 3
CmBOR1, 191
Crop productivity, 129–130

Cucumber *(C. sativus)*, 174
Cucurbita pepo, 54
Cytoskeleton, 55–57

D
Dhaincha plants *(Sesbania cannabina)*, 54
Dicotyledonous plants, 155–156

E
Ethylene, 95
Eucalyptus *(Eucalyptus camaldulensis)*, 106
Exogenous brassinolide (eBL) application, 280–281

F
Faba bean *(Vicia faba)*, 129–130
Fertilizers, 262–264, 264*t*
Foliar application, 31–32
Free amino acids, 107–108

G
Gas exchange, 108
Gene expression, 64
Gene regulation
  boron-deficient conditions, 193–194
  conditions of excessive boron, 194–200
Glutathione S-transferase (GST), 195
Glycosylinositol phosphorylceramides (GIPCs), 172
Grapevine, 34, 191
Gypsophila arrostil, 196

H
Helianthus annuus, 56–57
Henna tree, 301–302
High boron requiring1 (BOR1), 71–73
Highbush blueberry *(Vaccinium corymbosum)*, 106–107
Higher cell-wall pectin content, 274–277
Higher plants, boron in
  B deficiency, 12–16
  dynamics and transport, 9–10
  micronutrient, 148–149
  reproductive phase, 8–9
  toxicity, 16–19
  translocation and distribution, 10
  vegetative phase
    biochemical/hormonal functions of, 6–8
    structural functions of, 4–6
High soil salinity, 234–235
Homeostasis, 97–100
Hydrated minerals, 187

I
Indole acetic acid (IAA), 7–8

J
JA, 178–179
Jacquinia ruscifolia Jacq., 300
Jasmonic acid (JA), 195

K
Kiwifruit, 300–301

L
Late embryogenesis abundant protein 5 (LEA5), 197–198
Leaf abscission, 16–17
Leaf toxicity, 59
Lewis acid, 9–10, 149
LIM domain-containing protein, 197–198
Lipid peroxidation, 173
Long noncoding RNAs (lncRNAs), 217
Loquat (*Eriobotrya japonica* Lindl.), 32
Lowbush blueberry, 305–306
Lychee, 302
Lychee plants, 32
Lycopersicum esculentum, 54

M
Macronutrients, 147–148
Maize, 302
Major intrinsic protein (MIP), 137–138, 150, 188–189
Malondialdehyde (MDA), 241–242
Mango, 306
Membranes, 55–57
Meristem development, 71
Metabolic processes, 9–10

MF-inoculated plants, 3
Micro-RNA abundances, 73
Microsporogenesis, 112
miRNAs, 199
in post-transcriptional control, 215–217
Mitochondria, 130–132
Mitragyna parvifolia (Roxb.) Korth, 303
Monocotyledonous plants, 113
Moringa oleifera, 6
Mustard *(Brassica campestris)*, 132
Mycorrhiza, 3

N

NCED3 knockout plants, 101–102
Nicotiana tabacum plant, 134
Nitrogen (N), 13–15
Nitrogen metabolism, 134
NOD26-like intrinsic-type (NIP), 151, 196–197
Nodulin 26-like intrinsic proteins, 9–10, 97
Nonmetabolic processes, 9–10
NPK formulations, 37

O

Oilseed crops, 312–313
Olive, 303
Orange trees *(Citrus sinensis)*, 169–170
Oxidative damage, 93
Oxidative stress, 58, 175–176, 180, 239–242

P

Pathogenesis-related genes (PR), 195, 200
Pea *(Pisum sativum)*, 134
Peach, 307
Peach *(Prunus persica; Rosaceae)*, 297–298
Pear trees *(Pyrus pyrifolia)*, 32
Pectin methylesterase (PME), 136
Peroxidase (POD) activity, 241–242
Petunia and agapanthus, 307
pH, 2–3
Phenolic concentration, 274
Phenolics, 7
Phloem translocation, 171
Phosphorylation, 200
Phosphorylation intensities, 73
Photooxidation stress, 177–178
Photosynthesis, 132–133
Photosynthetic pigments, 174–175
Photosystem II (PSII) efficiency, 174
Physcomitrella patens, 154
Phytohormones, 73
boron–abscisic acid (ABA) interactions, 80–81
boron–auxin interactions, 76–79
boron–brassinosteroid interactions, 81
boron–cytokinin interactions, 79–80
boron–ethylene interactions, 80
boron–jasmonic acid interactions, 81–82
Picea meyeri (Pinaceae), 307–308
Pistachio *(Pistacia vera)*, 106, 303–304
Plant defense system, 239–242
Plant dry matter output, 147–148
Plant growth and development, 71
Plant growth traits, 235–236
Plant meristems, 71
Plant metabolism, 272–274
Plant physiology, 4
Plant responses
boron deficiency condition, 255–256, 257*t*
boron toxicity condition, 256, 257*t*
Plant tolerance, 157–158
Plasma membrane (PM), 130–132
Plasma membrane intrinsic proteins (PIPs) genes, 101
PLETHORA (PLT) transcription factor family, 73–74
Plum (*Prunus domestica* L.), 32
Pollen germinations, 16–17, 154
Pollen grains
biochemical, physiological morphological processes, 294
germination and pollen tube, 293
grains' fertility and viability, 293
morphogenetic process, 293
in vitro studies, 295–304, 296*t*
Almond *(Prunus amygdalus; Rosaceae)*, 297–298
apocynaceae family, 298–299
areca palm, 299
calabash tree, 299

Chinese fir, 299–300
henna tree, 301–302
Jacquinia ruscifolia Jacq., 300
kiwifruit, 300–301
lychee, 302
maize, 302
Mitragyna parvifolia (Roxb.) Korth, 303
olive, 303
peach *(Prunus persica; Rosaceae)*, 297–298
pistachio, 303–304
pomegranate, 304
in vivo studies
apple, 305
lowbush blueberry, 305–306
mango, 306
peach, 307
petunia and agapanthus, 307
Picea meyeri (Pinaceae), 307–308
Pollen–pistil interaction, 312
Pollen-stigma interaction, 312, 314
acid phosphatase, 317
boron stress impact on floral development, 318–320
esterase, 316
peroxidase, 317
Pollen tube formation, 136
Pollen tube growth, 16–17
Polyphenol oxidase activity (PPO), 58
Pomegranate, 304
Poplar, 212–213
Protochlorophyllide oxidoreductase, 197–198
Puccinellia distans, 196, 214–215
Purple-leafed sweet basil cultivars, 175–176

Q

Quince (*Cydonia oblonga* Mill), 32

R

Reactive oxygen species (ROS), 7, 234–235
Reproductive development, 71, 75–76
Rhamnogalacturonan II (RG-II), 4–5, 136–137, 149–150, 188
Rice *(Oryza sativa)*, 113, 213–214
Rizobial N_2 fixation, 7–8
Root apical meristem (RAM), 95–96
Root growth, 95, 272–274
Root system
allocation of biomass, 95
evolutionary development, 94
morphoanatomical mechanisms of, 96–97
nutrient uptake processes, 94
plastic development of, 94
Rotten Ear (RTE), 190–191

S

Saline irrigation water, 4
Secondary metabolism, 58
Seed yield, 312–313
Selaginella moellendorffii, 154
Shoot growth, 281–283
Soil B–Si compound formation, 237–238
Soils
in boron, 250–251
agricultural soils, 260–262
plants/varieties, responses of, 255–256
and foliar application, 31–32
in North Africa, 233–234
passive and active transportation in, 150
properties
effect on bioavailability of B, 257–258
and nutrients, 249–250
soil organic matter, 128
in South Australia, 233–234
in West Asia, 233–234
Soybean *(Glycine max)*, 134, 138–139
S. parvula, 110
Stress
biochemical and physiological changes, 104–108
biochemical response of crops
carbohydrate metabolism, 139
cell wall and membrane permeability, 136–137
reactive oxygen species and antioxidant system, 138–139
transporters, 137–138
on crop productivity, 129–130, 131*t*

Stress (*Continued*)
metabolites for, 192–193
physiological response of crops
carbon partitioning and source-sink relationship, 133–134
nitrogen metabolism, 134
nutrient interaction, 134–135
photosynthesis, 132–133
pollen tube formation, 136
transpiration, 135
reproductive phase, 110–113
on seed yield and quality, 320–322
Stress-related transcription factors, 200
Sunflower, 53
Superoxide dismutase (SOD), 239–241

T
TaBOR1.2, 191
TF binding sites (TFBSs), 207
TFBSs, 208
Thaumatin-like protein 1 (TLP1), 195, 197–198
Thiamin, 178–179
Tobacco *(Nicotiana tabacum)*, 104–105, 155
Tomato *(Solanum lycopersicum)*, 101
Toxicity
alleviation of effects, 283–285
long noncoding RNAs, 217
oxidative stress and plant defense system, 239–242
plants and boron nutrition, 149–150
silicon-induced alleviation of
plant growth traits, 235–236
silicon induced improvement, 242–243
Transcription factors (TFs)
in *Arabidopsis thaliana*, 209–210
in barley, 210–212
categories, 208
DNA-binding domain, 207
known functions of, 217–227
in poplar, 212–213
in *Puccinellia distans*, 214–215
in rice, 213–214
sequence and distribution analyses, 207
in wheat, 214
Transpiration, 135
Transporters, 137–138

U
Undissociated boric acid, 128–129
5'-UTR open reading frames (uORFs), 98–99

V
Vascular plants, 271–272
Vicia faba, 154–155
VvBOR1, 191

W
Wheat *(Triticum aestivum)*, 105–106, 214
White lupine *(Lupinus albus)*, 104–105
White mulberry *(Morus alba)*, 104–105
Wound-inducible protein kinase (WIPK), 193–194
WRKY6 mutation, 108–109
Wuschel-like homeobox5 (WOX5) transcription factor, 73–74

X
X intrinsic proteins (XIPs), 137–138

Z
Zea mays, 54–55
Zinnia elegans plants, 54
ZmRTE1, 190–191
Zucchini *(Cucurbita pepo)*, 174